Metric Measurement

Basic Prefixes

kilo- 10^3 $= 1000$
hecto- $10^2 = 100$
deka- $10^1 = 10$

deci- $10^{-1} = \dfrac{1}{10}$ $= .1$

centi- $10^{-2} = \dfrac{1}{100}$ $= .01$

milli- $10^{-3} = \dfrac{1}{1000}$ $= .001$

kilo means 1000 times basic unit
hecto means 100 times basic unit
deka means 10 times basic unit

deci means .1 times basic unit

centi means .01 times basic unit

milli means .001 times basic unit

Length

The meter is the basic unit.

1 kilometer = 1000 meters	1 km = 1000 m
1 hectometer = 100 meters	1 hm = 100 m
1 dekameter = 10 meters	1 dam = 10 m
1 decimeter $= \dfrac{1}{10}$ of a meter	1 dm = .1 m
1 centimeter $= \dfrac{1}{100}$ of a meter	1 cm = .01 m
1 millimeter $= \dfrac{1}{1000}$ of a meter	1 mm = .001 m

Some Approximate Comparisons between English and Metric Systems
(The symbol $\approx$ means "is approximately equal to".)

Length

1 inch $\approx$ 2.5 centimeters
1 foot $\approx$ 30.5 centimeters
1 yard $\approx$ .9 of a meter
1 mile $\approx$ 1.6 kilometers

1 centimeter $\approx$ .4 of an inch
1 centimeter $\approx$ .03 of a foot
1 meter $\approx$ 1.1 yards
1 kilometer $\approx$ .6 of a mile

Trigonometry

The Prindle, Weber & Schmidt Series in Mathematics

Althoen and Bumcrot, *Introduction to Discrete Mathematics*

Brown and Sherbert, *Introductory Linear Algebra with Applications*

Buchthal and Cameron, *Modern Abstract Algebra*

Burden and Faires, *Numerical Analysis,* Third Edition

Cullen, *Linear Algebra and Differential Equations*

Dobyns, Steinbach and Lunsford, *The Electronic Study Guide for Precalculus Algebra*

Eves, *In Mathematical Circles*

Eves, *Mathematical Circles Adieu*

Eves, *Mathematical Circles Revisited*

Eves, *Mathematical Circles Squared*

Eves, *Return to Mathematical Circles*

Fletcher and Patty, *Foundations of Higher Mathematics*

Geltner and Peterson, *Geometry for College Students*

Gilbert and Gilbert, *Elements of Modern Algebra,* Second Edition

Gobran, *Beginning Algebra,* Fourth Edition

Gobran, *College Algebra*

Gobran, *Intermediate Algebra,* Fourth Edition

Gordon, *Calculus and the Computer*

Hall, *Algebra for College Students*

Hall and Bennett, *College Algebra with Applications*

Hartfiel and Hobbs, *Elementary Linear Algebra*

Hunkins and Mugridge, *Applied Finite Mathematics,* Second Edition

Kaufmann, *Algebra for College Students,* Second Edition

Kaufmann, *Algebra with Trigonometry for College Students*

Kaufmann, *College Algebra*

Kaufmann, *College Algebra and Trigonometry*

Kaufmann, *Elementary Algebra for College Students,* Second Edition

Kaufmann, *Intermediate Algebra for College Students,* Second Edition

Kaufmann, *Precalculus*

Kaufmann, *Trigonometry*

Keisler, *Elementary Calculus: An Infinitesimal Approach,* Second Edition

Konvisser, *Elementary Linear Algebra with Applications*

Laufer, *Discrete Mathematics and Applied Modern Algebra*

Nicholson, *Linear Algebra with Applications*

Pasahow, *Mathematics for Electronics*

Powers, *Elementary Differential Equations*

Powers, *Elementary Differential Equations with Boundary Value Problems*

Powers, *Elementary Differential Equations with Linear Algebra*

Proga, *Arithmetic and Algebra*

Proga, *Basic Mathematics*, Second Edition

Radford, Vavra, and Rychlicki, *Introduction to Technical Mathematics*

Radford, Vavra, and Rychlicki, *Technical Mathematics with Calculus*

Rice and Strange, *Calculus and Analytic Geometry for Engineering Technology*

Rice and Strange, *College Algebra*, Third Edition

Rice and Strange, *Plane Trigonometry*, Fourth Edition

Rice and Strange, *Technical Mathematics*

Rice and Strange, *Technical Mathematics and Calculus*

Schelin and Bange, *Mathematical Analysis for Business and Economics*, Second Edition

Steinbach and Lunsford, *The Electronic Study Guide for Trigonometry*

Strnad, *Introductory Algebra*

Swokowski, *Algebra and Trigonometry with Analytic Geometry*, Sixth Edition

Swokowski, *Calculus with Analytic Geometry*, Second Alternate Edition

Swokowski, *Calculus with Analytic Geometry*, Fourth Edition

Swokowski, *Fundamentals of Algebra and Trigonometry*, Sixth Edition

Swokowski, *Fundamentals of College Algebra*, Sixth Edition

Swokowski, *Fundamentals of Trigonometry*, Sixth Edition

Swokowski, *Precalculus: Functions and Graphs*, Fifth Edition

Tan, *Applied Calculus*

Tan, *Applied Finite Mathematics*, Second Edition

Tan, *Calculus for the Managerial, Life, and Social Sciences*

Tan, *College Mathematics*, Second Edition

Venit and Bishop, *Elementary Linear Algebra*, Second Edition

Venit and Bishop, *Elementary Linear Algebra*, Alternate Second Edition

Willard, *Calculus and Its Applications*, Second Edition

Willerding, *A First Course in College Mathematics*, Fourth Edition

Wood and Capell, *Arithmetic*

Wood, Capell, and Hall, *Developmental Mathematics*, Third Edition

Wood, Capell, and Hall, *Intermediate Algebra*

Wood, Capell, and Hall, *Introductory Algebra*

Zill, *A First Course in Differential Equations with Applications*, Third Edition

Zill, *Calculus with Analytic Geometry*, Second Edition

Zill, *Differential Equations with Boundary-Value Problems*

Trigonometry

Jerome E. Kaufmann

Western Illinois University

PWS-KENT Publishing Company
Boston

PWS-KENT
Publishing Company

20 Park Plaza
Boston, Massachusetts 02116

PWS-KENT Publishing Company is a division of Wadsworth, Inc.

Library of Congress Cataloging-in-Publication Data

Kaufmann, Jerome E.
  Trigonometry.

  Includes index.
  1. Trigonometry. I. Title.
QA531.K3 1988   516.2′4   87–25801
ISBN 0–534–92106–X

Printed in the United States of America.
88 89 90 91 92—10 9 8 7 6 5 4 3 2 1

Production: Susan Graham
Composition: Polyglot Pte. Ltd.
Interior design: Susan Graham
Illustration: J&R Art Services
Cover design: Julie Gecha
Cover printer: New England Book Components
Text printer/binder: Arcata Graphics/Halliday

Cover art: *Cross Expressions* by George Snyder, 1983, 60″ × 60″ acrylic on canvas, used with permission of the artist.

Preface

Should the trigonometric functions be introduced as functions of angles or functions of real numbers? This is probably the number one issue relative to the teaching of trigonometry. For this type of course, my twenty-seven years of teaching experience suggest that *introducing* the trigonometric functions as functions of angles is easier and more meaningful for these students. Following such an introduction, the circular functions can be defined and used in a meaningful way.

Chapter 1 contains those review topics that are essential for the study of trigonometry. Basic graphing techniques are reviewed so that the graphing of trigonometric functions is a natural extension. Those geometric concepts needed in trigonometry are reviewed in Section 1.4. This includes work with degree measure, radian measure, Pythagorean Theorem, $30°-60°$ right triangles, and isosceles right triangles.

The focus of Chapter 2 is problem solving. The basic trigonometric functions are introduced and then immediately used to solve problems. In Section 2.6, vectors are given a geometric interpretation and used for problem solving.

Chapter 3 is primarily devoted to the graphing of trigonometric functions. Variations of all six basic trigonometric curves are covered in a carefully organized manner, consistent with the graphing techniques reviewed in Chapter 1. Take a good look at Chapter 3; I think that you will appreciate the wealth of graphing experiences available to the students.

Chapter 4 is a fairly standard treatment of trigonometric equations and identities. The proving of identities is introduced as an outgrowth of the simplification process. I have attempted to carefully guide the students through a development of the numerous sum, difference, multiple, and half-angle identities. This material is difficult; it does take time to absorb it.

Chapter 5, as indicated by the title, contains a variety of trigonometric topics. Graphing is again the key issue regarding the introduction of the polar coordinate system in Sections 5.4 and 5.5. Vectors are given an algebraic interpretation in Section 5.6.

Chapter 6 was included for sentimental reasons. Logarithms have long been associated with a trigonometry course. I felt that some instructors, if time permits, might want their students to see part of that association.

Other Special Features of the Book

1. I carefully constructed the problem sets on an even/odd basis; that is, all variations of skill-development exercises are contained in both the even- and odd-numbered problems. Thus, a meaningful assignment can be given using either the "evens" or the "odds" and a double dosage is available by assigning all of them.

2. Many of the problem sets contain a special section called **Miscellaneous Problems**. These problems encompass a variety of ideas. Some of them are proofs, some bring in supplementary topics and relationships, and some are just more difficult problems. All of them could be omitted without breaking the continuity pattern of the text; however, I feel that they do add another flexibility feature.

3. There is a **Review Problem Set** at the end of each chapter. These sets were designed to help students pull together the ideas presented in the chapter. For example, in Chapter 2, problems involving right triangles, oblique triangles for which the Law of Cosines applies, and oblique triangles for which the Law of Sines applies are presented on a section-by-section basis. Then the review problem set provides the students with the opportunity of deciding which idea applies to a particular problem.

4. I tried to make **Chapter Summaries** truly useful from a student's viewpoint. There is no standard format for these chapter summaries. At the end of each chapter I asked myself the question, "What is the most effective way of summarizing the big ideas of this chapter?" In Chapter 3, for example, I felt that it was effective to summarize the chapter in terms of the various graphing techniques that were introduced throughout the chapter.

5. In each of my texts I have tried to assign the calculator its rightful place in the study of mathematics—a *tool*, useful at times, unnecessary at other times. As I wrote this text, I found the calculator to be valuable throughout Chapters 2 and 6 but not needed very often in the other chapters. I would think that students using the text would react in a similar fashion. For those instructors who want their students to use the trigonometric tables, an appendix is included for that purpose.

Acknowledgments

I would like to take this opportunity to thank the following people who served as reviewers for this manuscript. Your comments and criticisms were extremely helpful and very much appreciated: Arthur Dull, Diablo Valley College; William Hinrichs, Rock Valley College; William Popejoy, University of Northern Colorado; John Snyder, Sinclair Community College; Ann Thorne, College of DuPage.

I am very grateful to the staff of **PWS-KENT** Publishing Company for their continuous cooperation and assistance throughout this project. I would also like to offer my special thanks to Susan Graham. She does an outstanding job of carrying out the details of preparation for publication in a dedicated and caring way.

Again I want to express my gratitude to my wife, Arlene, who continues to spend numerous hours proofreading and typing the manuscripts along with the answer books, solutions manuals, and accompanying sets of tests.

Jerome E. Kaufmann

Contents

CHAPTER 3
Graphing Trigonometric Functions

CHAPTER 4
Identities and Equations

CHAPTER 5
Topics in Trigonometry

Trigonometry

1 Some Prerequisites for Trigonometry

Courtesy: Cooper Industries, Inc. ©Tom Watson

This first chapter contains a brief review of some algebraic and geometric concepts that are used in the study of trigonometry. Most of these concepts will be familiar to you, but do take the time to be sure that you have them well under control since they are used frequently in subsequent chapters. We also call your attention to the inside front cover of this text, which contains some formulas and vocabulary from some of your previous mathematics courses.

Real Numbers and Coordinate Geometry

The following vocabulary is commonly used to classify different types of real numbers.

$1, 2, 3, 4, \ldots$	natural numbers; counting numbers; positive integers
$0, 1, 2, 3, \ldots$	whole numbers; nonnegative integers
$\ldots, -3, -2, -1$	negative integers
$\ldots, -3, -2, -1, 0$	nonpositive integers
$\ldots, -2, -1, 0, 1, 2, 3, \ldots$	integers

A **rational number** is defined as any number that can be expressed in the form a/b, where a and b are integers and b is not zero. The following are examples of rational numbers: $\frac{2}{3}$, $-\frac{3}{4}$, 6, -4, .3, $7\frac{1}{2}$, and 0.

A rational number can also be defined in terms of a decimal representation. Before doing so, let's briefly review the different possibilities for decimal representations. Decimals can be classified as **terminating, repeating**, or **nonrepeating**. Some examples of each are as follows.

$$\left. \begin{array}{l} .3 \\ .46 \\ .789 \\ .6234 \end{array} \right) \qquad \text{terminating decimals}$$

$$\left. \begin{array}{l} .6666\ldots \\ .141414\ldots \\ .694694694\ldots \\ .2317171717\ldots \\ .5417283283283\ldots \end{array} \right) \qquad \text{repeating decimals}$$

$$\left. \begin{array}{l} .276314583\ldots \\ .21411811161111\ldots \\ .673183329333\ldots \end{array} \right) \qquad \text{nonrepeating decimals}$$

A repeating decimal has a block of digits that repeats indefinitely. This repeating block of digits may be of any number of digits and may or may not begin immediately after the decimal point. A small horizontal bar is commonly used to indicate the repeat block. Thus, .6666... is written as $.\overline{6}$ and .2317171717... as $.23\overline{17}$.

In terms of decimals, a **rational number** is defined as a number that has either a terminating or repeating decimal representation. The following examples illustrate some rational numbers written in a/b form and in decimal form.

$$\frac{3}{4} = .75 \qquad \frac{3}{11} = .\overline{27} \qquad \frac{1}{8} = .125 \qquad \frac{1}{7} = .\overline{142857} \qquad \frac{1}{3} = .\overline{3}$$

An **irrational number** is defined as a number that cannot be expressed in a/b

form, where a and b are integers and b is not zero. Furthermore, an irrational number has a nonrepeating decimal representation. Some examples of irrational numbers and a partial decimal representation for each are as follows.

$$\sqrt{2} = 1.414213562373095\ldots \qquad \sqrt{3} = 1.73205080756887\ldots$$

$$\pi = 3.1415926538979\ldots$$

The irrational numbers π, $\sqrt{2}/2$, $\sqrt{3}$, $\sqrt{3}/2$, and $\sqrt{3}/3$ will be used frequently in our study of trigonometry. At times we will also use **rational approximations** for irrational numbers. For example, to the nearest tenth, $\sqrt{2}/2 = .7$.

The entire set of **real numbers** is composed of the rational numbers along with the irrationals. The following diagram can be used to summarize the various classifications of the real number system.

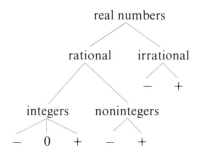

Any real number can be traced down through the diagram. For example,

7 is real, rational, an integer, and positive;

$-\frac{1}{2}$ is real, rational, a noninteger, and negative;

$\sqrt{7}$ is real, irrational, and positive;

0.59 is real, rational, a noninteger, and positive.

The Real Number Line and Absolute Value

It is often convenient to have a geometric representation of the set of real numbers in front of us, as indicated in Figure 1.1. Such a representation, called the **real number line**, indicates a one-to-one correspondence between the set of real numbers and the points on a line. That is to say, to each real number there corresponds one and only one point on the line, and to each point on the line there corresponds one and only one real number. The number that corresponds to a particular point on the line is called the **coordinate** of the point.

Figure 1.1

Many operations, relations, properties, and concepts pertaining to real numbers can be given a geometric interpretation on the number line. For example, the basic inequality relations have a geometric interpretation. The

Figure 1.2

statement $a > b$ (read as "a is greater than b") means that a is to the right of b and the statement $c < d$ (read as "c is less than d") means that c is to the left of d (Figure 1.2).

Sometimes we use an inequality statement such as $0 < x < 2\pi$ (read as "0 is less than x **and** x is less than 2π") to refer to the real numbers between 0 and 2π. Likewise, the statement $-\pi \leq x \leq \pi$ refers to all real numbers between $-\pi$ and π, inclusive.

The property $-(-x) = x$ can be "pictured" on the number line in a sequence of steps (Figure 1.3).

1. Choose a point having a coordinate of x.

2. Locate its opposite (written as $-x$) on the other side of zero.

3. Locate the opposite of $-x$ (written as $-(-x)$) on the other side of zero.

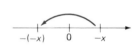

Figure 1.3

Therefore, we conclude that **the opposite of the opposite of any real number is the number itself**, and we symbolically express this by $-(-x) = x$.

> *Remark:* The symbol -1 can be read *negative one*, the *negative of one*, the *opposite of one*, or the *additive inverse of one*. The "opposite of" and "additive inverse of" terminology is especially meaningful when working with variables. For example, the symbol $-x$, read "the opposite of x" or "the additive inverse of x," emphasizes an important issue. Since x can be any real number, $-x$ (opposite of x) can be zero, positive, or negative. If x is positive, then $-x$ is negative. If x is negative, then $-x$ is positive. If x is zero, then $-x$ is zero.

The concept of **absolute value** can be interpreted on the number line. Geometrically, the absolute value of any real number is the distance between the number and zero on the number line. For example, the absolute value of 2 is 2, the absolute value of -3 is 3, and the absolute value of 0 is 0 (Figure 1.4). Symbolically, absolute value is denoted with vertical bars. Thus, we write $|2| = 2$, $|-3| = 3$, and $|0| = 0$.

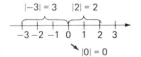

Figure 1.4

More formally, the concept of absolute value is defined as follows.

DEFINITION 1.1

For all real numbers a,

1. If $a \geq 0$, then $|a| = a$;
2. If $a < 0$, then $|a| = -a$.

According to Definition 1.1 we obtain

$$|6| = 6 \qquad \text{by applying part 1,}$$
$$|0| = 0 \qquad \text{by applying part 1,}$$
$$|-7| = -(-7) = 7 \qquad \text{by applying part 2.}$$

Notice that the absolute value of a positive number is the number itself, but the absolute value of a negative number is its opposite. Thus, the absolute value of any number, except 0, is positive and the absolute value of 0 is 0. Together, these facts also indicate that the absolute value of any real number is equal to the absolute value of its opposite. All of these ideas are summarized in the following properties.

Properties of Absolute Value

(a and b represent any real number.)

1. $|a| \geq 0$
2. $|a| = |-a|$
3. $|a - b| = |b - a|$ $a - b$ and $b - a$ are opposites of each other.

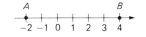

Figure 1.5

In Figure 1.5 we have indicated points A and B at -2 and 4, respectively. The distance between A and B is obviously 6 units and can be calculated by either using $|-2 - 4|$ or $|4 - (-2)|$. In general, if two points have coordinates of x_1 and x_2, the distance between the two points is determined by using either $|x_2 - x_1|$ or $|x_1 - x_2|$.

The Rectangular Coordinate System

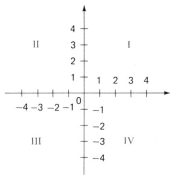

Figure 1.6

Consider two number lines, one vertical and one horizontal, perpendicular to each other at the point associated with zero on both lines (Figure 1.6). These number lines are referred to as the **horizontal and vertical axes,** or together as the **coordinate axes.** They partition the plane into four regions called **quadrants.** The quadrants are numbered counterclockwise from I through IV as indicated in Figure 1.6. The point of intersection of the two axes is called the **origin.**

It is now possible to set up a one-to-one correspondence between **ordered pairs of real numbers** and the points in a plane. To each ordered pair of real numbers there corresponds a unique point in the plane and to each point in the plane there corresponds a unique ordered pair of real numbers. We have illustrated a part of this correspondence in Figure 1.7. The ordered pair $(3, 2)$ means that point A is located three units to the right and two units up from the

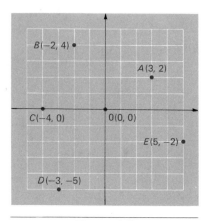

Figure 1.7

origin. (The ordered pair $(0, 0)$ is associated with the origin 0.) The ordered pair $(-3, -5)$ means that point D is located three units to the left and five units down from the origin.

In general, the real numbers a and b in an ordered pair, (a, b), are associated with a point; they are referred to as the **coordinates of the point**. The first number, a, called the **abscissa**, is the directed distance of the point from the vertical axis measured parallel to the horizontal axis. The second number, b, called the **ordinate**, is the directed distance of the point from the horizontal axis measured parallel to the vertical axis (Figure 1.8(a)). Thus, in the first quadrant all points have a positive abscissa and a positive ordinate. In the second quadrant all points have a negative abscissa and a positive ordinate. We have indicated the sign situations for all four quadrants in Figure 1.8(b). This system of associating points in a plane with pairs of real numbers is called the **rectangular coordinate system** or the **Cartesian coordinate system**.

Historically, the rectangular coordinate system provided the basis for the development of a branch of mathematics called **analytic geometry**, or what is often referred to as **coordinate geometry**. René Descartes, a French mathematician of the 17th century, was able to transform geometric problems into an algebraic setting and then to use the tools of algebra to solve the problems.

Basically, there are two kinds of problems in coordinate geometry:

1. Given an algebraic equation, find its geometric graph;

2. Given a set of conditions pertaining to a geometric figure, find its algebraic equation.

We will use the rectangular coordinate system extensively in our study of trigonometry.

For most problems in coordinate geometry it is customary to label the horizontal axis the **x-axis** and the vertical axis the **y-axis**. Then, ordered pairs

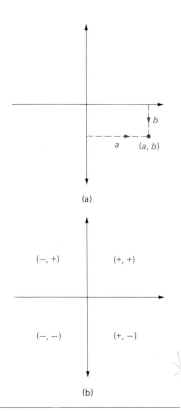

Figure 1.8

representing points in the **xy-plane** are of the form (x, y); that is, x is the first coordinate and y is the second coordinate.

Distance between Two Points

As we work with the rectangular coordinate system it is sometimes necessary to express the length of certain line segments. In other words, we need to be able to find the **distance between** two points. Using the Pythagorean Theorem, we can develop a general **distance formula** as follows.

Let $P_1(x_1, y_1)$ and $P_2(x_2, y_2)$ represent any two points in the xy-plane. Form a right triangle as indicated in Figure 1.9. The coordinates of the vertex of the right angle, point R, are (x_2, y_1). The length of $\overline{P_1 R}$ is $|x_2 - x_1|$ and the length of $\overline{RP_2}$ is $|y_2 - y_1|$. Letting d represent the length of $\overline{P_1 P_2}$ and applying the Pythagorean Theorem, we obtain

$$d^2 = |x_2 - x_1|^2 + |y_2 - y_1|^2.$$

Since $|a|^2 = a^2$, the distance formula can be stated as

$$d = \sqrt{(x_2 - x_1)^2 + (y_2 - y_1)^2}.$$

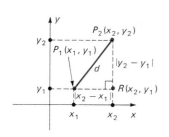

Figure 1.9

It makes no difference which point is called P_1 or P_2 when using the distance formula. Also, remember, if you forget the formula, don't panic, but merely form a right triangle as in Figure 1.9 and apply the Pythagorean Theorem. Now let's consider some examples illustrating the use of the distance formula.

EXAMPLE 1

Find the distance between $(-2, 5)$ and $(1, -1)$.

Solution Let $(-2, 5)$ be P_1 and $(1, -1)$ be P_2. Using the distance formula, we obtain

$$d = \sqrt{(x_2 - x_1)^2 + (y_2 - y_1)^2}$$
$$= \sqrt{(1 - (-2))^2 + (-1 - 5)^2}$$
$$= \sqrt{3^2 + (-6)^2} = \sqrt{9 + 36} = \sqrt{45} = 3\sqrt{5}.$$

The distance between the two points is $3\sqrt{5}$ units.

In Example 1, notice the simplicity of the approach when using the distance formula. No diagram was needed; we merely "plugged in" the values and did the computation. However, many times a figure is helpful in the analysis of the problem as we see in the next example.

EXAMPLE 2

Verify that the points $(-3, 6)$, $(3, 4)$, and $(1, -2)$ are vertices of an isosceles triangle. (An isosceles triangle has two sides of the same length.)

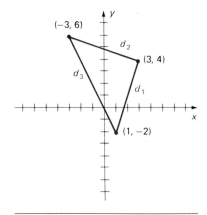

Figure 1.10

Solution Let's plot the points and draw the triangle (Figure 1.10). Using the distance formula, the lengths d_1, d_2, and d_3 can be found as follows.

$$d_1 = \sqrt{(3-1)^2 + (4-(-2))^2}$$
$$= \sqrt{4 + 36} = \sqrt{40} = 2\sqrt{10}$$
$$d_2 = \sqrt{(-3-3)^2 + (6-4)^2}$$
$$= \sqrt{36 + 4} = \sqrt{40} = 2\sqrt{10}$$
$$d_3 = \sqrt{(-3-1)^2 + (6-(-2))^2}$$
$$= \sqrt{16 + 64} = \sqrt{80} = 4\sqrt{5}$$

Since $d_1 = d_2$, it is an isosceles triangle.

Problem Set 1.1

For Problems 1–10, identify each statement as true or false.

1. Every rational number is a real number.
2. Every irrational number is a real number.
3. Every real number is a rational number.
4. If a number is real, then it is irrational.
5. Some irrational numbers are also rational numbers.
6. All integers are rational numbers.
7. The number zero is a rational number.
8. Zero is a positive integer.
9. Zero is a negative integer.
10. All whole numbers are integers.

For Problems 11–18, from the list 0, $\sqrt{5}$, $-\sqrt{2}$, $\frac{7}{8}$, $-\frac{10}{13}$, $7\frac{1}{8}$, 0.279, $0.4\overline{67}$, $-\pi$, -14, 46, and 6.75, identify each of the following.

11. The natural numbers
12. The whole numbers
13. The integers
14. The rational numbers
15. The irrational numbers
16. The nonnegative integers
17. The nonpositive integers
18. The real numbers

For Problems 19–24, find the distance on the real number line between two points whose coordinates are as indicated.

19. 17 and 35
20. -14 and 12
21. 18 and -21
22. -17 and -42
23. -56 and -27
24. 0 and -34

For Problems 25–28, evaluate each expression if x is a nonzero real number.

25. $\dfrac{|x|}{x}$

26. $\dfrac{x}{|x|}$

27. $\dfrac{|-x|}{-x}$

28. $|x| - |-x|$

Problems 29–34 pertain to a rectangular coordinate system.

29. In which quadrants do the coordinates of a point have the same sign?

30. In which quadrants do the coordinates of a point have opposite signs?

31. In which quadrants is the abscissa negative?

32. In which quadrants is the ordinate negative?

33. What is the value of the abscissa of any point on the vertical axis?

34. What is the value of the ordinate of any point on the horizontal axis?

For Problems 35–44, find the distance between each of the pairs of points. Express answers in simplest radical form.

35. $(2, 1)$ and $(10, 7)$ 36. $(-2, -1)$ and $(7, 11)$

37. $(-1, 3)$ and $(2, -2)$ 38. $(1, -1)$ and $(3, -4)$

39. $(-5, 2)$ and $(-1, 6)$ 40. $(6, -4)$ and $(9, -7)$

41. $(-2, -4)$ and $(4, 0)$ 42. $(-3, 3)$ and $(0, -3)$

43. $(-2, 3)$ and $(-2, -7)$ 44. $(1, -6)$ and $(-5, -6)$

45. Verify that the points $(0, 3)$, $(2, -3)$, and $(-4, -5)$ are vertices of an isosceles triangle.

46. Verify that the points $(-3, 1)$, $(5, 7)$, and $(8, 3)$ are vertices of a right triangle.

47. Verify that $(3, 1)$ is the midpoint of the line segment joining $(-2, 6)$ and $(8, -4)$.

48. Verify that the points $(7, 12)$ and $(11, 18)$ divide the line segment joining $(3, 6)$ and $(15, 24)$ into three segments of equal length.

For Problems 49–52, use your calculator to obtain a rational approximation, to the nearest hundredth, for each of the following.

49. $\dfrac{\sqrt{2}}{2}$ 50. $\sqrt{3}$ 51. $\dfrac{3}{2\sqrt{3}}$ 52. $\dfrac{1}{\sqrt{3}}$

1.2

Functions: Composite Functions

A function is a special kind of **relation**, so we will begin our discussion with a simple definition of a relation.

DEFINITION 1.2

> A **relation** is a set of ordered pairs.

Thus, a set of ordered pairs such as $\{(1, 2), (3, 7), (8, 14)\}$ is a relation. The set of all first components of the ordered pairs is the **domain** of the relation and the set of all second components is the **range** of the relation. The relation $\{(1, 2), (3, 7), (8, 14)\}$ has a domain of $\{1, 3, 8\}$ and a range of $\{2, 7, 14\}$.

The ordered pairs referred to in Definition 1.2 may be generated by various means, such as a graph or a chart. However, one of the most common ways of generating ordered pairs is by the use of equations. Since the solution set of an equation in two variables is a set of ordered pairs, such an equation describes a relation. Each of the following equations describes a relation between the variables x and y. We have listed *some* of the infinitely many ordered pairs (x, y) of each relation.

$$x^2 + y^2 = 4: (1, \sqrt{3}), (1, -\sqrt{3}), (0, 2), (0, -2) \tag{1}$$

$$y^2 = x^3: (0, 0), (1, 1)(1, -1), (4, 8), (4, -8) \tag{2}$$

$$y = x + 2: (0, 2), (1, 3), (2, 4), (-1, 1), (5, 7) \tag{3}$$

$$y = \frac{1}{x - 1}: (0, -1), (2, 1)\left(3, \frac{1}{2}\right), \left(-1, -\frac{1}{2}\right), \left(-2, -\frac{1}{3}\right) \tag{4}$$

$$y = x^2: (0, 0), (1, 1), (2, 4), (-1, 1), (-2, 4) \tag{5}$$

Now we direct your attention to the ordered pairs associated with equations (3), (4), and (5). Note that in each case no two ordered pairs have the same first component. Such a set of ordered pairs is called a **function**.

DEFINITION 1.3

A **function** is a relation in which no two ordered pairs have the same first component.

Stated another way, Definition 1.3 means that a function is a relation where each member of the domain is assigned **one and only one** member of the range. Thus, it is easy to determine that each of the following sets of ordered pairs is a function.

$$f = \{(x, y) \mid y = x + 2\}$$
$$g = \left\{(x, y) \mid y = \frac{1}{x - 1}\right\}$$
$$h = \{(x, y) \mid y = x^2\}$$

In each case there is one and only one value of y (an element of the range) associated with each value of x (an element of the domain).

Notice that we named the previous functions f, g, and h. It is common to name functions by means of a single letter and the letters f, g, and h are often used. We would suggest more meaningful labels when functions are used to portray real-world situations. For example, if a problem involves a profit function, then naming the function p or even P would seem natural.

The symbol for a function can be used along with a variable that represents an element in the domain to indicate the associated element in the range. For example, suppose that we have a function f specified in terms of the variable x. The symbol, $f(x)$, (read "f of x" or "the value of f at x") represents the element in the range associated with the element x from the domain. The function $f = \{(x, y) \mid y = x + 2\}$ can be written as $f = \{(x, f(x)) \mid f(x) = x + 2\}$ and this is usually shortened to read "f is the function determined by the equation $f(x) = x + 2$."

Remark: Be careful with the symbolism $f(x)$. As we stated above, it means the value of the function f at x. It does not mean f times x.

This **function notation** is very convenient when computing and expressing various values of the function. For example, the value of the function $f(x) = 3x - 5$ at $x = 1$ is

$$f(1) = 3(1) - 5 = -2.$$

Likewise, the functional values for $x = 2$, $x = -1$, and $x = 5$ are

$$f(2) = 3(2) - 5 = 1;$$
$$f(-1) = 3(-1) - 5 = -8; \text{ and}$$
$$f(5) = 3(5) - 5 = 10.$$

Thus, this function f contains the ordered pairs $(1, -2)$, $(2, 1)$, $(-1, -8)$, $(5, 10)$, and in general all ordered pairs of the form $(x, f(x))$, where $f(x) = 3x - 5$ and x is any real number.

It may be helpful for you to mentally picture the concept of a function in terms of a "function machine" as illustrated in Figure 1.11. Each time that a value of x is put into the machine, the equation $f(x) = x + 2$ is used to generate one and only one value for $f(x)$ to be ejected from the machine. For example, if 3 is put into this machine, then $f(3) = 3 + 2 = 5$, and 5 is ejected. Thus, the ordered pair $(3, 5)$ is one element of the function. Now let's look at some examples to help pull together some of the ideas about functions.

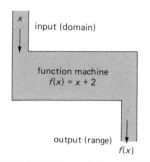

x
input (domain)

function machine
$f(x) = x + 2$

output (range)
$f(x)$

Figure 1.11

EXAMPLE 1

Determine whether the relation $\{(x, y) \mid y^2 = x\}$ is a function and specify its domain and range.

Solution Because $y^2 = x$ is equivalent to $y = \pm\sqrt{x}$, to each value of x there are assigned *two* values for y. Therefore, this relation is not a function. The expression $\sqrt{x}$ requires that x be nonnegative; therefore, the domain (D) is $D = \{x \mid x \geq 0\}$. To each nonnegative real number, the relation assigns two real numbers, $\sqrt{x}$ and $-\sqrt{x}$. Thus, the range (R) is

$$R = \{y \mid y \text{ is a real number}\}.$$

EXAMPLE 2

For the function $f(x) = x^2$,

(a) specify its domain,   (b) determine its range, and

(c) evaluate $f(-2)$, $f(0)$, and $f(4)$.

Solution

(a) Any real number can be squared; therefore, the domain (D) is

$$D = \{x \mid x \text{ is a real number}\}.$$

(b) Squaring a real number always produces a nonnegative result. Thus, the range (R) is

$$R = \{f(x) \mid f(x) \geq 0\}.$$

(c) $f(-2) = (-2)^2 = 4$
$f(0) = (0)^2 = 0$
$f(4) = (4)^2 = 16$

For our purposes in this text, if the domain of a function is not specifically indicated or determined by a real-world application, then we assume the domain to be all *real number* replacements for the variable, which produce *real number* functional values. Consider the following examples.

EXAMPLE 3

Specify the domain for each of the following.

(a) $f(x) = \dfrac{1}{x - 1}$ (b) $f(t) = \dfrac{1}{t^2 - 4}$ (c) $f(s) = \sqrt{s - 3}$

Solution

(a) We can replace x with any real number except 1, because 1 makes the denominator zero. Thus, the domain is given by

$$D = \{x \mid x \neq 1\}.$$

(b) We need to eliminate any value of t that will make the denominator zero. Thus, let's solve the equation $t^2 - 4 = 0$.

$$t^2 - 4 = 0$$
$$t^2 = 4$$
$$t = \pm 2$$

The domain is the set

$$D = \{t \mid t \neq -2 \quad \text{and} \quad t \neq 2\}.$$

(c) The radicand, $s - 3$, must be nonnegative.

$$s - 3 \geq 0$$
$$s \geq 3$$

The domain is the set

$$D = \{s \mid s \geq 3\}.$$

EXAMPLE 4

If $f(x) = -2x + 7$ and $g(x) = x^2 - 5x + 6$, find $f(3)$, $f(-4)$, $g(2)$, and $g(-1)$.

Solution

$$f(x) = -2x + 7 \qquad\qquad g(x) = x^2 - 5x + 6$$
$$f(3) = -2(3) + 7 = 1 \qquad\qquad g(2) = 2^2 - 5(2) + 6 = 0$$
$$f(-4) = -2(-4) + 7 = 15 \qquad\qquad g(-1) = (-1)^2 - 5(-1) + 6 = 12$$

(handwritten in left margin) difference quotient of a function =

$$\frac{f(a+h) - f(a)}{h}$$

In Example 4, notice that we are working with two different functions in the same problem. Thus, different names, f and g, are used.

EXAMPLE 5

If $f(x) = x^2 + 2x - 3$, find

$$\frac{f(a+h) - f(a)}{h}.$$

Solution

$$f(a+h) = (a+h)^2 + 2(a+h) - 3$$
$$= a^2 + 2ah + h^2 + 2a + 2h - 3;$$

and

$$f(a) = a^2 + 2a - 3.$$

Therefore,

$$f(a+h) - f(a) = (a^2 + 2ah + h^2 + 2a + 2h - 3) - (a^2 + 2a - 3)$$
$$= a^2 + 2ah + h^2 + 2a + 2h - 3 - a^2 - 2a + 3$$
$$= 2ah + h^2 + 2h$$

and

$$\frac{f(a+h) - f(a)}{h} = \frac{2ah + h^2 + 2h}{h}$$
$$= \frac{h(2a + h + 2)}{h}$$
$$= 2a + h + 2.$$

Functions and functional notation provide the basis for describing many real-world relationships. The next example illustrates this point.

EXAMPLE 6

Suppose a factory determines that the overhead for producing a quantity of a certain item is $500 and the cost for each item is $25. Express the total expenses as a function of the number of items produced and compute the expenses for producing 12, 25, 50, 75, and 100 items.

Solution Let n represent the number of items produced. Then $25n + 500$ represents the total expenses. Let's use E to represent the **expense function**, so that we have $E(n) = 25n + 500$, where n is a whole number, from which we obtain

$$E(12) = 25(12) + 500 = 800;$$
$$E(25) = 25(25) + 500 = 1125;$$
$$E(50) = 25(50) + 500 = 1750;$$
$$E(75) = 25(75) + 500 = 2375;$$
$$E(100) = 25(100) + 500 = 3000.$$

So the total expenses for producing 12, 25, 50, 75, and 100 items are $800, $1125, $1750, $2375, and $3000, respectively.

The Composition of Functions

The basic operations of addition, subtraction, multiplication, and division can be defined for functions. However, for our purposes in this text, there is an additional operation, called composition, that will be used a bit later.

DEFINITION 1.4

The **composition** of functions f and g is defined by

$$(f \circ g)(x) = f(g(x)),$$

for all x in the domain of g such that $g(x)$ is in the domain of f.

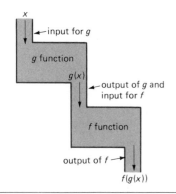

Figure 1.12

The left side, $(f \circ g)(x)$, of the equation in Definition 1.4 can be read as "the composition of f and g" and the right side, $f(g(x))$, can be read as "f of g of x." It may also be helpful for you to mentally picture Definition 1.4 as two function machines *hooked together* to produce another function (often called a **composite function**) as illustrated in Figure 1.12. Notice that what comes out of the function g is substituted into the function f. Thus, composition is sometimes called the substitution of functions.

Figure 1.12 also vividly illustrates the fact that $f \circ g$ is defined *for all x in the domain of g such that $g(x)$ is in the domain of f.* In other words, what comes out of g must be capable of being fed into f. Let's consider some examples.

EXAMPLE 7

If $f(x) = x^2$ and $g(x) = x - 3$, find $(f \circ g)(x)$ and determine its domain.

Solution Applying Definition 1.4 we obtain

$$(f \circ g)(x) = f(g(x))$$
$$= f(x - 3)$$
$$= (x - 3)^2.$$

Because g and f are both defined for all real numbers, so is $f \circ g$.

EXAMPLE 8

If $f(x) = \sqrt{x}$ and $g(x) = x - 4$, find $(f \circ g)(x)$ and determine its domain.

Solution Applying Definition 1.4 we obtain

$$(f \circ g)(x) = f(g(x))$$
$$= f(x - 4)$$
$$= \sqrt{x - 4}.$$

The domain of g is all real numbers but the domain of f is only the nonnegative real numbers. Thus $g(x)$, which is $x - 4$, has to be nonnegative. So

$$x - 4 \geq 0$$

$$x \geq 4,$$

and the domain of $f \circ g$ is $D = \{x \mid x \geq 4\}$.

Definition 1.4 with f and g interchanged defines the composition of g and f as $(g \circ f)(x) = g(f(x))$.

EXAMPLE 9

If $f(x) = x^2$ and $g(x) = x - 3$, find $(g \circ f)(x)$ and determine its domain.

Solution

$$(g \circ f)(x) = g(f(x))$$
$$= g(x^2)$$
$$= x^2 - 3.$$

Since f and g are both defined for all real numbers, the domain of $g \circ f$ is the set of all real numbers.

The results of Examples 7 and 9 demonstrate an important idea, namely, that the **composition of functions is not a commutative operation**. In other words, it is not true that $f \circ g = g \circ f$ for all functions f and g. However, as we will see later, there is a special class of functions where $f \circ g = g \circ f$.

EXAMPLE 10

If $f(x) = 2x + 3$ and $g(x) = \sqrt{x - 1}$, determine each of the following.

(a) $(f \circ g)(x)$ (b) $(g \circ f)(x)$ (c) $(f \circ g)(5)$

(d) $(g \circ f)(7)$

Solution

(a) $(f \circ g)(x) = f(g(x))$
$$= f(\sqrt{x - 1})$$
$$= 2\sqrt{x - 1} + 3$$

(b) $(g \circ f)(x) = g(f(x))$
$$= g(2x + 3)$$
$$= \sqrt{2x + 3 - 1}$$
$$= \sqrt{2x + 2}$$

(c) By using the composite function formed in part (a) we obtain
$$(f \circ g)(5) = 2\sqrt{5 - 1} + 3$$
$$= 2\sqrt{4} + 3$$
$$= 2(2) + 3 = 7.$$

(**d**) By using the composite function formed in part (b) we obtain

$$(g \circ f)(7) = \sqrt{2(7) + 2} = \sqrt{16} = 4.$$

Problem Set 1.2

For Problems 1–8, specify the domain and the range for each relation. Also indicate whether or not each relation is a function.

1. $\{(1, 4), (2, 6), (3, 12), (4, 17)\}$ 2. $\{(0, 0), (1, 1), (2, 8), (3, 27)\}$

3. $\{(0, 0), (1, 1), (1, -1), (-1, 1)\}$ 4. $\{(0, 2), (1, 2), (2, 2), (3, 2), (4, 2)\}$

5. $\{(1, 3), (1, 4), (1, -1), (1, -2)\}$

6. $\{(4, 2), (4, -2), (9, 3), (9, -3), (16, 4), (16, -4)\}$

7. $\{(x, y) \,|\, x + 2y = 4\}$

8. $\{(x, y) \,|\, y = x^3\}$

For Problems 9–22, specify the domain of each function.

9. $f(x) = x^2 + 2x - 1$ 10. $f(x) = 3x - 7$

11. $f(x) = \dfrac{1}{x}$ 12. $f(x) = \dfrac{3}{x + 6}$

13. $f(x) = \dfrac{3}{(x + 2)(x - 3)}$ 14. $f(x) = \dfrac{-4}{(x - 1)(x + 5)}$

15. $f(x) = \dfrac{1}{x^2 + 5x + 6}$ 16. $f(x) = \dfrac{2}{x^2 - 2x - 15}$

17. $f(x) = \dfrac{-3}{x^2 + 2x}$ 18. $f(x) = \dfrac{-5}{x^2 - 6x}$

19. $f(x) = \dfrac{-3}{x^2 - 9}$ 20. $f(x) = \dfrac{-4}{x^2 - 25}$

21. $f(x) = \sqrt{x + 1}$ 22. $f(x) = \sqrt{2x - 3}$

For Problems 23–30, find the indicated functional values.

23. If $f(x) = 3x + 6$, find $f(0)$, $f(2)$, $f(-3)$, and $f(a)$.

24. If $f(x) = -5x + 8$, find $f(-1)$, $f(-2)$, $f(3)$, and $f(5)$.

25. If $h(x) = -x^2 - 3$, find $h(1)$, $h(-1)$, $h(-3)$, and $h(5)$.

26. If $h(x) = -2x^2 - x + 4$, find $h(-2)$, $h(-3)$, $h(4)$, and $h(5)$.

27. If $f(x) = \sqrt{x - 1}$, find $f(1)$, $f(5)$, $f(13)$, and $f(26)$.

28. If $f(x) = \sqrt{2x + 1}$, find $f(3)$, $f(4)$, $f(10)$, and $f(12)$.

29. If $f(x) = 2x^2 - 7$ and $g(x) = x^2 + x - 1$, find $f(-2)$, $f(3)$, $g(-4)$, and $g(5)$.

30. If $f(x) = 5x^2 - 2x + 3$ and $g(x) = -x^2 + 4x - 5$, find $f(-2)$, $f(3)$, $g(-4)$, and $g(6)$.

For Problems 31–34, find $[f(a + h) - f(a)]/h$.

31. $f(x) = 3x + 2$ 32. $f(x) = -4x + 3$

33. $f(x) = x^2 - 6x + 4$ 34. $f(x) = 2x^2 + x - 1$

For Problems 35–48, determine $(f \circ g)(x)$ and $(g \circ f)(x)$ for each pair of functions. Also specify the domain for $f \circ g$ and $g \circ f$.

35. $f(x) = 2x$, $g(x) = 3x - 1$ 36. $f(x) = 4x + 1$, $g(x) = 3x$

37. $f(x) = 5x - 3,\quad g(x) = 2x + 1$

38. $f(x) = 3 - 2x,\quad g(x) = -4x$

39. $f(x) = 3x + 4,\quad g(x) = x^2 + 1$

40. $f(x) = 3,\quad g(x) = -3x^2 - 1$

41. $f(x) = 3x - 4,\quad g(x) = x^2 + 3x - 4$

42. $f(x) = 2x^2 - x - 1,\quad g(x) = x + 4$

43. $f(x) = \dfrac{1}{x},\quad g(x) = 2x + 7$

44. $f(x) = \dfrac{1}{x^2},\quad g(x) = x$

45. $f(x) = \sqrt{x - 2},\quad g(x) = 3x - 1$

46. $f(x) = \dfrac{1}{x},\quad g(x) = \dfrac{1}{x^2}$

47. $f(x) = \dfrac{1}{x - 1},\quad g(x) = \dfrac{2}{x}$

48. $f(x) = \dfrac{4}{x + 2},\quad g(x) = \dfrac{3}{2x}$

Your calculator may be of some help for Problems 49–54.

49. Suppose that the cost function for producing a certain item is given by $C(n) = 3n + 5$, where n represents the number of items produced. Compute $C(150)$, $C(500)$, $C(750)$, and $C(1500)$.

50. The height of a projectile fired vertically into the air (neglecting air resistance) at an initial velocity of 64 feet per second is a function of the time (t) and is given by the equation

$$h(t) = 64t - 16t^2.$$

Compute $h(1)$, $h(2)$, $h(3)$, and $h(4)$.

51. The profit function for selling n items is given by $P(n) = -n^2 + 500n - 61,500$. Compute $P(200)$, $P(230)$, $P(250)$, and $P(260)$.

52. A car rental agency charges $50 per day plus $0.32 a mile. Therefore, the daily charge for renting a car is a function of the number of miles traveled (m) and can be expressed as $C(m) = 50 + 0.32m$. Compute $C(75)$, $C(150)$, $C(225)$, and $C(650)$.

53. The equation $A(r) = \pi r^2$ expresses the area of a circular region as a function of the length of a radius (r). Use 3.14 as an approximation for π and compute $A(2)$, $A(3)$, $A(12)$, and $A(17)$.

54. The equation $I(r) = 500r$ expresses the amount of simple interest earned by an investment of $500 for one year as a function of the rate of interest (r). Compute $I(0.11)$, $I(0.12)$, $I(0.135)$, and $I(0.15)$.

1.3

Graphing Functions

As you continue to study mathematics, you will find that developing the ability to quickly sketch the graph of an equation is an important skill. No doubt you have been introduced to some graphing techniques in previous algebra courses. At this time we simply want to review some of those techniques, that apply to graphing functions, so that we can use them when graphing trigonometric functions.

The graph of any equation that determines a function is called **the graph of the function**. Therefore, if f is the function determined by $f(x) = 3x - 4$, then the graph of f is the set of all ordered pairs $(x, f(x))$ that satisfy $f(x) = 3x - 4$. Furthermore, for graphing purposes we will simply let $y = f(x)$ and then we can graph the ordered pairs (x, y) that satisfy $y = 3x - 4$. We will also use such phrases as "graph the function $y = 3x - 4$" instead of the more technically correct statement "graph the function determined by the equation $y = 3x - 4$."

Before tackling some specific graphing problems, let's review one graphing idea that is quite helpful and frequently easy to apply, that is, the idea of finding the **intercepts** of a graph. In general, the intercepts of a graph can be defined as follows.

> The x-coordinates of the points that a graph has in common with the x-axis are called the **x-intercepts** of the graph. (To compute the x-intercepts, let $y = 0$ and solve for x.)

> The y-coordinates of the points that a graph has in common with the y-axis are called the **y-intercepts** of the graph. (To compute the y-intercepts, let $x = 0$ and solve for y.)

Linear Functions

Any function that can be written in the form

$$f(x) = ax + b,$$

where a and b are real numbers, is called a linear function. The following are examples of linear functions.

$$f(x) = -2x - 4 \qquad f(x) = 5x + 8 \qquad f(x) = \tfrac{1}{2}x - \tfrac{2}{3}$$

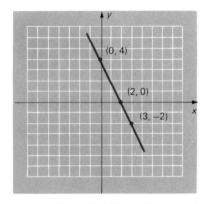

Figure 1.13

Graphing linear functions is easy because the graph of every linear function is a straight line. Therefore, all we need to do is to determine two points of the graph and draw the line determined by those two points. The two points involving the intercepts are usually easy to find, and it's generally a good idea to plot a third point to serve as a check.

EXAMPLE 1

Graph the linear function $y = -2x + 4$.

Solution

If $x = 0$, then $y = -2(0) + 4 = 4$; so the point $(0, 4)$ is on the graph.

If $y = 0$, then $0 = -2x + 4$ or $x = 2$; so the point $(2, 0)$ is on the graph.

If $x = 3$, then $y = -2(3) + 4 = -2$; so the point $(3, -2)$ is on the graph.

Plotting the points $(0, 4)$, $(2, 0)$, and $(3, -2)$ and connecting them with a straight line produces Figure 1.13.

x	y
0	0
1	2
−1	−2

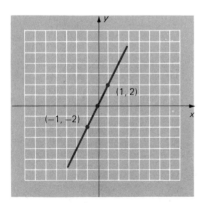

Figure 1.14

EXAMPLE 2

Graph the linear function $y = 2x$.

Solution   If $x = 0$, then $y = 2(0) = 0$; so the origin $(0, 0)$ is on the line. Since both intercepts are determined by $(0, 0)$, another point is necessary to determine the line. Then a third point should be found as a check point. The graph of $y = 2x$ is shown in Figure 1.14.

Quadratic Functions

Any function that can be written in the form

$$f(x) = ax^2 + bx + c,$$

where a, b, and c are real numbers and $a \neq 0$, is called a **quadratic function**. The graph of any quadratic function is a **parabola** and the vocabulary commonly associated with parabolas is indicated in Figure 1.15.

Graphing parabolas relies on being able to find the vertex, determining whether the parabola opens upward or downward, and locating two points on opposite sides of the axis of symmetry. It is also helpful to compare the parabolas produced by various types of equations such as $y = x^2 + k$, $y = ax^2$, $y = (x - h)^2$, and $y = a(x - h)^2 + k$ to the **basic parabola** produced by the equation $y = x^2$. The graph of $y = x^2$ is shown in Figure 1.16. Notice that the y-axis is the axis of symmetry for the graph of $y = x^2$. The graph is said to be **symmetrical with respect to the y-axis**. Now let's consider an equation of the form $y = x^2 + k$, where k is a constant.

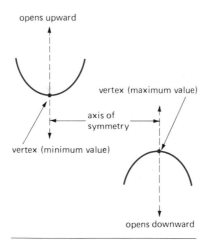

Figure 1.15

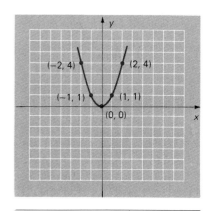

Figure 1.16

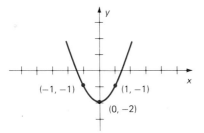

Figure 1.17

EXAMPLE 3

Graph the quadratic function $y = x^2 - 2$.

Solution It should be evident that y-values for $y = x^2 - 2$ are two less than corresponding y-values for $y = x^2$. For example, when $x = 1$ the equation $y = x^2$ produces $y = 1$, but the equation $y = x^2 - 2$ produces $y = -1$. Thus, the graph of $y = x^2 - 2$ is the same as the graph of $y = x^2$ except *moved down 2 units* as shown in Figure 1.17.

x	$y = x^2$	$y = 2x^2$
0	0	0
1	1	2
2	4	8
-1	1	2
-2	4	8

In general, the graph of a quadratic function of the form $y = x^2 + k$ is the same as the graph of $y = x^2$ except moved up or down k units depending on whether k is positive or negative.

Now let's consider some quadratic functions of the form $y = ax^2$, where a is a nonzero constant.

EXAMPLE 4

Graph $y = 2x^2$.

Solution The table of values to the left allows us to make some comparisons of y-values. Notice that the y-values for $y = 2x^2$ are *twice* the corresponding y-values for $y = x^2$. Thus, the parabola associated with $y = 2x^2$ has the same vertex (the origin) as the graph of $y = x^2$, but it is *narrower*, as shown in Figure 1.18.

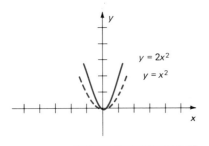

Figure 1.18

In Figure 1.19 the graphs of $y = \frac{1}{2}x^2$ and $y = -x^2$ are compared to the graph of $y = x^2$. Notice that $y = \frac{1}{2}x^2$ produces a parabola *wider* than the graph of $y = x^2$, and the graph of $y = -x^2$ is the x-axis reflection of the parabola $y = x^2$.

In general, the graph of a quadratic function of the form $y = ax^2$ has its vertex at the origin, and opens upward if a is positive and downward if a is negative. The parabola is narrower than the basic parabola if $|a| > 1$ and wider if $|a| < 1$.

Let's continue our investigation of quadratic functions by considering those of the form $y = (x - h)^2$, where h is a nonzero constant.

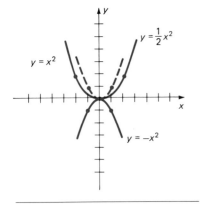

Figure 1.19

EXAMPLE 5

Graph $y = (x - 3)^2$.

Solution A fairly extensive table of values illustrates a pattern. Notice that $y = (x - 3)^2$ and $y = x^2$ take on the same y-values, but for different values of x. More specifically, if $y = x^2$ achieves a certain y-value at x equaling some constant, then $y = (x - 3)^2$ achieves that same y-value at x equaling the constant *plus three*. In other words, the graph of $y = (x - 3)^2$ is the graph of $y = x^2$ *moved three units to the right*, as shown in Figure 1.20.

x	$y = x^2$	$y = (x - 3)^2$
-1	1	16
0	0	9
1	1	4
2	4	1
3	9	0
4	16	1
5	25	4
6	36	9
7	49	16

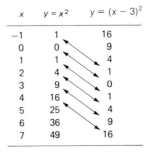

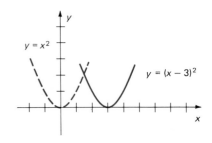

Figure 1.20

In general, the graph of a quadratic function of the form $y = (x - h)^2$ is the same as the graph of $y = x^2$ except moved to the right h units if h is positive; or moved to the left h units if h is negative.

The following diagram summarizes our work thus far for graphing quadratic functions.

$$y = x^2 + \textcircled{k} \longrightarrow \text{moves the parabola up or down}$$

$$y = x^2 \longleftrightarrow y = \textcircled{a}x^2 \longrightarrow \text{affects the "width" and which way the parabola opens}$$

basic parabola

$$y = (x - \textcircled{h})^2 \longrightarrow \text{moves the parabola right or left}$$

Now let's consider two examples that combine the previous ideas.

EXAMPLE 6

Graph $y = 3(x - 2)^2 + 1$.

Solution

$$y = 3(x - 2)^2 + 1$$

narrows the parabola and opens it upward | moves the parabola 2 units to the right | moves the parabola 1 unit up

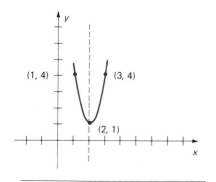

Figure 1.21

The vertex is at $(2, 1)$ and the line $x = 2$ is the axis of symmetry. If $x = 1$, then $y = 3(1 - 2)^2 + 1 = 4$. Thus, the point $(1, 4)$ is on the graph and so is its reflection across the line of symmetry, namely, the point $(3, 4)$. The parabola can now be drawn as in Figure 1.21.

EXAMPLE 7

Graph $y = -\frac{1}{2}(x + 1)^2 - 3$.

Solution

$$y = -\frac{1}{2}(x - (-1))^2 - 3$$

widens the parabola and opens it downward | moves the parabola 1 unit to the left | moves the parabola 3 units down

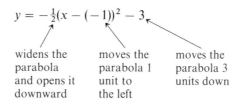

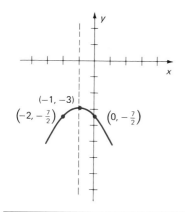

Figure 1.22

The vertex is at $(-1, -3)$ and the line $x = -1$ is the axis of symmetry. If $x = 0$, then $y = -\frac{1}{2}(0 + 1)^2 - 3 = -\frac{7}{2}$. So the point $(0, -\frac{7}{2})$ is on the graph and so is its reflection across the line of symmetry, namely, the point $(-2, -\frac{7}{2})$. The parabola is drawn in Figure 1.22.

The previous discussion regarding quadratic functions could be repeated over and over again relative to other basic functions. In fact, the vertical shifting of a parabola produced by equations of the form $y = x^2 + k$ and the horizontal shifting produced by equations of the form $y = (x - h)^2$ are common to all functions. For example, if we know what the graph of $y = x^3$ looks like, then the graph of $y = x^3 + 3$ is the same curve moved up three units and the graph of $y = (x + 2)^3$ is the same curve moved two units to the left. We will have you pursue this line of thinking a little further in the next set of problems.

Problem Set 1.3

For Problems 1–12, graph each of the linear functions.

1. $y = 2x - 4$ 	 2. $y = x + 3$ 	 3. $y = -x + 3$ 	 4. $y = -x - 1$
5. $y = -3x - 1$ 	 6. $y = 3x + 6$ 	 7. $y = -4x$ 	 8. $y = 3x$
9. $y = \frac{1}{2}x - 1$ 	 10. $y = \frac{2}{3}x + 2$ 	 11. $y = 2$ 	 12. $y = -3$

For Problems 13–30, graph each of the quadratic functions.

13. $y = x^2 + 1$ 	 14. $y = x^2 - 1$ square 	 15. $y = 3x^2$
16. $y = -2x^2$ 	 17. $y = -x^2 + 2$ 1st then 	 18. $y = -3x^2 - 1$
19. $y = -\frac{1}{2}x^2 - 3$ 	 20. $y = \frac{1}{2}x^2 + 2$ negate 	 21. $y = (x + 2)^2$
22. $y = 2(x - 1)^2$ 	 23. $y = -2(x - 3)^2$ 	 24. $y = -(x + 1)^2$
25. $y = (x - 2)^2 - 3$ 	 26. $y = (x + 2)^2 + 1$ 	 27. $y = 2(x + 1)^2 - 4$
28. $y = 2(x - 1)^2 + 3$ 	 29. $y = -3(x - 2)^2 - 1$ 	 30. $y = -3(x + 2)^2 - 3$

31. (a) Graph $y = \sqrt{x}$ by plotting enough points to determine the curve.
   (b) Sketch a graph of each of the following as variations of the basic square root curve from part (a).
      (i) $y = \sqrt{x} + 1$ 	 (ii) $y = 2\sqrt{x}$
      (iii) $y = -\sqrt{x}$ 	 (iv) $y = \sqrt{x - 2}$
      (v) $y = \sqrt{x + 1} - 2$ 	 (vi) $y = -\sqrt{x - 1} - 1$

32. (a) Graph $y = x^3$ by plotting enough points to determine the curve.
   (b) Sketch a graph of each of the following as variations of the basic curve from part (a).
      (i) $y = x^3 - 2$ 	 (ii) $y = \frac{1}{2}x^3$
      (iii) $y = -x^3$ 	 (iv) $y = (x - 3)^3$
      (v) $y = (x + 1)^3 + 2$ 	 (vi) $y = -(x - 2)^3 - 1$

1.4

Some Geometric Prerequisites for Trigonometry

From a geometric viewpoint, a **plane angle** is usually defined as the set of points consisting of two rays having a common endpoint. The common endpoint is called the **vertex** of the angle, and the rays are called the **sides** of the angle. In

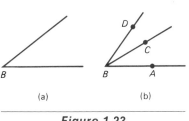

(a)   (b)

Figure 1.23

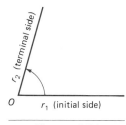

Figure 1.24

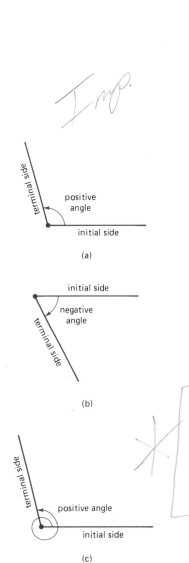

(a)

(b)

(c)

Figure 1.25

Figure 1.23(a), the angle can be named by its vertex only; thus, we can refer to $\angle B$. Figure 1.23(b) illustrates three angles having a common vertex. In such cases, points on the sides of the angles can be used to help name the angles. We can refer to $\angle DBC$, $\angle CBA$, and $\angle DBA$ in Figure 1.23(b).

In trigonometry it is more convenient to think of an angle as rotating a ray about its endpoint. More specifically, in Figure 1.24 let's begin with a ray r_1 and rotate it, in a plane, about its endpoint O to a position indicated by the ray r_2. We call r_1 the **initial side**, r_2 the **terminal side**, and O the **vertex** of the angle. If the rotation is in a counterclockwise direction, as indicated by an arrow, then the angle is a **positive angle**. If the rotation is in a clockwise direction, then a **negative angle** is formed. There is no restriction as to the amount of rotation (Figure 1.25). As indicated in parts (a) and (c) of Figure 1.25, different angles can have the same initial and terminal sides (the amount of rotation is different). Any two such angles are called **coterminal**.

The size of an angle (amount of rotation from initial side to terminal side) can be described using **degree measure**. The angle formed by rotating a complete counterclockwise revolution has a measure of 360 degrees, written $360°$. Thus, one degree ($1°$) is $\frac{1}{360}$ of a complete revolution. The diagrams at the base of the page illustrate angles of different degree measure and some commonly used terminology. Two acute angles are **complementary** if their sum is $90°$; one of the angles is referred to as the complement of the other. Two positive angles are **supplementary** if their sum is $180°$.

The degree system for angle measurement is quite similar to the hour-minute-second relationship of our time system. Each degree is divided into 60 parts, called **minutes**, and each minute is divided into 60 parts, called **seconds**. Thus, we can speak of an angle having a measure of 73 degrees, 12 minutes, and 36 seconds, and we write it as $73°12'36''$.

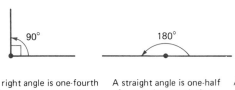

A right angle is one-fourth of a complete revolution, or $90°$.

A straight angle is one-half of a complete revolution, or $180°$.

An acute angle has a measure between $0°$ and $90°$.

An obtuse angle has a measure between $90°$ and $180°$.

This indicates an angle of measure $-225°$.

This indicates an angle of measure $420°$.

When using a calculator, fractional parts of a degree are written in decimal form; for example, an angle may have a measure of 73.21°. Some calculators are equipped with a special key sequence that will switch back and forth between the degree-minute-second form and the decimal form. If your calculator is not so equipped, you can proceed as follows.

From Degree-Minute-Second Form to Decimal Form:

$$73°12'36'' = 73° + \left(\frac{12}{60}\right)^° + \left(\frac{36}{3600}\right)^° \qquad \begin{array}{l}\text{If } 1° = 60' \text{ and} \\ 1' = 60'', \text{ then} \\ 1° = 3600''.\end{array}$$

$$= 73° + \left(\frac{1}{5}\right)^° + \left(\frac{1}{100}\right)^°$$

$$= 73° + (.2)^° + (.01)^°$$

$$= 73.21°$$

From Decimal Form to Degree-Minute-Second Form:

$$73.21° = 73° + .21(60')$$

$$= 73° + 12.6'$$

$$= 73° + 12' + .6(60'')$$

$$= 73° + 12' + 36''$$

$$= 73°12'36''$$

Radian Measure

The **radian** is another basic unit of angle measure that is used extensively in subsequent mathematics courses and in various mathematical applications in the physical sciences. To define a radian, let's first define the concept of a central angle. A **central angle of a circle** is an angle whose vertex is the center of the circle, as illustrated in Figure 1.26.

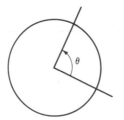

Lower case Greek letters are often used to denote angles. The symbol θ is the letter "theta."

Figure 1.26

Using the concept of a central angle, a radian can be defined as follows.

DEFINITION 1.5

One **radian** is the measure of the central angle of a circle where the sides of the angle intercept an arc equal in length to the radius of the circle (Figure 1.27).

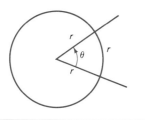

Figure 1.27

Therefore, in Figure 1.27, θ is an angle of measure 1 radian. Furthermore, since the circumference of a circle is given by $C = 2\pi r$, and each arc of length r determines an angle of one radian, there are $2\pi r/r = 2\pi$ radians in one complete revolution. Thus,

$$2\pi \text{ radians} = 360°$$

or, equivalently,

$$\pi \text{ radians} = 180°.$$

So we have the following two basic relationships between degree and radian measure.

$$1 \text{ radian} = \frac{180}{\pi} \text{ degrees}$$

and

$$1 \text{ degree} = \frac{\pi}{180} \text{ radians}$$

Remark: By using 3.1416 as an approximation for π, we can determine that 1 radian is approximately $\frac{180}{3.1416} = 57.3$ degrees. This relationship need not be memorized, but it may strengthen your perception of the size of one radian.

Sometimes it is necessary to switch back and forth between degree and radian measure. This creates no great difficulty, as illustrated by the following examples.

EXAMPLE 1

Change 150° to radians.

Solution Since 1 degree $= \pi/180$ radians,

$$150 \text{ degrees} = 150\left(\frac{\pi}{180}\right) \text{ radians}$$

$$= \frac{5\pi}{6} \text{ radians.} \quad = 2.61 \text{ radians}$$

EXAMPLE 2

Change $3\pi/4$ radians to degrees.

Solution Since 1 radian $= 180/\pi$ degrees,

$$\frac{3\pi}{4} \text{ radians} = \frac{3\pi}{4}\left(\frac{180}{\pi}\right) \text{ degrees} \quad \text{or } 3/4 = .75$$

$$= 135 \text{ degrees.} \quad \text{so } .75(180) = 135°$$

Some calculators have a $\boxed{d \leftrightarrow r}$ key that can be used for direct conversion between degrees and radians. If not, you will need to use $\pi/180$ to convert degrees to radians and $180/\pi$ to convert radians to degrees. Therefore,

$$5 \text{ radians} = 5\left(\frac{180}{\pi}\right) = 286.5 \text{ degrees, to the nearest tenth of a degree,}$$

and

$$127.4 \text{ degrees} = 127.4\left(\frac{\pi}{180}\right) = 2.2 \text{ radians, to the nearest tenth of a radian.}$$

Even though the process of switching between degree and radian measure is quite simple, it would be advantageous for you to know a few basic conversions as listed in the following chart.

DEGREES	30°	45°	60°	90°	180°	270°	360°
RADIANS	$\dfrac{\pi}{6}$	$\dfrac{\pi}{4}$	$\dfrac{\pi}{3}$	$\dfrac{\pi}{2}$	π	$\dfrac{3\pi}{2}$	2π

Having these few conversions at your fingertips also provides a basis for expressing the radian measure of many other angles. For example, since $300° = 5(60°)$ and $60° = \pi/3$ radians, then $300° = 5(\pi/3) = 5\pi/3$ radians.

Arc Length

Consider, as in Figure 1.28, a circle of radius r and a central angle θ, measured in **radians**, intercepting an arc of length s. Because the arc length s compares to the total circumference of $2\pi r$ as θ radians compare to a complete revolution of 2π radians, we have the proportion

$$\frac{s}{2\pi r} = \frac{\theta}{2\pi}.$$

Solving for s produces

$$(s)(2\pi) = (2\pi r)(\theta)$$

$$s = r\theta.$$

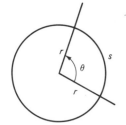

Figure 1.28

EXAMPLE 3

Find the length of the arc intercepted by a central angle of $\pi/6$ radians if a radius of the circle is 11 inches long.

Solution Using $s = r\theta$, we obtain

$$s = r\theta$$

$$= 11\left(\frac{\pi}{6}\right)$$

$$= 5.8 \text{ inches, to the nearest tenth of an inch.}$$

EXAMPLE 4

How high will the weight in the accompanying figure be lifted if the drum is rotated through an angle of 70°?

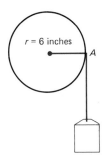

Solution First, we need to change 70° to radians.

$$70° = 70\left(\frac{\pi}{180}\right) = \frac{7\pi}{18} \text{ radians.}$$

Therefore, point A will move

$$s = r\theta = 6\left(\frac{7\pi}{18}\right) = 7.3 \text{ inches, to the nearest tenth of an inch.}$$

Thus, the weight will be lifted approximately 7.3 inches.

Triangles

The concepts of linear and angular measurement provide the basis for classifying various geometric figures. Triangles are often classified as follows. It is possible for triangles to fit both classification schemes. For example, an **isosceles right triangle** is a right triangle having two sides of the same length.

An acute triangle has three acute angles.

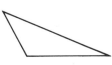

An obtuse triangle has one obtuse angle.

A right triangle has one right angle.

A scalene triangle has no two sides of the same length.

An equilateral triangle has three sides of the same length.

An isosceles triangle has two sides of the same length.

Classification by angles

Classification by sides

In Section 1.1 we used the Pythagorean Theorem to develop the distance formula for coordinate geometry. At this time, let's state the theorem and review some other applications of it.

Pythagorean Theorem

If for a right triangle, a and b are measures of the legs and c is the measure of the hypotenuse, then

$$a^2 + b^2 = c^2.$$

leg b hypotenuse c a leg

EXAMPLE 5

A 50-foot rope hangs from the top of a flagpole. When pulled taut, the rope reaches a point on the ground 18 feet from the base of the pole. Find the height of the pole to the nearest tenth of a foot.

Solution Let's sketch a figure and record the given information. Using the Pythagorean Theorem, we can solve for p as follows.

$$p^2 + 18^2 = 50^2$$

$$p^2 + 324 = 2500$$

$$p^2 = 2176$$

$$p = \pm\sqrt{2176} = \pm 46.6$$

50 feet p
18 feet

p represents the height of the flagpole.

The negative solution is disregarded and the height of the flagpole, to the nearest tenth of a foot, is 46.6 feet.

EXAMPLE 6

Find the length of each leg of an isosceles right triangle having a hypotenuse with a measure of 6 centimeters.

Solution Because it is an **isosceles right triangle**, both legs are of the same length. Thus, let's use x to represent the length of each leg. Now, applying the Pythagorean Theorem, we can proceed as follows.

$$x^2 + x^2 = 6^2$$

$$2x^2 = 36$$

$$x^2 = 18$$

$$x = \pm\sqrt{18} = \pm 3\sqrt{2}$$

Therefore, each leg is $3\sqrt{2}$ centimeters long.

Another very useful property comes directly from an equilateral triangle. Consider the equilateral triangle in Figure 1.29(a) having sides of length s. We know from elementary geometry that an angle bisector of an equilateral triangle is also a perpendicular bisector of the opposite side. Therefore, in Figure 1.29(b),

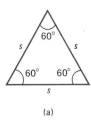

(a)

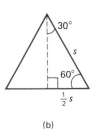

(b)

Figure 1.29

the angle bisector (dashed line segment) divides the equilateral triangle into two right triangles having acute angles of 30° and 60°. The following general property can be stated: **In a 30°–60° right triangle, the length of the side opposite the 30° angle is equal to one-half of the length of the hypotenuse.**

EXAMPLE 7

Suppose that a 20-foot ladder is leaning against a building and makes an angle of 60° with the ground. How far up on the building does the top of the ladder reach? Express your answer to the nearest tenth of a foot.

Solution The accompanying figure depicts the situation.

$$h^2 + 10^2 = 20^2$$

$$h^2 + 100 = 400$$

$$h^2 = 300$$

$$h = \pm\sqrt{300} = \pm 17.3$$

The top of the ladder touches the building approximately 17.3 feet from the ground.

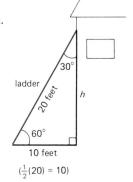

The following two right triangles are used frequently in subsequent trigonometry sections. Notice that in each triangle the hypotenuse is 1 unit long. We can determine the lengths of the legs for each triangle as we did in Examples 6 and 7. (Remember the dimensions of these right triangles!)

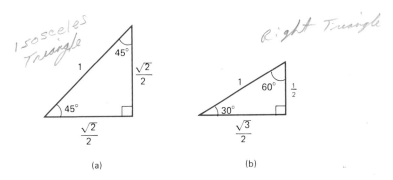

(a)   (b)

Problem Set 1.4

1. Find the complement of an angle of measure 38°.

2. Find the supplement of an angle of measure 72°.

3. The difference between two complementary angles is 40°. How large is each angle?

4. Find the measure of an angle that is $10°$ smaller than its complement.

5. One of two supplementary angles is 8 times as large as the other. How large is each angle?

6. The measures of two supplementary angles are in the ratio of 2 to 7. How large is each angle?

In Problems 7–18, if the measurement is given in degree-minute-second form, change it to decimal form to the nearest one-hundredth of a degree. If the measurement is given in decimal form, change it to degree-minute-second form.

7. $14°30'$	8. $62°15'$	9. $22.3°$	10. $114.6°$
11. $8°45'18''$	12. $34°50'30''$	13. $45.32°$	14. $132.15°$
15. $150°10'$	16. $94°45'$	17. $9.13°$	18. $73.47°$

In Problems 19–30, change each angle to radians. Do not use a calculator.

19. $10°$	20. $15°$	21. $80°$	22. $120°$
23. $150°$	24. $210°$	25. $225°$	26. $300°$
27. $-30°$	28. $-330°$	29. $-570°$	30. $480°$

In Problems 31–42, each angle is expressed in radians. Change each angle to degrees without using a calculator.

31. $\dfrac{\pi}{9}$	32. $\dfrac{5\pi}{18}$	33. $\dfrac{13\pi}{18}$	34. $\dfrac{7\pi}{12}$
35. $\dfrac{4\pi}{3}$	36. $\dfrac{7\pi}{4}$	37. $\dfrac{13\pi}{6}$	38. $\dfrac{17\pi}{6}$
39. $-\dfrac{\pi}{4}$	40. $-\dfrac{5\pi}{9}$	41. $-\dfrac{7\pi}{6}$	42. $-\dfrac{7\pi}{3}$

In Problems 43–48, each angle is expressed in radians. Use your calculator and change each angle to the nearest tenth of a degree.

43. 2	44. 3	45. 7
46. 4.1	47. -4	48. -6.2

In Problems 49–54, use your calculator to help change each angle to the nearest tenth of a radian.

49. $27°$	50. $212°$	51. $14.5°$
52. $141.8°$	53. $-251.6°$	54. $-373.4°$

55. Find, to the nearest tenth of an inch, the length of the arc intercepted by a central angle of $2\pi/3$ radians if a radius of the circle is 22 inches long.

56. Find, to the nearest tenth of a meter, the length of the arc intercepted by a central angle of $130°$ if a radius of the circle is 8 meters long.

57. Find, to the nearest tenth of a centimeter, the length of a radius of a circle if a central angle of $80°$ intercepts an arc of 25 centimeters.

58. Find, to the nearest tenth of a foot, the length of a radius of a circle if a central angle of $3\pi/5$ radians intercepts an arc of 12 feet.

59. Find, to the nearest tenth of a degree, the measure of a central angle that intercepts an arc of 7.1 centimeters if a radius of the circle is 3.2 centimeters long.

60. Referring to the accompanying diagram, how much will the weight be lifted if the drum is rotated through an angle of $150°$? Express the result to the nearest tenth of an inch.

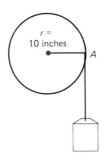

61. Referring to the diagram in Problem 60, through what angle (to the nearest tenth of a degree) must the drum be rotated to raise the weight 6 feet?

62. Two pulleys are connected with a belt as indicated in the following diagram. If the smaller pulley makes a complete revolution, through what size angle does the larger pulley turn? Express the result to the nearest tenth of a degree.

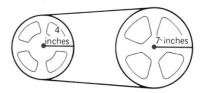

63. Referring to the diagram in Problem 62, through what angle does the smaller pulley turn while the larger pulley makes a complete revolution?

64. A gas gauge on an automobile is illustrated in the accompanying diagram. The scale from E to F is 3.5 inches long and it is an arc of a circle having a radius of 1.5 inches. What angle will the needle make between the empty and full readings? Express the angle to the nearest tenth of a degree.

65. A speedometer on an automobile is illustrated in the accompanying diagram. An angle of 110° is made as the needle moves from the 55-miles-per-hour position to the 0-miles-per-hour position. The radius of the circle is 2 inches. Find, to the nearest tenth of an inch, the length of the arc from the 55-reading to the 0-reading.

66. The accompanying diagram depicts the back-wheel drive-chain apparatus of a bicycle. How far will the bicycle move forward for each complete revolution of the drive sprocket? Express the result to the nearest inch.

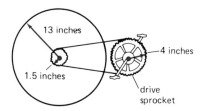

67. Referring to the diagram in Problem 66, how much rotation of the drive sprocket is needed to move the bicycle forward 50 feet? Express your result to the nearest tenth of a revolution.

For Problems 68–75, we are using a right triangle labeled as follows.

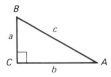

Express lengths of sides in simplest radical form.

68. Find c if $a = 5$ inches and $b = 12$ inches.

69. Find b if $c = 25$ feet and $a = 24$ feet.

70. Find b if $c = 12$ meters and $a = 6$ meters.

71. Find a if $c = 8$ centimeters and $b = 4$ centimeters.

72. Find b and c if $A = 30°$ and $a = 3$ yards.

73. Find a and b if $B = 60°$ and $c = 12$ yards.

74. Find c if $A = 45°$ and $a = 4$ meters.

75. Find a and b if $B = 45°$ and $c = 10$ meters.

76. The length of the hypotenuse of an isosceles right triangle is 8 meters. Find, to the nearest tenth of a meter, the length of each leg.

77. The length of the hypotenuse of an isosceles right triangle is 11 centimeters. Find, to the nearest tenth of a centimeter, the length of each leg.

78. An 18-foot ladder resting against a house reaches a windowsill 16 feet above the ground. How far is the foot of the ladder from the foundation of the house? Express your answer to the nearest tenth of a foot.

79. A 42-foot guy-wire makes an angle of 60° with the ground and it is attached to a telephone pole. Find the distance from the base of the pole to the point on the pole where the wire is attached. Express your answer to the nearest tenth of a foot.

80. A rectangular plot of ground measures 18 meters by 24 meters. Find the distance, to the nearest meter, from one corner of the plot to the diagonally opposite corner.

81. Consecutive bases of a square-shaped baseball diamond are 90 feet apart. Find the

distance, to the nearest foot, from first base diagonally across the diamond to third base.

82. A diagonal of a square parking lot is 50 meters. Find, to the nearest meter, the length of a side of the lot.

83. Find the length of an altitude of an equilateral triangle if each side of the triangle is 6 centimeters long. Express your answer in simplest radical form.

84. Suppose that we are given a cube with edges of length 12 centimeters. Find the length of a diagonal from a lower corner to the diagonally opposite upper corner. Express your answer to the nearest tenth of a centimeter.

85. Suppose that we are given a rectangular box with a length of 8 centimeters, a width of 6 centimeters, and a height of 4 centimeters. Find the length of a diagonal from a lower corner to the diagonally opposite upper corner. Express your answer to the nearest tenth of a centimeter.

Chapter Summary

Let's review Chapter 1 in terms of specific skills that you will need in subsequent chapters.

You should be able to identify natural numbers, counting numbers, positive integers, whole numbers, nonnegative integers, negative integers, nonpositive integers, integers, rational numbers, irrational numbers, and real numbers.

You should be able to use the distance formula, $d = \sqrt{(x_2 - x_1)^2 + (y_2 - y_1)^2}$, to find the distance between any two points in a rectangular coordinate system.

You should be able to distinguish between a relation and a function.

You should be able to use the function notation. For example, if $f(x) = 5x - 1$, then $f(3) = 5(3) - 1 = 15 - 1 = 14$.

You should be able to determine the domain of a function.

You should be able to determine the composition of two functions and determine the domain of the resulting composite function.

You should be able to graph any linear function.

You should be able to graph any quadratic function of the form $y = a(x - h)^2 + k$.

You should be able to define each of the following geometric concepts:

plane angle,
positive angle,
negative angle,
coterminal angles,
right angle,
straight angle,
acute angle,
obtuse angle,
complementary angles,
supplementary angles,

central angle,
radian,
acute triangle,
obtuse triangle,
right triangle,
scalene triangle,
isosceles triangle,
equilateral triangle,
isosceles right triangle.

You should be able to change back and forth between degree and radian measure.

You should be able to apply the arc length formula $s = r\theta$.

You should be able to solve problems using the Pythagorean Theorem.

Chapter 1 Review Problem Set

For Problems 1–8, from the list, 0, $\sqrt{7}$, $-\sqrt{3}$, $\frac{5}{6}$, $-\frac{1}{2}$, $-\frac{\sqrt{2}}{2}$, 0.14, $0.\overline{12}$, 2π, -1, 1, and 10, identify each of the following.

1. The natural numbers 1, 10
2. The whole numbers 0, 1, 10
3. The integers 0, -1, 1, 10
4. The nonpositive integers 0, -1
5. The rational numbers
6. The irrational numbers
7. The nonnegative integers
8. The counting numbers

For Problems 9–12, find the distance between each of the pairs of points.

9. $(-2, 3)$ and $(4, -5)$
10. $(-1, -3)$ and $(-2, -4)$
11. $(-2, -3)$ and $(2, 2)$
12. $(4, -5)$ and $(-1, 2)$

For Problems 13–16, specify the domain of each function.

13. $f(x) = \dfrac{4}{x - 3}$
14. $f(x) = \dfrac{-2}{x^2 - 16}$

15. $f(x) = \dfrac{1}{x^2 - 2x}$
16. $f(x) = \sqrt{4x - 3}$

For Problems 17–20, find the indicated functional values.

17. If $f(x) = -5x - 2$, find $f(-2)$, $f(-3)$, and $f(4)$.
18. If $f(x) = 2x^2 - x - 4$, find $f(-1)$, $f(1)$, and $f(3)$.
19. If $f(x) = -x^2 + 2x - 5$, find $f(-2)$, $f(2)$, and $f(5)$.
20. If $f(x) = 3x^2 - 2x + 8$, find $\dfrac{f(a + h) - f(a)}{h}$.

For Problems 21–24, determine $(f \circ g)(x)$ and $(g \circ f)(x)$ for each pair of functions. Also specify the domain for $f \circ g$ and $g \circ f$.

21. $f(x) = -3x$, $g(x) = 4x + 5$
22. $f(x) = 2x + 1$, $g(x) = x^2 - 1$

23. $f(x) = \sqrt{x + 1}$, $g(x) = 5x - 2$
24. $f(x) = \dfrac{1}{x + 1}$, $g(x) = \dfrac{2}{x - 3}$

For Problems 25–28, graph each of the functions.

25. $y = -3x - 6$
26. $y = -2x^2 + 3$
27. $y = -(x - 2)^2 + 4$
28. $y = 2(x + 3)^2 - 1$

29. The measures of two complementary angles are in the ratio of 7 to 11. How large is each angle?

30. Change $35°17'$ and $82°15'36''$ to decimal form and express each of them to the nearest one-hundredth of a degree.

31. Change $93.35°$ and $163.27°$ to degree-minute-second form.

32. Without using a calculator, change each of the following to radians.
    (a) 420°   (b) 570°   (c) −45°

33. Without using a calculator, change each of the following to degrees.
    (a) $\dfrac{7\pi}{6}$ (b) $-\dfrac{4\pi}{3}$ (c) $\dfrac{17\pi}{4}$

34. Find, to the nearest tenth of a centimeter, the length of the arc intercepted by a central angle of $4\pi/3$ radians if the radius of the circle is 17 centimeters long.

35. Find, to the nearest tenth of an inch, the length of the arc intercepted by a central angle of 130° if a radius of the circle is 14 inches long.

36. A rectangular floor measures 22 feet by 13 feet. Find the distance, to the nearest tenth of a foot, from one corner of the floor to the diagonally opposite corner.

37. A diagonal of a square plot of ground is 46 meters long. Find, to the nearest tenth of a meter, the length of a side of the square plot.

38. A 50-foot guy-wire makes an angle of 60° with the ground and is attached to a telephone pole. Find the distance from the base of the pole to the point on the pole where the guy-wire is attached. Express your answer to the nearest tenth of a foot.

2 Trigonometric Functions and Problem Solving

Courtesy, RCA OR ©Wm. Floyd Holdman

The word trigonometry was derived from two Greek words meaning "measurement of triangles". Historically, the development of trigonometry began with the study of the various relationships that exist between the angles and sides of triangles. This aspect of trigonometry has many applications in surveying, navigation, carpentry, and the various branches of engineering.

Originally, the trigonometric functions were restricted to having domains of angles. However, a more modern viewpoint allows for the domains to be the set of real numbers independent of any angle relationship. This viewpoint has resulted in a larger variety of applications for the trigonometric functions in such areas as light, sound, and electrical wave theories.

Our approach to trigonometry in this text will follow the historical route. That is to say, we will first introduce the trigonometric functions in terms of angles and then extend their domains to real numbers in general.

2.1

Trigonometric Functions

point of intersection point
of lines — highest point

If a rectangular coordinate system is introduced, then the **standard position** of an angle is obtained by taking the vertex at the origin and letting the initial side coincide with the positive side of the x-axis. The angles in Figure 2.1 are each in standard position. Each angle is named by using a Greek letter positioned next to the curved arrow. Thus, we can refer to angles θ(theta), ϕ(phi), α(alpha), β(beta), ψ(psi), and ω(omega). Angle θ is called a **first-quadrant angle** because its terminal side lies in the first quadrant. Angles ϕ, α, and β are **second-quadrant, third-quadrant**, and **fourth-quadrant** angles, respectively. If the terminal side of an angle in standard position coincides with a coordinate axis, then the angle is called a **quadrantal angle** as indicated by ψ and ω in Figure 2.1. Also note that θ, ϕ, and ψ are positive angles, whereas α, β, and ω are negative angles.

Figure 2.1

DEFINITION 2.1
Six Basic Trigonometric Functions

Let θ be an angle in standard position; let $P(x, y)$ be any point (except the origin) on the terminal side of θ (Figure 2.2). The six trigonometric functions are:

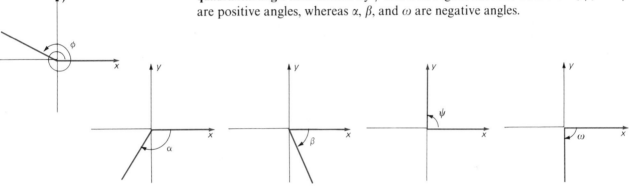

$$\sin \theta = \frac{y}{r} \text{ (read as "sine theta")}$$

$$\cos \theta = \frac{x}{r} \text{ (read as "cosine theta")}$$

$$\tan \theta = \frac{y}{x} \text{ (read as "tangent theta")}$$

$$\csc \theta = \frac{r}{y} \text{ (read as "cosecant theta")}$$

$$\sec \theta = \frac{r}{x} \text{ (read as "secant theta")}$$

$$\cot \theta = \frac{x}{y} \text{ (read as "cotangent theta")}$$

Figure 2.2

These functions are basic to all of trigonometry; **memorize them**.

In Definition 2.1, r is the distance between the origin and point P; **it is always a positive number** and it is determined by $r = \sqrt{x^2 + y^2}$. Recall that a function assigns to each member of a set (called the **domain**) a unique member of another set (called the **range**). The domain of each of the six trigonometric functions is a set of angles, and Definition 2.1 assigns to each angle (with a few exceptions) a real number determined by the ratios y/r, x/r, y/x, r/y, r/x and x/y. (The fact that a *unique* number is assigned to each angle will be demonstrated a bit later.) Because division by zero is not permitted, $\tan \theta$ and $\sec \theta$ cannot be defined for $x = 0$, and $\csc \theta$ and $\cot \theta$ cannot be defined for $y = 0$. Furthermore, notice that $\csc \theta$, $\sec \theta$, and $\cot \theta$ are the **reciprocals** of $\sin \theta$, $\cos \theta$, and $\tan \theta$, respectively. That is to say,

$$\csc \theta = \frac{1}{\sin \theta}, \qquad \sin \theta \neq 0$$

$$\sec \theta = \frac{1}{\cos \theta}, \qquad \cos \theta \neq 0$$

$$\cot \theta = \frac{1}{\tan \theta}, \qquad \tan \theta \neq 0.$$

Another very useful relationship that follows directly from Definition 2.1 is

$$\sin^2 \theta + \cos^2 \theta = 1.$$

The notation $\sin^2 \theta$ (usually read as "sine squared of theta") means $(\sin \theta)^2$; therefore, $\sin^2 \theta = (\sin \theta)(\sin \theta)$. This relationship can be verified as follows.

$$\sin^2 \theta + \cos^2 \theta = \left(\frac{y}{r}\right)^2 + \left(\frac{x}{r}\right)^2$$

$$= \frac{y^2}{r^2} + \frac{x^2}{r^2}$$

$$= \frac{y^2 + x^2}{r^2} = \frac{r^2}{r^2} = 1$$

The reciprocal relationships and the property $\sin^2 \theta + \cos^2 \theta = 1$ are called trigonometric identities. More identities will be discussed in Chapter 4.

EXAMPLE 1

Find the values of the six trigonometric functions of angle θ if θ is in standard position and the point $(-3, 4)$ is on the terminal side of θ.

Solution Figure 2.3 shows θ and the point $(-3, 4)$ on the terminal side of θ. Using $r = \sqrt{x^2 + y^2}$, we obtain

$$r = \sqrt{(-3)^2 + 4^2}$$

$$= \sqrt{9 + 16} = \sqrt{25} = 5.$$

Now using $x = -3$, $y = 4$, and $r = 5$, the values of the six trigonometric functions of θ can be determined.

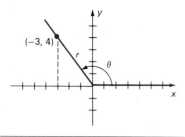

Figure 2.3

$$\sin \theta = \frac{y}{r} = \frac{4}{5} \qquad\qquad \cos \theta = \frac{x}{r} = \frac{-3}{5} = -\frac{3}{5}$$

$$\tan \theta = \frac{y}{x} = \frac{4}{-3} = -\frac{4}{3} \qquad\qquad \csc \theta = \frac{r}{y} = \frac{5}{4}$$

$$\sec \theta = \frac{r}{x} = \frac{5}{-3} = -\frac{5}{3} \qquad\qquad \cot \theta = \frac{x}{y} = \frac{-3}{4} = -\frac{3}{4}$$

It is important to realize that *any point* (other than the origin) on the terminal side of an angle in standard position can be used to determine the trigonometric functions of the angle. This fact is based on a property of similar triangles illustrated in Figure 2.4. Triangles OQP and $OQ'P'$ are similar triangles, and corresponding sides of similar triangles are proportional. That is to say, the ratios of corresponding sides are equal. For example, $y/r = y'/r'$ and therefore either of the ratios can be used to determine $\sin \theta$. Similar arguments are true for the remaining five trigonometric functions.

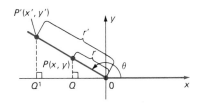

Figure 2.4

EXAMPLE 2

Find $\sin \theta$ and $\cos \theta$ if the terminal side of θ lies on the line $y = 2x$ in the third quadrant.

Solution First, let's sketch the line $y = 2x$ (Figure 2.5). The point $(-2, -4)$ can be used to sketch the line and also as a point on the terminal side of θ. Therefore,

$$\begin{aligned} r &= \sqrt{(-2)^2 + (-4)^2} \\ &= \sqrt{4 + 16} \\ &= \sqrt{20} \\ &= 2\sqrt{5}. \end{aligned}$$

Now the values for $\sin \theta$ and $\cos \theta$ can be determined.

$$\sin \theta = \frac{y}{r}$$

$$= \frac{-4}{2\sqrt{5}} = -\frac{2}{\sqrt{5}} = -\frac{2\sqrt{5}}{5}$$

$$\cos \theta = \frac{x}{r}$$

$$= \frac{-2}{2\sqrt{5}} = -\frac{1}{\sqrt{5}} = -\frac{\sqrt{5}}{5}$$

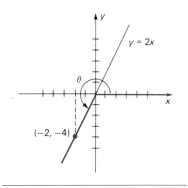

Figure 2.5

EXAMPLE 3

Find $\sin \theta$, $\cos \theta$, and $\tan \theta$ for $\theta = 30°$.

Solution Let's choose a point P on the terminal side of θ so that $r = 1$ (Figure 2.6). Because $\theta = 30°$, the right triangle indicated is a 30°-60° right triangle.

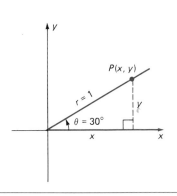

Figure 2.6

Therefore, $y = \frac{1}{2}$ because it is the side opposite the 30° angle. Then by remembering the special 30°-60° right triangle illustrated in Section 1.4 or by applying the Pythagorean Theorem, we can determine that $x = \sqrt{3}/2$. Therefore, using $r = 1$, $x = \sqrt{3}/2$, and $y = \frac{1}{2}$, we obtain

$$\sin 30° = \frac{y}{r} = \frac{\frac{1}{2}}{1} = \frac{1}{2},$$

$$\cos 30° = \frac{x}{r} = \frac{\frac{\sqrt{3}}{2}}{1} = \frac{\sqrt{3}}{2},$$

and

$$\tan 30° = \frac{y}{x} = \frac{\frac{1}{2}}{\frac{\sqrt{3}}{2}} = \frac{1}{\sqrt{3}} = \frac{\sqrt{3}}{3}.$$

Before considering another example, let's agree on some symbolism. It is customary to omit writing the word "radian" when using radian measure. For example, an angle θ of radian measure π is usually written as $\theta = \pi$ instead of $\theta = \pi$ radians. However, if the measure is stated in degrees, then be sure to use the degree symbol; in other words, an angle θ of measure 70 degrees is written as $\theta = 70°$ and not $\theta = 70$.

EXAMPLE 4

Find $\sin\theta$, $\cos\theta$, and $\tan\theta$ for $\theta = \pi/4$.

Solution Remember that $\pi/4$ radians equals 45°. Therefore, in Figure 2.7 we sketched a 45° angle and chose a point P on the terminal side so that $r = 1$. Since $\theta = 45°$, the indicated right triangle is an isosceles right triangle with $x = y$. Then by remembering the special isosceles right triangle illustrated in Section 1.4 or by applying the Pythagorean Theorem, we can determine that $x = y = \sqrt{2}/2$. (The

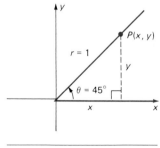

Figure 2.7

values of x and y are both positive since P is in the first quadrant.) Thus, the coordinates of point P are $(\sqrt{2}/2, \sqrt{2}/2)$ and we can determine the trigonometric functional values of $\pi/4$.

$$\sin\frac{\pi}{4} = \frac{y}{r} = \frac{\frac{\sqrt{2}}{2}}{1} = \frac{\sqrt{2}}{2}$$

$$\cos\frac{\pi}{4} = \frac{x}{r} = \frac{\frac{\sqrt{2}}{2}}{1} = \frac{\sqrt{2}}{2}$$

$$\tan\frac{\pi}{4} = \frac{y}{x} = \frac{\frac{\sqrt{2}}{2}}{\frac{\sqrt{2}}{2}} = 1$$

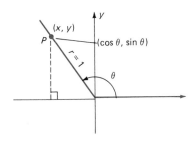

Figure 2.8

In Examples 3 and 4 notice the convenience of choosing $r = 1$. In general, for any angle θ, by choosing $r = 1$ we obtain

$$\sin\theta = \frac{y}{r} = \frac{y}{1} = y \quad \text{and} \quad \cos\theta = \frac{x}{r} = \frac{x}{1} = x.$$

In other words, if P is a point one unit from the origin on the terminal side of θ, then the coordinates of point P are $(\cos\theta, \sin\theta)$ as indicated in Figure 2.8. Furthermore, the tangent function can be expressed as

$$\tan\theta = \frac{y}{x} = \frac{\sin\theta}{\cos\theta}, \quad \text{for } \cos\theta \neq 0.$$

EXAMPLE 5

Find $\sin\theta$, $\cos\theta$, and $\tan\theta$ for $\theta = 240°$.

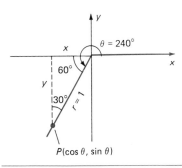

Figure 2.9

Solution In Figure 2.9 we sketched a $240°$ angle and indicated a point P on the terminal side so that $r = 1$. The indicated right triangle is a $30°$-$60°$ right triangle with the $30°$ angle at P. Therefore, $x = -(1/2)$ and $y = -(\sqrt{3}/2)$ (x and y are both negative because P is in the third quadrant) and the coordinates of P are $(-1/2, -\sqrt{3}/2)$. Thus, we obtain

$$\sin 240° = y = -\frac{\sqrt{3}}{2},$$

$$\cos 240° = x = -\frac{1}{2},$$

and

$$\tan 240° = \frac{\sin 240°}{\cos 240°} = \frac{-\frac{\sqrt{3}}{2}}{-\frac{1}{2}} = \sqrt{3}.$$

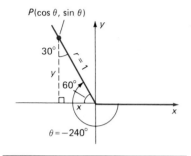

Figure 2.10

EXAMPLE 6

Find $\sin\theta$, $\cos\theta$, and $\tan\theta$ for $\theta = -240°$.

Solution   In Figure 2.10 we sketched an angle of $-240°$ and indicated a point P on the terminal side so that $r = 1$. The indicated right triangle is a $30°$-$60°$ right triangle with the $30°$ angle at P. Therefore, $x = -(1/2)$ and $y = \sqrt{3}/2$ and we can express the functional values as follows.

$$\sin(-240°) = y = \frac{\sqrt{3}}{2}$$

$$\cos(-240°) = x = -\frac{1}{2}$$

$$\tan(-240°) = \frac{\sin(-240°)}{\cos(-240°)} = \frac{\dfrac{\sqrt{3}}{2}}{-\dfrac{1}{2}} = -\sqrt{3}$$

Compare your results for Example 5 and Example 6. Note that $\sin 240° = -(\sqrt{3}/2)$ and $\sin(-240°) = \sqrt{3}/2$. In other words, $\sin(-240°) = -\sin 240°$. Likewise, observe that $\cos(-240°) = \cos 240°$ and $\tan(-240°) = -\tan 240°$. More will be said about some general relationships of this type in the next section.

If θ is a quadrantal angle (terminal side lies on an axis), it is easy to determine the values of the trigonometric functions. The next example illustrates this point.

EXAMPLE 7

Find the values of the six trigonometric functions of θ if $\theta = \pi/2$.

Solution   First, remember that $\pi/2 = 90°$. Then, as indicated in Figure 2.11, let's choose a point P on the terminal side of a $90°$ angle so that $r = 1$. The coordinates of point P are $(0, 1)$. Therefore,

$$\sin\frac{\pi}{2} = y = 1,$$

$$\cos\frac{\pi}{2} = x = 0,$$

$$\tan\frac{\pi}{2} = \frac{\sin\dfrac{\pi}{2}}{\cos\dfrac{\pi}{2}} = \frac{1}{0} \quad \text{(undefined)},$$

$$\csc\frac{\pi}{2} = \frac{1}{\sin\dfrac{\pi}{2}} = \frac{1}{1} = 1,$$

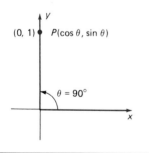

Figure 2.11

$$\sec\frac{\pi}{2} = \frac{1}{\cos\frac{\pi}{2}} = \frac{1}{0} \quad \text{(undefined)},$$

and

$$\cot\frac{\pi}{2} = \frac{x}{y} = \frac{0}{1} = 0. \quad \left(\text{The reciprocal relationship } \cot\theta = \frac{1}{\tan\theta} \right.$$

$$\left. \text{cannot be used here because } \tan\frac{\pi}{2} \text{ is} \right.$$

$$\left. \text{undefined.} \vphantom{\Big)} \right)$$

Problem Set 2.1

For Problems 1–8, point P is on the terminal side of θ and θ is a positive angle less than $360°$ in standard position. Draw θ, and determine the values of the six trigonometric functions of θ.

1. $P(3, -4)$ 2. $P(-3, -4)$ 3. $P(-5, 12)$ 4. $P(12, 5)$
5. $P(1, -1)$ 6. $P(-1, -1)$ 7. $P(-2, -3)$ 8. $P(3, -2)$

For Problems 9–16, point P is on the terminal side of θ and $0° > \theta > -360°$ in standard position. Draw θ, and determine the values of the six trigonometric functions of θ.

9. $P(2, 4)$ 10. $P(1, -3)$ 11. $P(3, -1)$ 12. $P(-2, 2)$
13. $P(0, 2)$ 14. $P(-1, 0)$ 15. $P(0, -1)$ 16. $P(4, 4)$

For Problems 17–34, determine $\sin\theta$, $\cos\theta$, and $\tan\theta$.

17. $\theta = 60°$ 18. $\theta = 150°$ 19. $\theta = \frac{3}{4}\pi$ 20. $\theta = \frac{7\pi}{6}$

21. $\theta = 300°$ 22. $\theta = 330°$ 23. $\theta = -\frac{\pi}{4}$ 24. $\theta = -\frac{\pi}{3}$

25. $\theta = -30°$ 26. $\theta = -210°$ 27. $\theta = 225°$ 28. $\theta = 315°$
29. $\theta = 390°$ 30. $\theta = 480°$ 31. $\theta = 585°$ 32. $\theta = 660°$

33. $\theta = \frac{23\pi}{6}$ 34. $\theta = \frac{11\pi}{4}$

35. Complete the following table.

θ	θ IN RADIANS	SIN θ	COS θ	TAN θ	CSC θ	SEC θ	COT θ
0°							
30°							
45°							
60°							
90°							
180°							
270°							

36. Find $\sin \theta$ if the terminal side of θ lies on the line $y = x$ in the third quadrant.

37. Find $\cos \theta$ if the terminal side of θ lies on the line $y = -x$ in the second quadrant.

38. Find $\tan \theta$ if the terminal side of θ lies on the line $y = -2x$ in the fourth quadrant.

39. Find $\sin \theta$ if the terminal side of θ lies on the line $y = 3x$ in the first quadrant.

40. If $\sin \theta = -\frac{4}{5}$ and the terminal side of θ is in the fourth quadrant, find $\cos \theta$ and $\tan \theta$.

41. If $\cos \theta = -\frac{4}{5}$ and the terminal side of θ is in the third quadrant, find $\sin \theta$ and $\cot \theta$.

42. If $\tan \theta = -\frac{5}{12}$ and the terminal side of θ is in the second quadrant, find $\sin \theta$ and $\cos \theta$.

43. If $\tan \theta = \frac{7}{24}$ and the terminal side of θ is in the first quadrant, find $\sin \theta$ and $\sec \theta$.

44. In which quadrant(s) must the terminal side of θ lie if $\sin \theta$ and $\tan \theta$ are to have the same sign?

45. In which quadrant(s) must the terminal side of θ lie if $\sin \theta$ is negative and $\cos \theta$ is positive?

46. In which quadrant(s) must the terminal side of θ lie if $\sin \theta$, $\cos \theta$, and $\tan \theta$ are all to have the same sign?

47. In which quadrant(s) must the terminal side of θ lie if $\sin \theta$ and $\cos \theta$ have opposite signs?

For Problems 48–53, determine θ if θ is a positive angle less than $360°$ satisfying the stated conditions.

48. $\tan \theta = 1$ and $\sin \theta$ is negative

49. $\cos = \frac{1}{2}$ and $\tan \theta$ is positive

50. $\sin \theta = \frac{\sqrt{3}}{2}$ and $\cos \theta$ is negative

51. $\cos \theta = -\frac{\sqrt{3}}{2}$ and $\sin \theta$ is negative

52. $\cos \theta = -\frac{1}{2}$ and $\tan \theta$ is positive

53. $\sin \theta = -1$ and $\cos \theta = 0$

2.2

Trigonometric Functions of Any Angle

Let's begin by summarizing some ideas from the previous section and its problem set. It is easy to determine the signs (positive or negative) of the trigonometric functions in each of the quadrants. For example, using Definition 2.1 with $r = 1$, we know that $\sin \theta = y$, and therefore $\sin \theta$ is positive in quadrants I and II and negative in quadrants III and IV. Furthermore, because $\csc \theta$ is the reciprocal of $\sin \theta$, its signs will agree with $\sin \theta$. The following chart summarizes the signs of all six trigonometric functions in the four quadrants.

QUADRANT CONTAINING θ	POSITIVE FUNCTIONS	NEGATIVE FUNCTIONS
I	All	None
II	sin, csc	cos, sec, tan, cot
III	tan, cot	sin, csc, cos, sec
IV	cos, sec	sin, csc, tan, cot

Table 2.1 summarizes the trigonometric functional values of some special angles that we have worked with thus far. It will be *very helpful* for you to have these values at your fingertips.

Table 2.1

θ	θ IN RADIANS	SIN θ	COS θ	TAN θ
0°	0	0	1	0
30°	$\dfrac{\pi}{6}$	$\dfrac{1}{2}$	$\dfrac{\sqrt{3}}{2}$	$\dfrac{\sqrt{3}}{3}$
45°	$\dfrac{\pi}{4}$	$\dfrac{\sqrt{2}}{2}$	$\dfrac{\sqrt{2}}{2}$	1
60°	$\dfrac{\pi}{3}$	$\dfrac{\sqrt{3}}{2}$	$\dfrac{1}{2}$	$\sqrt{3}$
90°	$\dfrac{\pi}{2}$	1	0	undefined
180°	π	0	-1	0
270°	$\dfrac{3\pi}{2}$	-1	0	undefined

Coterminal Angles

Recall that coterminal angles have the same initial and terminal sides. Therefore, angles in standard position that have the same terminal side are coterminal. Coterminal angles differ from each other by a multiple of 360° or of 2π radians. For example, angles of 120° and 480° are coterminal as are angles of π and 5π. The angles that are coterminal with any angle θ can be represented by $\theta + 360°n$ or $\theta + 2\pi n$, where n is any integer. Furthermore, it follows from the definitions of the trigonometric functions that **corresponding functions of coterminal angles are equal**. Therefore, using some values from Table 2.1 the following statements can be made.

$$\sin 390° = \sin 30° = \frac{1}{2} \quad \text{because 390° and 30° are coterminal.}$$

$$\tan 780° = \tan 60° = \sqrt{3} \quad \text{because 780° and 60° are coterminal.}$$

$$\cos \frac{9\pi}{4} = \cos \frac{\pi}{4} = \frac{\sqrt{2}}{2} \quad \text{because } \frac{9\pi}{4} \text{ and } \frac{\pi}{4} \text{ are coterminal.}$$

$$\sin(-300°) = \sin 60° = \frac{\sqrt{3}}{2} \quad \text{because } -300° \text{ and } 60° \text{ are coterminal.}$$

Reference Angle

The concept of a **reference angle** can be defined as follows.

DEFINITION 2.2

> Let θ be any angle in standard position with its terminal side in one of the four quadrants. The **reference angle** associated with θ (we will call it θ') is the **acute angle** formed by the terminal side of θ and the x-axis.

In Figure 2.12 we have indicated the reference angle θ' for the four different situations in which the terminal side of θ lies in Quadrants I, II, III, or IV.

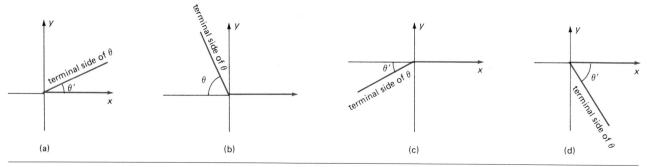

Figure 2.12

From our work in the previous section the following fact becomes evident: **The trigonometric functions of any angle θ are equal to those of the reference angle associated with θ, except possibly for the sign. The sign can be determined by considering the quadrant in which the terminal side of θ lies.** Let's consider an example.

EXAMPLE 1

Find $\cos \theta$ if $\theta = 225°$.

Solution In Figure 2.13 we have sketched $\theta = 225°$ and indicated its reference angle $\theta' = 45°$. Since the terminal side of θ lies in the third quadrant, $\cos \theta$ is negative. Therefore,

$$\cos 225° = -\cos 45° = -\frac{\sqrt{2}}{2}.$$

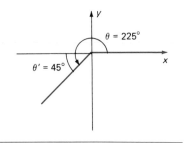

Figure 2.13

Using the information from Table 2.1 and our knowledge of coterminal angles, reference angles, signs of the trigonometric functions in each of the quadrants, and the reciprocal relationships, we can determine the six trigonometric functions of many special angles.

EXAMPLE 2

Find the six trigonometric functions of θ for $\theta = 510°$.

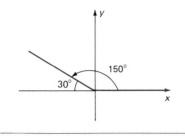

Figure 2.14

Solution An angle of 510° is coterminal with an angle of $510° - 360° = 150°$. The reference angle associated with a 150° angle is a 30° angle (Figure 2.14). Because θ is a second-quadrant angle, $\sin \theta$ is positive, $\cos \theta$ is negative, and $\tan \theta$ is negative. Therefore, we obtain

$$\sin 510° = \sin 30° = \frac{1}{2}$$

$$\cos 510° = -\cos 30° = -\frac{\sqrt{3}}{2}$$

$$\tan 510° = -\tan 30° = -\frac{\sqrt{3}}{3}.$$

Using the reciprocal relationships we obtain

$$\csc 510° = \frac{1}{\sin 510°} = \frac{1}{\frac{1}{2}} = 2$$

$$\sec 510° = \frac{1}{\cos 510°} = \frac{1}{-\frac{\sqrt{3}}{2}} = -\frac{2\sqrt{3}}{3}$$

$$\cot 510° = \frac{1}{\tan 510°} = \frac{1}{-\frac{\sqrt{3}}{3}} = -\sqrt{3}.$$

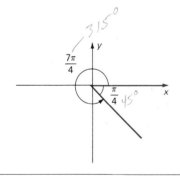

Figure 2.15

EXAMPLE 3

Find $\sin \theta$ for $\theta = 15\pi/4$.

Solution An angle of $15\pi/4$ radians is coterminal with an angle of $(15\pi/4) - 2\pi = 7\pi/4$. The reference angle associated with an angle of $7\pi/4$ radians is an angle of $\pi/4$ radians (Figure 2.15). Because θ is a fourth-quadrant angle, $\sin \theta$ is negative. Therefore,

$$\sin \frac{15\pi}{4} = -\sin \frac{\pi}{4} = -\frac{\sqrt{2}}{2}.$$

Remark: For a problem such as Example 3, you may find it easier to begin by switching from radians to degrees. That's alright, but it is advantageous in later sections to feel comfortable working with radian measure.

EXAMPLE 4

Find $\cos \theta$ for $\theta = -480°$.

Solution An angle of $-480°$ is coterminal with an angle of $-480° + 720° = 240°$. The reference angle associated with an angle of 240° is a 60° angle. (Try to

$720 - 480 = 240°$

mentally picture that without drawing a figure.) Because θ is a third-quadrant angle, $\cos \theta$ is negative. Therefore,

$$\cos(-480°) = -\cos 60° = -\tfrac{1}{2}.$$

Trigonometric Functions of Any Angle

Until now, we have been finding **exact values** for the trigonometric functions of some **special angles**. Now suppose we need a value for sin 23°. An approximate value could be found by drawing, in standard position, a 23° angle with a protractor (Figure 2.16). The ordinate of point P, measured in terms of the unit used for r, is an approximate value for sin 23°. Obviously, this approach would yield a very crude approximation, but it does emphasize one of the meanings of the trigonometric functions. For our purposes, better approximations can be found more efficiently by using a table of trigonometric values or a calculator.

If you are going to use a table, then turn to Appendix A in the back of the book. You will find a table of trigonometric values and a brief discussion regarding the use of the table.

The following examples illustrate the use of a calculator to find trigonometric functional values.

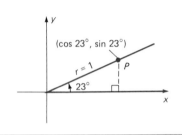

Figure 2.16

EXAMPLE 5

Use a calculator to find the value of (**a**) sin 23°, (**b**) cos 212°, (**c**) tan($-114.2°$), and (**d**) tan 90°.

Solutions   First, be sure that your calculator is set for degree measure. Your calculator manual will indicate that procedure. Many calculators are automatically set for degree measure when turned on.

(**a**) To find sin 23°, enter the number 23 and press the $\boxed{\text{SIN}}$ key. The display, to seven decimal places, should read .3907311. Therefore, to the nearest ten-thousandth, sin 23° = .3907.

(**b**) To find cos 212°, enter the number 212 and press the $\boxed{\text{COS}}$ key. The display, to seven decimal places, should read $-.8480481$. Therefore, to the nearest ten-thousandth, cos 212° = $-.8480$.

(**c**) To evaluate tan($-114.2°$), enter the number 114.2, press the $\boxed{+/-}$ key, and then press the $\boxed{\text{TAN}}$ key. The display, to seven decimal places, will read 2.2251009. Therefore, to the nearest ten-thousandth, tan($-114.2°$) = 2.2251.

(**d**) To attempt to evaluate tan 90°, enter the number 90 and press the $\boxed{\text{TAN}}$ key. The display will either blink 9's or give an "error" message. Either way, it is telling us that tan 90° is undefined.

The trigonometric functions csc θ, sec θ, and cot θ can be evaluated by using the reciprocal relationships, as the next example illustrates.

EXAMPLE 6

Use a calculator to find csc θ, sec θ, and cot θ for $\theta = 57°$.

Solution Because $\csc 57° = 1/\sin 57°$, we can enter 57, press the $\boxed{\text{SIN}}$ key, and then press the $\boxed{1/x}$ key. This will yield, to the nearest ten-thousandth, $\csc 57° = 1.1924$. In a like manner, because $\sec 57° = 1/\cos 57°$, we can enter 57, press the $\boxed{\text{COS}}$ key, and then press the $\boxed{1/x}$ key. To the nearest ten-thousandth, we will obtain $\sec 57° = 1.8361$.

Similarly, because $\cot 57° = 1/\tan 57°$, we can enter 57, press the $\boxed{\text{TAN}}$ key, and then press the $\boxed{1/x}$ key. To the nearest ten-thousandth, we will obtain $\cot 57° = .6494$.

From Examples 5 and 6 it is evident that using a calculator to evaluate the six trigonometric functions of any angle is very easy. However, we do suggest that you organize your use of the calculator to minimize the chances of making a human error, such as pressing the wrong key. Along this line, we have two specific suggestions to offer at this time. Suppose we want to evaluate $\sin(-23°)$. First, before pressing any keys on your calculator, mentally picture the terminal side of a $-23°$ angle being in the fourth quadrant. Therefore, $\sin(-23°)$ must be a negative number. Secondly, after entering -23, check your display to be sure you have entered the correct number. Now by pressing the $\boxed{\text{SIN}}$ key, you should be sure of the result $\sin(-23°) = -.3907$, to the nearest ten-thousandth. We will use the calculator and involve radian measure in the next chapter.

In Example 6 we used the reciprocal relationships to help determine some trigonometric values. The relationship $\sin^2 \theta + \cos^2 \theta = 1$ can also be used at times for that purpose. For example, suppose that θ is a second-quadrant angle and $\sin \theta = \frac{1}{2}$. We can determine the value for $\cos \theta$ from $\sin^2 \theta + \cos^2 \theta = 1$.

$$\left(\frac{1}{2}\right)^2 + \cos^2 \theta = 1$$

$$\frac{1}{4} + \cos^2 \theta = 1$$

$$\cos^2 \theta = \frac{3}{4}$$

$$\cos \theta = -\frac{\sqrt{3}}{2} \qquad \text{$\cos \theta$ is negative since θ is a second-quadrant angle.}$$

Some additional relationships are suggested by Figure 2.17. Points P and P' are located one unit from the origin on the terminal sides of θ and $-\theta$, respectively. The points P and P' are x-axis reflections of each other. Therefore, the coordinates of P' are $(x, -y)$. Then the following relationships can be observed.

$$\sin(-\theta) = -y = -\sin \theta$$
$$\cos(-\theta) = x = \cos \theta$$

$$\tan(-\theta) = \frac{-y}{x} = -\frac{y}{x} = -\tan \theta$$

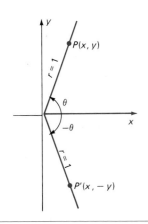

Figure 2.17

These properties allow us to make statements such as the following.

$$\sin(-30°) = -\sin 30° = -\frac{1}{2}$$

$$\cos(-30°) = \cos 30° = \frac{\sqrt{3}}{2}$$

$$\tan(-30°) = -\tan 30° = -\frac{\sqrt{3}}{3}$$

Problem Set 2.2

For Problems 1–8, find the quadrant that contains the terminal side of θ if the given conditions are true.

1. $\sin \theta > 0$ and $\cos \theta > 0$
2. $\sin \theta < 0$ and $\cos \theta < 0$
3. $\sin \theta < 0$ and $\cos \theta > 0$
4. $\tan \theta < 0$ and $\cos \theta > 0$
5. $\sin \theta < 0$ and $\cot \theta > 0$
6. $\sec \theta < 0$ and $\tan \theta < 0$
7. $\csc \theta > 0$ and $\cot \theta < 0$
8. $\cos \theta > 0$ and $\cot \theta < 0$

For Problems 9–16, find α such that $0° < \alpha < 360°$ and is coterminal with θ.

9. $\theta = 510°$
10. $\theta = 570°$
11. $\theta = 960°$
12. $\theta = 750°$
13. $\theta = -60°$
14. $\theta = -210°$
15. $\theta = -480°$
16. $\theta = -660°$

For Problems 17–22, find α such that $0 < \alpha < 2\pi$ and is coterminal with θ.

17. $\theta = \frac{7\pi}{2}$
18. $\theta = \frac{11\pi}{4}$
19. $\theta = \frac{31\pi}{6}$

20. $\theta = \frac{17\pi}{3}$
21. $\theta = -\frac{5\pi}{4}$
22. $\theta = -\frac{2\pi}{3}$

For Problems 23–34, find the reference angle θ' for each of the given values of θ.

23. $\theta = 265°$
24. $\theta = 285.3°$
25. $\theta = 431.8°$
26. $\theta = 510°$

27. $\theta = -73°$
28. $\theta = -190°$
29. $\theta = \frac{5\pi}{4}$
30. $\theta = \frac{11\pi}{6}$

31. $\theta = \frac{8\pi}{3}$
32. $\theta = \frac{13\pi}{4}$
33. $\theta = -\frac{4\pi}{3}$
34. $\theta = -\frac{5\pi}{6}$

For Problems 35–66, find exact values. Do not use a calculator or a table.

35. $\sin 120°$
36. $\cos 150°$
37. $\cos 210°$
38. $\sin 210°$
39. $\tan 300°$
40. $\tan 315°$
41. $\csc 135°$
42. $\sec 240°$
43. $\sec 420°$
44. $\csc 480°$
45. $\sin(-150°)$
46. $\cos(-210°)$
47. $\cos(-300°)$
48. $\sin(-390°)$
49. $\cot(-930°)$
50. $\cot(-480°)$
51. $\sin 630°$
52. $\cos 540°$
53. $\cos 315°$
54. $\sin(-315°)$

55. $\sin \frac{2\pi}{3}$
56. $\cos \frac{3\pi}{4}$
57. $\tan \frac{4\pi}{3}$
58. $\cot \frac{5\pi}{3}$

59. $\cos \frac{11\pi}{4}$
60. $\sin \frac{13\pi}{4}$
61. $\cot \frac{13\pi}{3}$
62. $\tan \frac{31\pi}{6}$

63. $\sin\left(-\dfrac{7\pi}{6}\right)$ 64. $\cos\left(-\dfrac{5\pi}{3}\right)$ 65. $\tan\left(-\dfrac{3\pi}{2}\right)$ 66. $\tan(-3\pi)$

For Problems 67–80, use your calculator (or the table in Appendix A) to find approximate values. Express the values to the nearest ten-thousandth.

67. $\sin 75°$ 68. $\cos 80°$ 69. $\tan 256°$ 70. $\tan 171.4°$

71. $\sin 59.4°$ 72. $\cos 117.6°$ 73. $\cos(-156°)$ 74. $\sin(-43.7°)$

75. $\sec 15.1°$ 76. $\csc 114.9°$ 77. $\csc(-14.7°)$ 78. $\cot 214.3°$

79. $\cot 328°$ 80. $\sec 412.3°$

For Problems 81–88, use as necessary the reciprocal relationships, $\sin^2\theta + \cos^2\theta = 1$, $\sin(-\theta) = -\sin\theta$, $\cos(-\theta) = \cos\theta$, and $\tan(-\theta) = -\tan\theta$ to find the required values.

81. If $\sin\theta = \dfrac{\sqrt{3}}{2}$ and θ is a second-quadrant angle, find $\cos\theta$.

82. If $\cos\theta = -\dfrac{1}{2}$ and θ is a third-quadrant angle, find $\sin\theta$.

83. If $\cos\theta = -\dfrac{\sqrt{3}}{2}$, find $\sec\theta$.

84. If $\sin\theta = -\dfrac{1}{2}$, find $\csc\theta$.

85. If $\sin\theta = .1080$, find $\sin(-\theta)$.

86. If $\cos\theta = .2062$, find $\cos(-\theta)$.

87. If $\tan\theta = 1.897$, find $\tan(-\theta)$.

88. If $\sin\theta = \frac{3}{4}$ and θ is a second-quadrant angle, find $\tan\theta$.

Miscellaneous Problems

For Problems 89–94, use your calculator to find approximate values to the nearest ten-thousandth.

89. $\sin 117°6'$ 90. $\cos 234°12'$ 91. $\tan(-114°48')$

92. $\sec 221°36'$ 93. $\csc 317°54'$ 94. $\cot 373°24'$

For Problems 95–100, verify each of the statements using a calculator or the table in Appendix A.

95. $\sin 25° = \cos 65°$ 96. $\cos 72° = \sin 18°$

97. $\sec 10° = \csc 80°$ 98. $\csc 14.3° = \sec 75.7°$

99. $\tan 47° = \cot 43°$ 100. $\cot 25.7° = \tan 64.3°$

2.3

Right Triangle Trigonometry

In these next three sections we will be involved in the problem solving aspect of trigonometry. More specifically, we will be using right triangles and triangles in general to solve problems that deal with a variety of applications. In this context

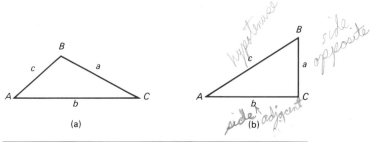

Figure 2.18

point of intersection of lines or surfaces / highest point

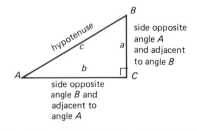

Figure 2.19

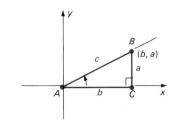

Figure 2.20

it is customary to designate the vertices and corresponding angles of a triangle by the capital letters A, B, and C. Then the sides opposite the angles A, B, and C are designated by the lowercase letters a, b, and c, respectively (Figure 2.18(a)). If triangle ACB is a right triangle, then we will use C to denote the right angle (Figure 2.18(b)). Furthermore, the sides of a right triangle are often referred to in terms of the acute angles. For example, side a is called the **side opposite** angle A, side b is called the **side adjacent** to angle A, and side C is called the **hypotenuse**. This terminology is summarized in Figure 2.19.

In Section 2.1 we defined the six basic trigonometric functions in terms of an angle θ in standard position on a Cartesian coordinate system. Now let's consider a special case of this definition, namely, the situation when $\theta = A$ is an acute angle of a right triangle as indicated in Figure 2.20. The six trigonometric functions can be defined as follows:

$$\sin A = \frac{a}{c} = \frac{\text{side opposite } A}{\text{hypotenuse}}, \qquad \csc A = \frac{c}{a} = \frac{\text{hypotenuse}}{\text{side opposite } A}$$

$$\cos A = \frac{b}{c} = \frac{\text{side adjacent } A}{\text{hypotenuse}}, \qquad \sec A = \frac{c}{b} = \frac{\text{hypotenuse}}{\text{side adjacent } A}$$

$$\tan A = \frac{a}{b} = \frac{\text{side opposite } A}{\text{side adjacent } A}, \qquad \cot A = \frac{b}{a} = \frac{\text{side adjacent } A}{\text{side opposite } A}$$

We could also place right triangle ACB so that angle B is in standard position and the six trigonometric functions of angle B could be defined in a similar manner. For example, $\sin B = b/c = (\text{side opposite } B)/\text{hypotenuse}$. Furthermore, once we have the definitions in terms of the ratios of sides of the right triangle, we no longer need to consider the coordinate system. Thus, in right triangle ACB, where C is the right angle, the following statements can be made.

$$\sin A = \frac{a}{c} = \cos B \qquad \tan A = \frac{a}{b} = \cot B \qquad \sec A = \frac{c}{b} = \csc B$$

$$\cos A = \frac{b}{c} = \sin B \qquad \cot A = \frac{b}{a} = \tan B \qquad \csc A = \frac{c}{a} = \sec B$$

Note also that A and B are complementary angles. Thus, the sine and cosine functions are called **cofunctions**. The tangent and cotangent functions are cofunctions, as are the secant and cosecant functions. In general, if θ is an acute angle, any trigonometric function of θ is equal to the cofunction of the complement of θ. The above relationships can be restated in terms of any acute angle θ as follows:

$$\sin \theta = \cos(90° - \theta) \quad \text{and} \quad \cos \theta = \sin(90° - \theta)$$

$$\tan \theta = \cot(90° - \theta) \quad \text{and} \quad \cot \theta = \tan(90° - \theta)$$

$$\sec \theta = \csc(90° - \theta) \quad \text{and} \quad \csc \theta = \sec(90° - \theta)$$

We refer to a triangle as having six parts—the three sides and the three angles. The phrase "solving a triangle" refers to finding the values of all six parts. Solving a right triangle can be analyzed as follows.

1. If an acute angle and the length of one of the sides is known, then the other acute angle is the complement of the one given. The lengths of the other sides can be determined from equations involving the appropriate trigonometric functions.

2. If the lengths of two sides are given, the third side can be found by using the Pythagorean Theorem. The two acute angles can be determined from equations involving the appropriate trigonometric functions.

Let's consider two examples of solving right triangles.

EXAMPLE 1

Solve the right triangle in Figure 2.21.

Solution Because A and B are complementary angles, $B = 90° - 34° = 56°$. Using the tangent function, a can be determined as follows.

$$\tan 34° = \frac{a}{8.1}$$

$$a = 8.1 \tan 34° = 5.5 \text{ centimeters, to the nearest tenth}$$

Using the cosine function, c can be determined as follows.

$$\cos 34° = \frac{8.1}{c}$$

$$c \cos 34° = 8.1$$

$$c = \frac{8.1}{\cos 34°} = 9.8 \text{ centimeters, to the nearest tenth}$$

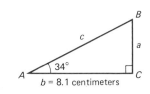

Figure 2.21

In Example 1 we could also have found c by using the secant function.

$$\sec 34° = \frac{c}{8.1}$$

$$c = 8.1 \sec 34° = 9.8 \text{ centimeters}$$

Before considering the next right-triangle problem, we need to expand our use of the calculator. We need to be able to find the measure of an acute angle

when given a trigonometric functional value of the angle. Consider the following examples.

EXAMPLE 2

Find the measure of an acute angle θ when given (a) $\sin \theta = .2706$, (b) $\cos \theta = .9449$, (c) $\tan \theta = 3.8947$, and (d) $\csc \theta = 1.249$.

Solution First, be sure that your calculator is in the degree mode.

(a) Enter the number .2706 and press the $\boxed{\text{INV}}$ and $\boxed{\text{SIN}}$ keys in that order. Your result should be $\theta = 15.7°$ to the nearest tenth of a degree.

(b) Enter the number .9449 and press the $\boxed{\text{INV}}$ and $\boxed{\text{COS}}$ keys in that order. Your result should be $\theta = 19.1°$, to the nearest tenth of a degree.

(c) Enter the number 3.8947 and press the $\boxed{\text{INV}}$ and $\boxed{\text{TAN}}$ keys in that order. Your result should be $\theta = 75.6°$, to the nearest tenth of a degree.

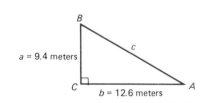

(d) Enter the number 1.249 and press the $\boxed{1/x}$, $\boxed{\text{INV}}$, and $\boxed{\text{SIN}}$ keys in that order. Your result should be $\theta = 53.2°$. (Pressing the $\boxed{1/x}$ key first after entering the number produces the value of $\sin \theta$.)

Remark: The $\boxed{\text{INV}}$ key refers to the concept of an inverse function. The inverse trigonometric functions are specifically defined and discussed in the next chapter. Some calculators may have an $\boxed{\text{ARC}}$ key instead of an $\boxed{\text{INV}}$ key.

Now let's return to solving some problems that deal with right triangles.

EXAMPLE 3

Solve right triangle ACB for which $C = 90°$, $a = 9.4$ meters, and $b = 12.6$ meters.

Solution Let's sketch the figure and record the known facts (Figure 2.22). Using the Pythagorean Theorem, c can be obtained as follows.

$$c = \sqrt{a^2 + b^2} = \sqrt{(9.4)^2 + (12.6)^2}$$
$$= \sqrt{247.12} = 15.7 \text{ meters, to the nearest tenth of a meter}$$

Angles A and B can be found as follows.

$$\tan A = \frac{9.4}{12.6} = .7460$$

$$A = 36.7°, \text{ to the nearest tenth of a degree}$$

$$\tan B = \frac{12.6}{9.4} = 1.3404$$

$$B = 53.3°, \text{ to the nearest tenth of a degree}$$

Figure 2.22

B

a = 9.4 meters

c

C   *b* = 12.6 meters   *A*

Notice in Example 3 that after finding angle A to be 36.7°, we did not subtract this value from 90° to find angle B. In so doing, an error in calculating angle A would produce a corresponding error in angle *B*. Instead, we would suggest finding *A* and *B* from the given information and then using the complementary relationship for checking purposes.

> ## *Applications*

Right-triangle trigonometry has a variety of applications. Let's consider a few examples.

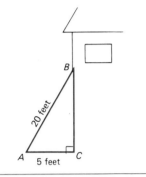

Figure 2.23

EXAMPLE 4

Suppose that a 20-foot ladder is placed against a building so that its lower end is 5 feet from the base of the building (Figure 2.23). What angle does the ladder make with the ground?

Solution Using the cosine function, angle *A* can be determined as follows.

$$\cos A = \frac{5}{20} = .2500$$

$$A = 75.5°, \text{ to the nearest tenth of a degree}$$

In Figure 2.24 we have indicated some terminology that is commonly used in **line of sight** problems. Angles of elevation and depression are measured with reference to a horizontal line. If the object being sighted is above the observer, then the angle formed by the line of sight and the horizontal line is called an **angle of elevation** (Figure 2.24(a)). If the object being sighted is below the observer, then the angle formed by the line of sight and the horizontal line is called an **angle of depression** (Figure 2.24(b)).

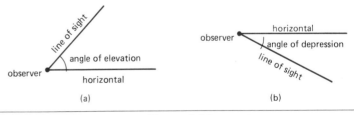

Figure 2.24

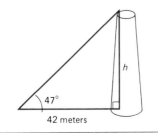

Figure 2.25

EXAMPLE 5

At a point 42 meters from the base of a smokestack, the angle of elevation of the top of the stack is 47°. Find the height of the smokestack to the nearest meter.

Solution Let's sketch a figure and record the given information (Figure 2.25). Letting *h* represent the height of the stack and using the tangent function, we can find *h* as follows.

$$\tan 47° = \frac{h}{42}$$

$$h = 42 \tan 47° = 45, \text{ to the nearest meter}$$

The height of the smokestack is approximately 45 meters.

EXAMPLE 6

From the top of a building 350 feet tall, the angle of depression of a special landmark is 71.5°. How far from the base of the building is the landmark?

Solution Figure 2.26 depicts the situation described in the problem. Angle θ and the given angle of depression are alternate interior angles; thus, $\theta = 71.5°$. Using the tangent function of θ we can find x as follows.

$$\tan 71.5° = \frac{350}{x}$$

$$x \tan 71.5° = 350$$

$$x = \frac{350}{\tan 71.5°} = 117 \text{ feet, to the nearest foot}$$

The landmark is approximately 117 feet from the base of the building.

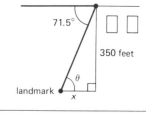

Figure 2.26

EXAMPLE 7

A TV tower stands on the top of the One Shell Plaza building in Houston. From a point 3000 feet from the base of the building, the angle of elevation of the top of the TV tower is 18.4°. If the building itself is 714 feet tall, how tall is the TV tower?

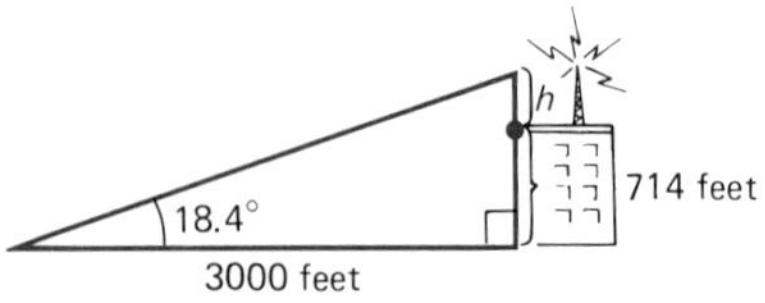

Solution Figure 2.27 depicts the situation described in the problem. Using the tangent function we can proceed as follows.

$$\tan 18.4° = \frac{h + 714}{3000}$$

$$h + 714 = 3000 \tan 18.4°$$

$$h = 3000 \tan 18.4° - 714 = 284 \text{ feet, to the nearest foot}$$

The TV tower is approximately 284 feet tall.

Figure 2.27

Our final example of this section illustrates a situation where two overlapping right triangles are used to generate a system of two equations and two unknowns.

EXAMPLE 8

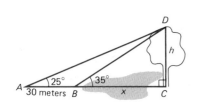

Figure 2.28

In Figure 2.28 a tree is located on the opposite side of a pond from points A and B. From point B, the angle of elevation to the top of the tree is 35°. From point A, the angle of elevation to the top of the tree is 25°. If points A and B are 30 meters apart, find the height of the tree to the nearest tenth of a meter.

Solution Let x represent the distance between B and C. Using right triangle BCD we have

$$\tan 35° = \frac{h}{x} \quad \text{where } h = x \tan 35°.$$

Using right triangle ACD we have $\tan 25° = h/(x + 30)$ where $h = (x + 30)\tan 25°$. Equating the two values for h and solving for x produces

$$x \tan 35° = (x + 30)\tan 25°$$

$$x \tan 35° = x \tan 25° + 30 \tan 25°$$

$$x \tan 35° - x \tan 25° = 30 \tan 25°$$

$$x(\tan 35° - \tan 25°) = 30 \tan 25°$$

$$x = \frac{30 \tan 25°}{\tan 35° - \tan 25°}$$

$$= 59.81, \text{ to the nearest hundredth.}$$

Substituting 59.81 for x in $h = x \tan 35°$ produces

$$h = 59.81 \tan 35° = 41.9, \text{ to the nearest tenth.}$$

The tree is approximately 41.9 meters tall.

Problem Set 2.3

For Problems 1–14, find an acute angle θ to the nearest tenth of a degree, satisfying the given conditions.

1. $\sin \theta = .2233$ 2. $\cos \theta = .7902$ 3. $\cos \theta = .4051$ 4. $\sin \theta = .8281$

5. $\tan \theta = .3365$ 6. $\tan \theta = 1.235$ 7. $\cot \theta = .5704$ 8. $\cot \theta = 6.940$

9. $\sin \theta = .5182$ 10. $\cos \theta = .7768$ 11. $\tan \theta = 6.400$ 12. $\tan \theta = .4937$

13. $\sec \theta = 3.103$ 14. $\csc \theta = 1.354$

In Problems 15–24, we are referring to a right triangle labeled as follows.

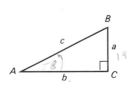

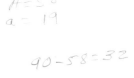

Solve each of the right triangles expressing lengths of sides to the nearest unit, and angles to the nearest degree.

15. $A = 37°$ and $b = 14$ 16. $A = 58°$ and $a = 19$ 17. $B = 23°$ and $b = 12$

18. $B = 42°$ and $a = 9$ 19. $A = 67°$ and $c = 26$ 20. $B = 19°$ and $c = 34$

21. $a = 5$ and $b = 12$ 22. $a = 24$ and $c = 25$ 23. $b = 12$ and $c = 29$

24. $a = 18$ and $b = 14$

25. A 30-foot ladder, leaning against the side of a building, makes a 50° angle with the ground. How far up on the building does the top of the ladder reach? Express your answer to the nearest tenth of a foot.

26. For safety purposes, a manufacturer recommends that the maximum angle made by a ladder with the ground as it leans against a building be 70°. What is the maximum height, to the nearest tenth of a foot, that the top of a 24-foot ladder can reach? How far, to the nearest tenth of a foot, is the bottom of the ladder from the base of the building?

27. From a point 50 meters from the base of a fir tree, the angle of elevation to the top of the tree is 61.5°. Find the height of the tree to the nearest tenth of a meter.

28. Bill is standing on top of a 175-foot cliff overlooking a lake. The measurement of the angle of depression to a boat on the lake is 29°. How far is the boat from the base of the cliff? Express your answer to the nearest foot.

29. From a point 2156 feet from the base of the Sears Tower in Chicago, Illinois the angle of elevation to the top of the building is 34°. Find, to the nearest foot, the height of the Sears Tower.

30. A radar station is tracking a missile. The angle of elevation is 22.6° and the line of sight distance is 36.8 kilometers (see the accompanying figure). Find the altitude and the horizontal range of the missile to the nearest tenth of a kilometer.

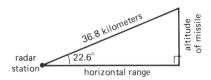

31. A person wishing to know the width of a river walks 50 yards downstream from a point directly across from a tree on the opposite bank. The angle between the river bank and the line of sight to the tree at this point is 40°. Find the width of the river to the nearest yard.

32. A diagonal of a rectangle is 17 centimeters long and makes an angle of 27° with a side of the rectangle. Find the length and width of the rectangle to the nearest centimeter.

33. Use the information in the following illustration to compute, to the nearest foot, the height of the Gateway Arch in St. Louis, Missouri.

34. From the top of the Ala Moana Hotel in Honolulu, Hawaii the angle of depression of a landmark on the ground is 57.3°. If the height of the hotel is 390 feet, determine, to the nearest foot, the distance that the landmark is from the hotel.

35. An upward moving escalator, 35 feet long, has an angle of elevation of 34°. What is the vertical distance between the floors traveled by the escalator? Express your answer to the nearest tenth of a foot.

36. The lengths of the three sides of an isosceles triangle are 18 centimeters, 18 centimeters, and 12 centimeters. Find the measure of each of the three angles to the nearest tenth of a degree.

37. A TV tower stands on top of the Empire State building. From a point 1150 feet from the base of the building, the angle of elevation of the top of the TV tower is 52°. If the building itself is 1250 feet tall, find the height of the TV tower to the nearest foot.

38. In the accompanying illustration find the length of x to the nearest tenth of a meter.

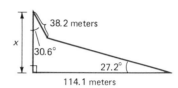

39. The length of each blade of a pair of shears from the pivot to the point is 8 inches. Find, to the nearest tenth of a degree, what angle the blades make with each other when the points of the open shears are 6 inches apart.

40. Two people 200 feet apart are in line with the base of a tower. The angle of elevation of the top of the tower from one person is 30° and from the other person is 60°. How far is the tower from each person? (Since the angles are 30° and 60°, try doing this problem without your calculator.)

41. Two buildings are separated by an alley that is 15 feet wide. From a second floor window in one of the buildings, it can be determined that the angle of elevation of the top of the other building is 75° and the angle of depression of the bottom of the building is 50°. Find the height, to the nearest foot, of the building for which these measurements have been taken.

42. Two buildings are separated by an alley. From a window 66 feet above the ground in one of the buildings, it can be observed that the angle of elevation of the top of the other building is 52°, and the angle of depression of the bottom of the building is 68°. Find the height, to the nearest foot, of the building for which these angle measurements have been taken.

43. A TV antenna sits on top of a building. From a point on the ground at a distance of 75 feet from the base of the building, a person determines the angle of elevation of the bottom of the antenna to be 48°, and the angle of elevation of the top of the antenna to be 54°. Find the height of the antenna to the nearest foot.

44. Find h, to the nearest tenth of a meter, in the accompanying illustration.

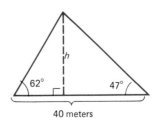

45. The right circular cone in the accompanying figure has a radius of 3 feet and a volume of 50 cubic feet. Find the measure of angle θ, to the nearest tenth of a degree. The volume of a right circular cone is given by the formula $V = \frac{1}{3}\pi r^2 h$.

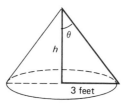

46. A jet takes off at a 15° angle traveling 220 feet per second. Find, to the nearest second, how long it takes to reach an altitude of 12,000 feet.

47. Suppose that a spacelab is circling the earth at an altitude of 400 miles as shown in the accompanying figure. The angle θ in the figure is 65.5°. Use this information to estimate the radius of the earth to the nearest mile.

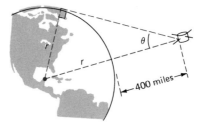

48. The length of a guy-wire to a pole is 60 feet as indicated in the accompanying figure. It makes an angle of 72° with the ground. How high above the ground is it attached to the pole? Express the answer to the nearest tenth of a foot.

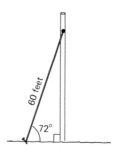

49. The rectangular box in the accompanying diagram is 20 inches long, 12 inches wide, and 10 inches deep. Find the angle θ, to the nearest tenth of a degree, formed by a diagonal of the base and a diagonal of the box.

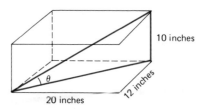

50. A draw bridge is 160 feet long. As shown in the accompanying figure, the two sections of the bridge can be lifted upward to an angle of $40°$. If the water level is 20 feet below the bridge, find the distance d between the end of a section and the water level when the bridge is fully open. Express the answer to the nearest foot.

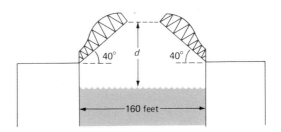

2.4

Solving Oblique Triangles: Law of Cosines

In the previous section we solved a variety of problems using the trigonometric functions of the acute angles of right triangles. Now we want to expand our problem solving capabilities to include any kind of triangle. Any triangle that is not a right triangle is called an **oblique triangle**.

In elementary geometry you studied the concept of congruence as it pertains to geometric figures. Two geometric figures are said to be congruent if they have exactly the same shape and size; that is to say, they can be made to coincide. You also discovered that there are certain conditions that determine the **congruence of triangles**. For example, the SAS condition states that if two sides and the included angle of one triangle are equal in measure, respectively, to two sides and the included angle of another triangle, then the triangles are congruent. In other words, knowing the lengths of two sides and the measure of the included angle determines the exact shape and size of the triangle. Let's work through an example to illustrate this idea.

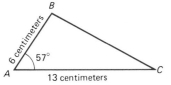

Figure 2.29

EXAMPLE 1

Using the information given in Figure 2.29 find the length of $\overline{BC}$.

Solution By drawing a line segment $\overline{BD}$ perpendicular to $\overline{AC}$, we can form two right triangles as indicated in Figure 2.30. From triangle ADB, the values of x and h can be found as follows.

$$\cos 57° = \frac{x}{6} \qquad\qquad \sin 57° = \frac{h}{6}$$

$$x = 6\cos 57° = 3.3 \qquad h = 6\sin 57° = 5.0$$

Now applying the Pythagorean Theorem to triangle CDB we obtain

$$\begin{aligned} a^2 &= h^2 + (13 - x)^2 \\ &= (5.0)^2 + (9.7)^2 \qquad (13 - x = 13 - 3.3 = 9.7) \\ &= 25.0 + 94.09 = 119.09. \end{aligned}$$

Figure 2.30

Thus,

$$a = \sqrt{119.09} = 11 \text{ centimeters, to the nearest centimeter.}$$

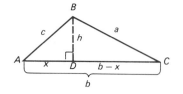

Figure 2.31

Without carrying out the details, it should be evident that the measures of angles B and C in Figure 2.30 are also determined by the given information. In other words, knowing the lengths of two sides and the measure of the angle included by those two sides does determine the remaining parts of the triangle.

Our experience in solving the previous problem should give us some direction as to how to develop a general formula for solving such problems. Let's consider a triangle labeled as in Figure 2.31. By drawing $\overline{BD}$ perpendicular to $\overline{AC}$, we can form two right triangles. From triangle ADB we obtain

$$\cos A = \frac{x}{c} \quad \text{and} \quad \sin A = \frac{h}{c}$$
$$x = c \cos A \qquad h = c \sin A.$$

Using the Pythagorean Theorem and triangle CDB produces

$$a^2 = h^2 + (b - x)^2$$
$$= h^2 + b^2 - 2bx + x^2.$$

Now we can substitute $c \cos A$ for x and $c \sin A$ for h and simplify.

$$a^2 = h^2 + b^2 - 2bx + x^2$$
$$= (c \sin A)^2 + b^2 - 2b(c \cos A) + (c \cos A)^2$$
$$= c^2 \sin^2 A + b^2 - 2bc \cos A + c^2 \cos^2 A$$
$$= c^2 \sin^2 A + c^2 \cos^2 A + b^2 - 2bc \cos A$$
$$= c^2 (\sin^2 A + \cos^2 A) + b^2 - 2bc \cos A \qquad \text{(Remember that } \sin^2 A + \cos^2 A = 1.\text{)}$$
$$= c^2 + b^2 - 2bc \cos A$$

A similar type of development could be used to show that $b^2 = a^2 + c^2 - 2ac \cos B$ and $c^2 = a^2 + b^2 - 2ab \cos C$. These three relationships are referred to as the **Law of Cosines** and can be formally stated as follows.

Law of Cosines

In any triangle ABC having sides of length a, b, and c, the following relationships are true.

$$a^2 = b^2 + c^2 - 2bc \cos A$$
$$b^2 = a^2 + c^2 - 2ac \cos B$$
$$c^2 = a^2 + b^2 - 2ab \cos C$$

Remark: You should realize that the development based on Figure 2.31 assumes angle A to be an acute angle. A similar type of development does follow if A is an obtuse angle, and we will have you carry out the details in the next set of problems.

Using the appropriate part of the Law of Cosines, problems like Example 1 are easy to solve. Let's consider another example of that type.

EXAMPLE 2

Using the information given in Figure 2.32, find the value of c to the nearest tenth of a meter.

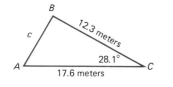

Figure 2.32

Solution Using $c^2 = a^2 + b^2 - 2ab \cos C$ we obtain

$$c^2 = (12.3)^2 + (17.6)^2 - 2(12.3)(17.6)\cos 28.1°$$
$$= 79.12.$$

Therefore,

$$c = \sqrt{79.12} = 8.9 \text{ meters, to the nearest tenth of a meter.}$$

The SSS Condition from Elementary Geometry

Referring back again to elementary geometry, you may recall the SSS property of congruence. It stated that if the lengths of three sides of one triangle are equal to the lengths of three sides of another triangle, then the triangles are congruent. In other words, knowing the lengths of three sides of a triangle determines that triangle's exact shape and size. Furthermore, from the Law of Cosines we see that $\cos A$, $\cos B$, and $\cos C$ can each be expressed in terms of the lengths of the sides as follows.

$$\cos A = \frac{b^2 + c^2 - a^2}{2bc} \qquad \cos B = \frac{a^2 + c^2 - b^2}{2ac} \qquad \cos C = \frac{a^2 + b^2 - c^2}{2ab}$$

Therefore, it becomes evident that the Law of Cosines can also be used to find the size of the angles of a triangle when given the lengths of the three sides. Let's consider an example.

EXAMPLE 3

Find the measure of each angle of a triangle having sides of length 9 feet, 15 feet, and 19 feet.

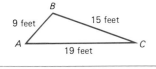

Figure 2.33

Solution Let's sketch and label a triangle to organize our use of the Law of Cosines (Figure 2.33).

$$\cos A = \frac{b^2 + c^2 - a^2}{2bc}$$

$a = 15$
$b = 19$
$c = 9$

$$= \frac{19^2 + 9^2 - 15^2}{2(19)(9)} = \frac{217}{342} = .6345, \text{ to the nearest ten-thousandth}$$

Therefore, $A = 50.6°$, to the nearest tenth of a degree.

$$\cos B = \frac{a^2 + c^2 - b^2}{2ac}$$

$$= \frac{15^2 + 9^2 - 19^2}{2(15)(9)} = -\frac{55}{270} = -.2037$$

Therefore, $B = 101.8°$, to the nearest tenth of a degree.

$$\cos C = \frac{a^2 + b^2 - c^2}{2ab}$$

$$= \frac{15^2 + 19^2 - 9^2}{2(15)(19)} = \frac{505}{570} = .8860$$

Therefore, $C = 27.6°$, to the nearest tenth of a degree. As a partial check, we see that $A + B + C = 50.6° + 101.8° + 27.6° = 180°$.

EXAMPLE 4

A vertical pole 50 feet tall stands on a hillside that makes an angle of $25°$ with the horizontal. A guy-wire is attached to the top of the pole and to a point on the hillside 35 feet down from the base of the pole. Find the length of the guy-wire.

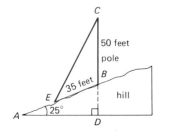

Figure 2.34

Solution First, let's sketch and record the data from the problem (Figure 2.34). From the figure we see that $\angle ABD = 90° - 25° = 65°$. Therefore, $\angle CBE = 180° - 65° = 115°$. Now using triangle EBC and the Law of Cosines, we can find the length of the guy-wire (EC).

$$(EC)^2 = (35)^2 + (50)^2 - 2(35)(50)\cos 115°$$
$$= 1225 + 2500 - (-1479.16)$$
$$= 5204.16, \text{ to the nearest hundredth}$$

Therefore,

$$EC = \sqrt{5204.16} = 72.1 \text{ feet, to the nearest tenth of a foot.}$$

EXAMPLE 5

A triangular plot of ground has sides of lengths 300 yards, 275 yards, and 250 yards. Find the measure, to the nearest tenth of a degree, of the smallest angle of the triangle.

Solution Let's sketch the triangular plot and record the given information. Since the smallest angle will be opposite the smallest side, we are looking for the angle θ opposite the side that is 250 yards long. Therefore,

$$\cos \theta = \frac{(300)^2 + (275)^2 - (250)^2}{2(300)(275)}$$

$$= \frac{103125}{165000} = .625$$

and

$$\theta = 51.3°, \text{ to the nearest tenth of a degree.}$$

One final comment about the material in this section needs to be made. Keep in mind that the Law of Cosines was developed by partitioning an oblique triangle into two right triangles and then using our knowledge of right triangles.

Therefore, if we should forget the Law of Cosines, all is not lost. We can solve a specific problem by forming right triangles and using our knowledge of right triangle trigonometry as we did in Example 1 of this section.

Problem Set 2.4

Each of the Problems 1–10 refers to triangle ABC. Express measures of angles to the nearest tenth of a degree and lengths of sides to the nearest tenth of a unit.

1. If $b = 8$ centimeters, $c = 12$ centimeters, and $A = 53°$, find a.

2. If $c = 13$ meters, $a = 10$ meters, and $B = 22°$, find b.

3. If $a = 11.6$ feet, $b = 5.1$ feet, and $C = 85°$, find c.

4. If $b = 21.4$ yards, $c = 15.1$ yards, and $A = 74°$, find a.

5. If $a = 27$ centimeters, $c = 21$ centimeters, and $B = 112°$, find b.

6. If $a = 14$ centimeters, $b = 18$ centimeters, and $c = 12$ centimeters, find A.

7. If $a = 17$ feet, $b = 25$ feet, and $c = 17$ feet, find B.

8. If $a = 14.6$ meters, $b = 11.2$ meters, and $c = 4.1$ meters, find C.

9. If $a = 8.3$ centimeters, $b = 16.4$ centimeters, and $c = 11.8$ centimeters, find A.

10. If $a = 7.2$ feet, $b = 11.4$ feet, and $c = 5.1$ feet, find B.

11. In triangle ABC, if $a = 7$ yards, $b = 10$ yards, and $c = 4$ yards, find A, B, and C to the nearest tenth of a degree.

12. In triangle ABC, if $a = 7.1$ centimeters, $b = 17.8$ centimeters, and $c = 12.5$ centimeters, find A, B, and C to the nearest tenth of a degree.

13. In triangle ABC, if $a = 41$ feet, $c = 32$ feet, and $B = 100°$, find b to the nearest tenth of a foot and A and C to the nearest tenth of a degree.

14. In triangle ABC, if $a = 8.1$ inches, $b = 14.3$ inches, and $C = 12°$, find c to the nearest tenth of an inch, and A and B to the nearest tenth of a degree.

15. In triangle ACB, if $a = 24$ centimeters, $b = 7$ centimeters, and $c = 25$ centimeters, find C.

16. Suppose that the lengths of the sides of a triangle are given as $a = 7.3$ inches, $b = 14.1$ inches, and $c = 5.2$ inches. Using our calculator to find the size of angle A, we keep getting an error message. Explain the reason for this.

17. The diagonals of a parallelogram are of length 30 centimeters and 22 centimeters, and intersect at an angle of $67°$. Find the lengths of the sides of the parallelogram to the nearest tenth of a centimeter. (Remember that the diagonals of a parallelogram bisect each other.)

18. A triangular plot of ground measures 50 meters by 65 meters by 80 meters. Find, to the nearest tenth of a degree, the size of the angle opposite the longest side.

19. City A is located 14 miles due north of City B. City C is located 17 miles from City B on a line that is $13°$ north of west from B. Find the distance between City A and City C, to the nearest mile.

20. A plane travels 250 miles due east after takeoff, then adjusts its course $15°$ southward and flies another 175 miles. How far is it from the point of departure? Express your answer to the nearest tenth of a mile.

21. To measure the length of a pond, a surveyor determines the measurements indicated on the accompanying diagram. Find the length of the pond to the nearest meter.

Problems 22–24 refer to baseball fields consisting of the 90-foot square diamond and the surrounding outfield as indicated in the following diagram.

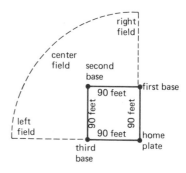

22. The pitcher's mound on a baseball diamond is located 60.5 feet from home plate on the diagonal connecting home plate to second base. Find the distance, to the nearest tenth of a foot, between the pitcher's mound and first base.

23. Suppose that the center fielder is standing in center field 375 feet from home plate. How far is it from where he is standing to third base? Express your answer to the nearest foot.

24. Suppose the left fielder is standing 320 feet from home plate and the line segment that connects him to home plate bisects the line segment connecting second base and third base. How far is he from second base? Express your answer to the nearest foot.

25. The following diagram shows a rooftop with some indicated measurements. Find, to the nearest tenth of a degree, the size of angle θ.

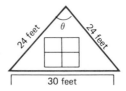

26. A solar collector is attached to a roof as indicated in the following diagram. Find the length of the collector to the nearest tenth of a foot.

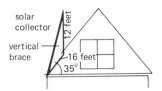

27. Two airplanes leave an airport at the same time, one going west at 375 miles per hour and the other going northeast at 425 miles per hour. How far apart are they 2 hours after leaving? Express your answer to the nearest mile.

28. Two points A and B are on opposite sides of a building. In order to find the distance between the points, Bob chooses a point C that is 200 feet from A and 175 feet from B, and then determines that angle ACB has a measure of $110°$. Find, to the nearest foot, the distance between A and B.

29. The rectangular box in the accompanying figure is 20 inches long, 12 inches wide, and 10 inches deep. Find the angle θ, to the nearest tenth of a degree, formed by a diagonal of the base and a diagonal of the $12''$-by-$10''$ side.

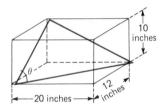

30. A parallelogram has sides of lengths 25 centimeters and 40 centimeters and one of the angles has measure $62°$. Find the length of each diagonal to the nearest tenth of a centimeter.

31. Verify that $a^2 = b^2 + c^2 - 2bc \cos A$ for a triangle in which A is an obtuse angle.

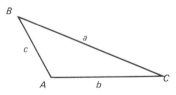

2.5

Law of Sines

Again, from elementary geometry you may recall the ASA property of congruence. It states that if two angles and the included side of one triangle are equal in measure, respectively, to two angles and the included side of another triangle, then the triangles are congruent. That is to say, knowing the measures of two angles and the included side determines the exact shape and size of the triangle. Furthermore, knowing the size of two angles of a triangle determines the size of the third angle. Therefore, the ASA property can also be stated as an AAS property.

In the previous section we found that the Law of Cosines can be used to solve triangles when the given information fits the SAS or SSS properties. However, the Law of Cosines is not useful in the ASA or AAS situations. Instead we need another set of properties referred to as the **Law of Sines**.

Let's consider triangle ACB in Figure 2.35 with $\overline{BD}$ drawn perpendicular to $\overline{AC}$. Using right triangle ADB we have

$$\sin A = \frac{h}{c}$$

$$h = c \sin A.$$

Using right triangle CDB, we obtain

$$\sin C = \frac{h}{a}$$

$$h = a \sin C.$$

Equating the two expressions for h produces

$$a \sin C = c \sin A,$$

which can be written as

$$\frac{a}{\sin A} = \frac{c}{\sin C}.$$

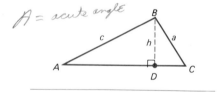

A = acute angle

Figure 2.35

Returning to triangle ACB in Figure 2.35 we could also draw a line segment from vertex C perpendicular to $\overline{AB}$. A similar line of reasoning would produce the relationship

$$\frac{a}{\sin A} = \frac{b}{\sin B}.$$

Therefore, the Law of Sines can be stated as follows.

Law of Sines

In any triangle ABC, having sides of length a, b, and c, the following relationships are true.

$$\frac{a}{\sin A} = \frac{b}{\sin B} = \frac{c}{\sin C}$$

Remark: The previous development of the Law of Sines was based on Figure 2.35, which has angle A as an acute angle. A similar type of development does follow if A is an obtuse angle. We will have you carry out the details of this in the next set of problems.

Let's illustrate some uses of the Law of Sines.

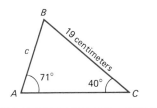

Figure 2.36

EXAMPLE 1

Suppose that in triangle ACB, $A = 71°$, $C = 40°$, and $a = 19$ centimeters. Find c to the nearest tenth of a centimeter.

Solution Let's sketch a triangle and record the given information (Figure 2.36). Using $a/\sin A = c/\sin C$ from the Law of Sines we can substitute 19 for a, 71° for A, 40° for C, and solve for c.

$$\frac{a}{\sin A} = \frac{c}{\sin C}$$

$$\frac{19}{\sin 71°} = \frac{c}{\sin 40°}$$

$$c \sin 71° = 19 \sin 40°$$

$$c = \frac{19 \sin 40°}{\sin 71°} = 12.9 \text{ centimeters, to the nearest tenth}$$
$$\text{of a centimeter}$$

EXAMPLE 2

Two points A and B are on opposite sides of a river. Point C is located 350 feet from A on the same side of the river as A. In triangle ACB, $C = 52°$ and $A = 67°$. Find the distance between A and B to the nearest foot. (See Figure 2.37.)

Solution Because $A + B + C = 180°$, $A = 67°$, and $C = 52°$, we know that

$$67° + B + 52° = 180°$$

$$B + 119° = 180°$$

$$B = 61°.$$

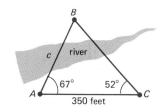

Figure 2.37

Now using $\dfrac{b}{\sin B} = \dfrac{c}{\sin C}$, we obtain

$$\frac{350}{\sin 61°} = \frac{c}{\sin 52°}$$

$$c \sin 61° = 350 \sin 52°$$

$$c = \frac{350 \sin 52°}{\sin 61°} = 315 \text{ feet, to the nearest foot.}$$

The next example illustrates a situation where both the Law of Cosines and the Law of Sines can be used to find the size of an angle, *but* one approach may be preferable to the other. Let's solve the problem both ways to illustrate our point.

Figure 2.38

EXAMPLE 3

In Figure 2.38, $A = 23.1°$, $a = 14$ yards, $b = 21$ yards, and $c = 8$ yards. Find B to the nearest tenth of a degree.

Solution A Using the Law of Cosines we obtain

$$\cos B = \frac{a^2 + c^2 - b^2}{2ac}$$

$$= \frac{14^2 + 8^2 - 21^2}{2(14)(8)}$$

$$= -\frac{181}{224} = -.8080, \text{ to the nearest ten-thousandth.}$$

Therefore,

$$B = 143.9°, \text{ to the nearest tenth of a degree.}$$

Solution B Using the Law of Sines we obtain

$$\frac{a}{\sin A} = \frac{b}{\sin B}$$

$$\frac{14}{\sin 23.1°} = \frac{21}{\sin B}$$

$$14 \sin B = 21 \sin 23.1°$$

$$\sin B = \frac{21 \sin 23.1°}{14}$$

$$= .5885.$$

Using our calculator we obtain $B = 36.1°$, to the nearest tenth of a degree. But remember that the sine function is also positive in the second quadrant. Therefore, B might be an obtuse angle with a measure of $180° - 36.1° = 143.9°$. Note that the Law of Sines does not indicate which value of B should be used for a particular problem. However, in this problem our sketch of the triangle in Figure 2.38 clearly indicates that B is an obtuse angle and thus we need to use $B = 143.9°$.

Relative to Solution B, we should realize that situations do arise where a rough sketch of the triangle may not be sufficient to decide which value of an angle determined by the Law of Sines is to be used. Therefore, in general we suggest that you use the Law of Cosines whenever possible.

The SSA Situation (The Ambiguous Case)

Thus far, the Law of Cosines and the Law of Sines have provided us with methods for solving triangles when the given information fits the SAS, SSS, ASA, or AAS patterns. But what happens if the given information fits the SSA pattern? In other words, how do we solve a triangle if we know the lengths of two sides and the measure of an angle opposite one of those two sides? This question is a little more difficult to answer because the given information of **the lengths of two sides and the measure of an angle opposite one of those two sides** may determine two triangles, one triangle, or no triangle at all. Let's examine why this is true.

Consider triangle ACB assuming that we are given A, a, and c. The following situations may exist.

Suppose A is an acute angle.

If $h < a < c$, then two possible triangles exist.

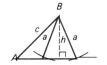

If $a > c$, then one triangle exists.

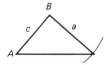

If $a = h$, then one triangle exists.

If $a < h$, then no triangle is determined.

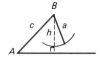

Suppose A is an obtuse angle.

If $a > c$, then one triangle exists.

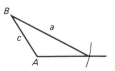

If $a < c$, then no triangle is determined.

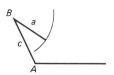

Fortunately, it is not necessary to memorize all of the possibilities in the previous list. Instead, each problem can be analyzed on an individual basis. Frequently, by making a careful sketch and using some common sense, we can determine which situation exists. The important issue is that we are alert to the fact that when the given information is of the SSA pattern, then various possibilities might exist. Let's analyze a few problems.

EXAMPLE 4

Suppose we are given $A = 57°$, $a = 17$ feet, and $c = 14$ feet. How many triangles exist satisfying these conditions? Find C for each such triangle.

Solution Let's make a careful sketch (Figure 2.39). Because $17 > 14$ there is only one possible triangle determined. We can use the Law of Sines to determine C.

$$\frac{a}{\sin A} = \frac{c}{\sin C}$$

$$\frac{17}{\sin 57°} = \frac{14}{\sin C}$$

$$17 \sin C = 14 \sin 57°$$

$$\sin C = \frac{14 \sin 57°}{17}$$

$$= .6907, \text{ to the nearest ten-thousandth}$$

Therefore,

$$C = 43.7° \quad \text{or} \quad C = 136.3°, \text{ to the nearest tenth of a degree.}$$

Either from the figure or from the fact that $57° + 136.3° > 180°$, we know that the answer of $136.3°$ must be discarded and $C = 43.7°$.

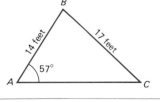

Figure 2.39

EXAMPLE 5

Suppose we are given $A = 43.2°$, $a = 7.7$ meters, and $c = 9.1$ meters. How many triangles exist satisfying these conditions? Find C for each such triangle.

Solution As we attempt to sketch a triangle for this situation (Figure 2.40) we might not be able to tell whether two, one, or no triangles exist. Therefore, let's find an approximate value for h.

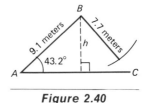

Figure 2.40

$$\sin 43.2° = \frac{h}{9.1}$$

$$h = 9.1 \sin 43.2° = 6.2 \text{ meters, to the nearest tenth of a meter}$$

Now, because $h < a < c$, we know that two triangles exist as indicated in Figure 2.41.

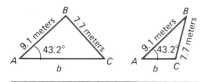

Figure 2.41

Using the Law of Sines we can find the two possible values for C as follows.

$$\frac{a}{\sin A} = \frac{c}{\sin C}$$

$$\frac{7.7}{\sin 43.2°} = \frac{9.1}{\sin C}$$

$$7.7 \sin C = 9.1 \sin 43.2°$$

$$\sin C = \frac{9.1 \sin 43.2°}{7.7} = .8090$$

Therefore,

$$C = 54.0° \quad \text{or} \quad C = 180° - 54.0° = 126.0°, \text{to the nearest tenth}$$
$$\text{of a degree.}$$

Each of the two values for angle C in Example 5 produces a different value for the length of side b as indicated by the two triangles in Figure 2.41. If $C = 54.0°$, then $B = 180° - (43.2° + 54.0°) = 82.8°$. Then b can be determined using the Law of Sines.

$$\frac{b}{\sin 82.8°} = \frac{7.7}{\sin 43.2°}$$

$$b \sin 43.2° = 7.7 \sin 82.8°$$

$$b = \frac{7.7 \sin 82.8°}{\sin 43.2°}$$

$$b = 11.2 \text{ meters, to the nearest tenth of a meter}$$

If $C = 126.0°$, then $B = 180° - (43.2° + 126.0°) = 10.8°$. Then b can be determined as follows.

$$\frac{b}{\sin 10.8°} = \frac{7.7}{\sin 43.2°}$$

$$b \sin 43.2° = 7.7 \sin 10.8°$$

$$b = \frac{7.7 \sin 10.8°}{\sin 43.2°}$$

$$b = 2.1 \text{ meters, to the nearest tenth of a meter}$$

EXAMPLE 6

Suppose we are given $A = 68°$, $a = 22$ inches, and $c = 25$ inches. How many triangles exist satisfying these conditions? Find C for each such triangle.

Solution As we attempt to sketch a triangle for the situation (Figure 2.42), we might not be able to tell whether two, one, or no triangles exist. Therefore, let's find an approximate value for h.

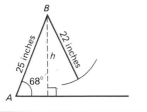

Figure 2.42

$$\sin 68° = \frac{h}{25}$$

$$h = 25 \sin 68° = 23.2 \text{ inches}$$

Because $a < h$, no triangle exists satisfying these conditions.

In Example 6, had we attempted to find c using the Law of Sines, the following situation would have arisen.

$$\frac{a}{\sin A} = \frac{c}{\sin C}$$

$$\frac{22}{\sin 68°} = \frac{25}{\sin C}$$

$$22 \sin C = 25 \sin 68°$$

$$\sin C = \frac{25 \sin 68°}{22} = 1.053$$

At this stage, we should recognize that because the sine of an angle cannot exceed 1, the set of conditions given in this problem does not determine a triangle.

Problem Set 2.5

Each of the Problems 1–10 refers to triangle ABC. Express measures of angles to the nearest tenth of a degree and lengths of sides to the nearest tenth of a unit.

1. If $A = 64°$, $C = 47°$, and $a = 17$ centimeters, find c.
2. If $A = 28°$, $C = 61°$, and $a = 6$ feet, find c.
3. If $A = 20.4°$, $B = 31.2°$, and $b = 25$ meters, find a.
4. If $B = 115°$, $C = 32°$, and $c = 6.1$ yards, find b.
5. If $A = 41°$, $C = 37°$, and $a = 14$ centimeters, find B; then find b.
6. If $A = 26.3°$, $B = 94.5°$, and $a = 9.2$ feet, find C; then find c.
7. If $A = 132°$, $C = 17°$, and $a = 75$ miles, find B; then find b.
8. If $A = 34.1°$, $a = 23$ feet, $b = 35$ feet, and $c = 17$ feet, find B and C.
9. If $B = 71.7°$, $a = 15$ miles, $b = 17$ miles, and $c = 14$ miles, find A and C.
10. If $C = 136°$, $a = 6$ kilometers, $b = 8$ kilometers, and $c = 13$ kilometers, find A and B.
11. In triangle ABC, if $A = 54°$, $C = 33°$, and $a = 28$ feet, find b and c to the nearest tenth of a foot.
12. In triangle ABC, if $A = 144°$, $B = 19°$, and $b = 12$ yards, find a and c to the nearest tenth of a yard.

In Problems 13–20, first decide whether two, one, or no triangles are determined by the given information. If one or two triangles are determined, calculate all possible values for the indicated part.

13. $A = 59°$, $a = 14$ centimeters, $c = 9$ centimeters; find C to the nearest tenth of a degree.
14. $A = 17°$, $a = 7$ feet, $c = 22$ feet; find C to the nearest tenth of a degree.
15. $A = 53.1°$, $a = 10$ meters, $c = 14$ meters; find C to the nearest tenth of a degree.

16. $A = 119°$, $a = 25$ yards, $c = 12$ yards; find C to the nearest tenth of a degree.

17. $A = 28°$, $a = 19$ miles, $c = 32$ miles; find B and C to the nearest tenth of a degree, and find b to the nearest tenth of a mile.

18. $A = 55°$, $a = 31$ feet, $c = 34$ feet; find B and C to the nearest tenth of a degree.

19. $A = 30°$, $a = 21$ centimeters, $c = 42$ centimeters; find C to the nearest degree and b to the nearest tenth of a centimeter.

20. $A = 124°$, $a = 21$ yards, $c = 27$ yards; find C to the nearest tenth of a degree and b to the nearest tenth of a yard.

21. Verify $a/\sin A = c/\sin C$ for a triangle in which A is an obtuse angle.

22. Two points A and B are located on opposite sides of a river. Point C is located 275 meters from B and on the same side of the river as B. In triangle ABC, $B = 71°$ and $C = 46°$. Find the distance between A and B to the nearest meter.

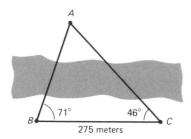

23. In the following illustration, a surveyor is standing at point A wanting to locate point C such that A, B, and C are on a straight line. However, his line of sight toward C is blocked by a building. Therefore, he determines angle ABP to be $140°$, the distance from B to P to be 75 feet, and angle BPC to be $58°$. Find the distance from P to C to the nearest tenth of a foot.

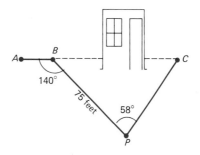

24. One end of a 20-foot plank is placed on the ground at a point 8 feet from the start of a

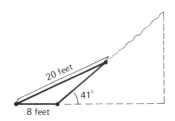

41° incline. The other end of the plank rests on the incline. How far up the incline does the plank extend? Express the answer to the nearest tenth of a foot.

25. A 200-foot TV antenna stands on the top of a hill that has an incline of 23° with the horizontal. How far down the hill will a 150-foot support cable extend if it is attached halfway up the antenna? Express the answer to the nearest foot.

26. A building 55 feet tall is on top of a hill that has an incline of 19°. A surveyor, standing at a point on the hillside, determines that the angle of elevation to the top of the building is 42°. (Remember that an angle of elevation is measured with reference to a horizontal line.) How far is the surveyor from the bottom of the building? Express the answer to the nearest tenth of a foot.

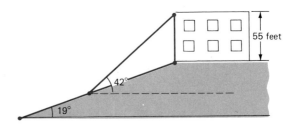

27. A triangular lot faces two streets that meet at an angle of 82°. The two sides of the lot facing the streets are each 175 feet long. Find the length of the third side to the nearest tenth of a foot.

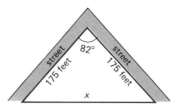

28. Two people 75 feet apart are in line with the base of a tower. The angle of elevation of the top of the tower from one person is 41.2° and from the other person is 32.6°. Find the height of the tower to the nearest tenth of a foot.

29. A triangular lot is bounded by three straight streets, Washington, Jefferson, and Monroe. Washington and Monroe Streets intersect at an angle of 65°. The side of the lot along Washington is 250 meters long and the side along Jefferson is 235 meters long. Find the length of the side of the lot along Monroe Street. Express the length to the nearest meter.

30. Two landmarks, A and B, are on the same side of a river and 750 yards apart. A surveyor stands at point C on the opposite side of the river, thus forming triangle ABC. If C = 57.2° and the distance between A and C is 600 yards, find the distance between B and C to the nearest yard.

31. A solar collector that is 10 feet wide is attached to a roof as indicated in the following diagram. The roof makes an angle of 35° with the horizontal and the collector makes an angle of 50° with the horizontal. Find the length of the vertical brace to the nearest tenth of a foot.

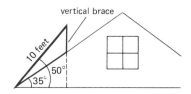

32. An office building is located at the top of a hill. From the base of the hill to the top of the office building, the angle of elevation is 40°. From a point 150 feet from the base of the hill, the angle of elevation to the top of the building is 32°. The hill rises at a 25° angle with the horizontal. Find the height of the office building to the nearest tenth of a foot.

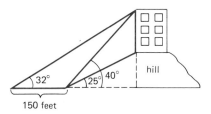

Miscellaneous Problems

33. Consider the following triangle.

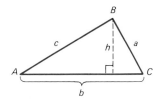

The area (K) of the triangle is given by $K = \frac{1}{2}bh$. Furthermore, $\sin C = \frac{h}{a}$ or $h = a \sin C$. Therefore, substituting $a \sin C$ for h in the formula $K = \frac{1}{2}bh$ produces

$$K = \frac{1}{2}ab \sin C.$$

This formula can be used to find the area of a triangle if the measurements of two sides and the included angle are known. Use the formula to help solve the following problems.

(a) Find the area of a triangle if two sides are 18 meters and 24 meters long, and the included angle has a measure of 47°. Express the area to the nearest square meter.

(b) Two sides of a triangular plot of ground are 75 yards and 90 yards long, and the angle included by those two sides is 65°. Find the area of the plot to the nearest square yard.

(c) The gable end of a house is shown in the accompanying figure. Find the area of the shaded triangle, to the nearest square foot.

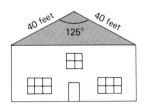

34. From the law of sines we know that $\dfrac{a}{\sin A} = \dfrac{b}{\sin B}$ or $b = \dfrac{a \sin B}{\sin A}$. Substituting this expression for b in $K = \dfrac{1}{2} ab \sin C$ produces

$$K = \frac{a^2 \sin B \sin C}{2 \sin A}.$$

This formula can be used to find the area of a triangle if the measurements of two angles and the included side are known. Use it to help solve the following problems.

(a) Find the area of a triangle if two angles have measures of $43°$ and $61°$, and the included side is 14 centimeters long. Express the area to the nearest square centimeter.

(b) A triangular plot of ground has two angles that measure $50°$ each. The side included between the two angles is 80 feet long. Find the area of the plot to the nearest square foot.

(c) The dimensions of a triangle are as indicated in the following figure. Find the area to the nearest square inch.

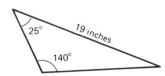

35. There is yet another formula, commonly referred to as Heron's Formula, that can be used to find the area of a triangle if the lengths of the three sides are known. We will not show the development of the formula but simply state it and have you use it to solve some problems. The area of a triangle is given by

$$K = \sqrt{s(s - a)(s - b)(s - c)},$$

where a, b, and c are the lengths of the three sides and $s = \frac{1}{2}(a + b + c)$.

(a) Find the area, to the nearest square centimeter, of a triangle that measures 14 centimeters by 16 centimeters by 18 centimeters.

(b) Find the area, to the nearest square yard, of a triangular plot of ground that measures 45 yards, by 60 yards, by 75 yards.

(c) Find the area of an equilateral triangle, each of whose sides is 18 inches long. Express the area to the nearest square inch.

2.6

Vectors

Figure 2.43

In the previous three sections we solved numerous problems involving the lengths of sides of triangles. We did not specify any direction in which the measurements were made; in other words, the sides of the triangles had magnitude (length), but no direction. Many quantities, however, can be described completely by specifying both magnitude and direction. For example, we speak of a car traveling north at 55 miles per hour or an airplane flying east at 350 miles per hour. Another example is force that is exerted with a given magnitude and in a given direction—30 pounds exerted at a 30° angle to the horizontal. Quantities having both magnitude and direction are called **vector quantities**.

Geometrically, vector quantities can be represented by directed line segments called **vectors**. In Figure 2.43, a vector has been drawn from point A (called the **initial point**) to the point B (called the **terminal point**) and is denoted by $\overrightarrow{AB}$. Boldface letters are also commonly used to name vectors. For example, vectors **u** and **v** are also represented in Figure 2.43. (Since boldface letters are hard to write, you might use $\vec{u}$ and $\vec{v}$.) The length of a directed line segment represents the **magnitude** of the vector and is denoted by $\|\overrightarrow{AB}\|$ or $\|\mathbf{v}\|$. Vectors that have the same magnitude and same direction are considered **equal vectors**. In Figure 2.43, **u** and **v** are equal vectors and we write **u** = **v**. The vector quantities (car traveling north at 55 miles per hour, airplane flying east at 350 miles per hour, and a force of 30 pounds acting at a 30° angle to the horizontal) are represented in Figure 2.44.

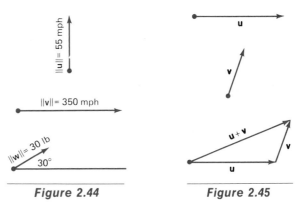

Figure 2.44   Figure 2.45

Vector Addition

The **sum** (also called the **resultant**) of two vectors **u** and **v** is pictured in Figure 2.45. The initial point of **v** is placed at the terminal point of **u** and the vector from the initial point of **u** to the terminal point of **v** is the sum **u** + **v**.

Another way to describe the addition of vectors is known as the **parallelogram law**. Consider two vectors $\overrightarrow{AB}$ and $\overrightarrow{AC}$ having a common initial

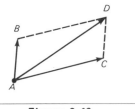

Figure 2.46

point as in Figure 2.46. Complete the parallelogram $ACDB$. Since $\overrightarrow{AB} = \overrightarrow{CD}$, the diagonal vector $\overrightarrow{AD}$ represents the sum of $\overrightarrow{AC}$ and $\overrightarrow{AB}$. So we can write $\overrightarrow{AC} + \overrightarrow{AB} = \overrightarrow{AD}$. (If $\overrightarrow{AC}$ and $\overrightarrow{AB}$ are perpendicular, then $\overrightarrow{AD}$ will be the diagonal of a rectangle.) If $\overrightarrow{AC}$ and $\overrightarrow{AB}$ represent two physical forces acting on some object at A, then it can be shown by physical experiments that $\overrightarrow{AD}$ represents the **resultant force**, that is, the single force that produces the same effect as the two combined forces. In other words, vector addition is consistent with the action of the vector quantities being added.

Scalar Multiplication and Vector Subtraction

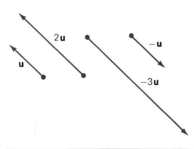

Figure 2.47

If $\mathbf{u}$ is a vector, then $2\mathbf{u}$ is the vector in the same direction as $\mathbf{u}$ but twice as long; $-3\mathbf{u}$ is three times as long as $\mathbf{u}$ but in the opposite direction (Figure 2.47). In general, if k is a real number (called a **scalar**) and $\mathbf{u}$ is any vector, then $k\mathbf{u}$ (called the **scalar multiple** of $\mathbf{u}$) is a vector whose magnitude is $|k|$ times the magnitude of $\mathbf{u}$ and whose direction is the same as $\mathbf{u}$ if $k > 0$ and opposite that of $\mathbf{u}$ if $k < 0$. In particular, $(-1)\mathbf{u}$ (usually written as $-\mathbf{u}$) has the same magnitude as $\mathbf{u}$ but is directed in the opposite direction of $\mathbf{u}$, as indicated in Figure 2.47. The vector $-\mathbf{u}$ is called the **opposite of** or the **negative of $\mathbf{u}$** and when added to $\mathbf{u}$ produces the **zero vector** denoted by $\mathbf{0}$. So we can write $\mathbf{u} + (-\mathbf{u}) = \mathbf{0}$. (The zero vector is interpreted as a point having a magnitude of 0 and no direction.) The zero vector is the identity element for vector addition, that is, $\mathbf{u} + \mathbf{0} = \mathbf{0} + \mathbf{u} = \mathbf{u}$. Finally, vector subtraction is defined by $\mathbf{u} - \mathbf{v} = \mathbf{u} + (-\mathbf{v})$.

EXAMPLE 1

Using the following two vectors, draw $3\mathbf{u} + 2\mathbf{v}$ and $\mathbf{u} - 3\mathbf{v}$.

Solution Place $3\mathbf{u}$ (a vector in the same direction as $\mathbf{u}$ but three times as long) and $2\mathbf{v}$ (a vector in the same direction as $\mathbf{v}$ but twice as long) with their initial points in common. Then complete the parallelogram and draw $3\mathbf{u} + 2\mathbf{v}$. Place $\mathbf{u}$ and $-3\mathbf{v}$

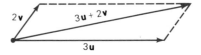

(a vector in the opposite direction of $\mathbf{v}$ and three times as long) with their initial points in common. Then complete the parallelogram and draw $\mathbf{u} - 3\mathbf{v}$.

Vectors and Problem Solving

Vectors can be used to solve a large variety of problems from diverse areas. In this brief introduction to vectors, we want to show you a few such applications. Some of these problems can be solved using only right triangle trigonometry, but others require the use of the Law of Cosines and the Law of Sines as applied to oblique triangles.

PROBLEM 1

Suppose that an airplane flying at 300 miles per hour is headed due north but is blown off of its course by the wind that is blowing at 30 miles per hour from the west. By what angle is the plane blown off of its intended path and what is its actual ground speed?

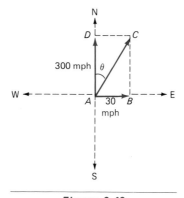

Figure 2.48

Solution The velocity of the plane and the wind velocity can be represented by two vectors, as shown in Figure 2.48. The actual path of the plane is represented by the resultant $\overrightarrow{AC}$, the diagonal of the rectangle. Angle θ is the angle that the plane is blown off of its intended course. Because DC is also 30, we can find θ by using the tangent function.

$$\tan \theta = \frac{30}{300} = .1$$

Therefore,

$\theta = 5.7°$ to the nearest tenth of a degree.

The magnitude of $\overrightarrow{AC}$ represents the ground speed of the plane.

$$\|\overrightarrow{AC}\|^2 = 300^2 + 30^2 = 90{,}900$$

Therefore,

$\|\overrightarrow{AC}\| = 301.5$ miles per hour.

Instead of adding two vectors to form a resultant vector, it is sometimes necessary to reverse the process and to find two vectors whose sum is a given vector. The two vectors that we find are called **components** of the given vector. The components are especially easy to find if they are to be perpendicular.

PROBLEM 2

Using a rope to pull a sled with a child on it, a man applies a force of 90 pounds. If the rope makes a 40° angle with the horizontal (Figure 2.49), find the horizontal force that tends to move the sled along the ground and the vertical force that tends to lift the sled vertically.

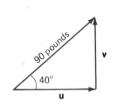

Figure 2.49

Solution   The vectors **u** and **v** in Figure 2.49 represent the horizontal and vertical components, respectively, of the given force of 90 pounds.

$$\cos 40° = \frac{\|\mathbf{u}\|}{90}$$

$$\|\mathbf{u}\| = 90 \cos 40° = 68.9, \text{ to the nearest tenth of a pound}$$

$$\sin 40° = \frac{\|\mathbf{v}\|}{90}$$

$$\|\mathbf{v}\| = 90 \sin 40° = 57.9, \text{ to the nearest tenth of a pound}$$

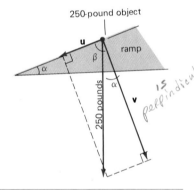

250-pound object

ramp

Figure 2.50

Suppose that a spherical object weighing 250 pounds is placed on an inclined ramp, as indicated in Figure 2.50. Gravity pulls directly downward with a force of 250 pounds. Part of this force (represented by **u**) tends to pull the object down the ramp, and another part (represented by **v**) presses the object against the ramp on a line perpendicular to the ramp. The angle of the ramp, α, is also the angle between two of the vectors since they are both complementary to angle β. Therefore, if the angle of the ramp is known, then the 250-pound resultant vector, which is a diagonal of the rectangle, can be *resolved into two components*—one directed down the ramp and the other directed perpendicular to the ramp. Let's consider a specific problem.

PROBLEM 3

A 250-pound spherical lead ball is placed on a ramp that has an incline of 22° with the horizontal. How much force is pulling down the ramp and how much force is pressing on a line perpendicular against the ramp?

Solution   We can use Figure 2.50 with $\alpha = 22°$.

$$\cos 22° = \frac{\|\mathbf{v}\|}{250}$$

$$\|\mathbf{v}\| = 250 \cos 22°$$
$$= 231.8, \text{ to the nearest tenth} \quad 231.8 \text{ lbs.}$$

$$\sin 22° = \frac{\|\mathbf{u}\|}{250}$$

$$\|\mathbf{u}\| = 250 \sin 22°$$
$$= 93.7, \text{ to the nearest tenth}$$

$$\left(\sec x - \tan x\right)\left(\frac{1}{\sin x} + 1\right)$$

$$\frac{\sec x}{\sin x} \cdot \frac{\tan x}{\sin x} + \frac{\sec x}{1} - \frac{\tan x}{1}$$

(annotations: $\frac{\sin}{\cos}$, $\frac{\sin}{\cos}$, "Foil")

~~sec~~

$$\frac{1}{\sin x \cos x} - \frac{\sin x}{\cos x}$$

~~2cos x~~ $\boxed{\cot x}$

$$\frac{1 - \sin^2 x}{\sin x \cos x} = \frac{\cos^2 x}{\sin x \cos x}$$

(0 -1 2 3 -1<)

$$\frac{\sec x}{\sin x} + \frac{\sec x}{1} - \frac{\tan x}{\sin x} - \frac{\tan x}{1}$$

$$\frac{\frac{1}{\cos}}{\sec x}{\sin x} - \tan x \Bigg| \qquad \frac{1 - \sin^2 x}{\sin x \cos x}$$

$$\frac{\frac{1}{\cos}}{\sin}$$

S - Satisfactory; only issued in courses numbered 050 and below.

U - Unsatisfactory; only issued in courses numbered 050 and below.

DF - Deferred grade; only issued in Math and English courses that are self-paced. The DF grade indicates that the student is making normal progress but has been unable to complete the required work. The student must re-enroll in the course and complete the required work the following semester (Spring and Summer session excluded).

5. An asterisk next to a name indicates that the student is receiving veterans benefits. A veteran receiving an 'N' grade must have the last recorded date of attendance written after the name. A 'W' grade must be accompanied by a dated drop card. This is to comply with VA regulations and veterans benefits.

6. Return only the top copy to the Student Records Office. The carbon copy is for your records or give it to your Instructional Coordinator.

7. After grades are mailed to students (approximately 3-4 days) an audit sheet will be distributed to you to verify and/or correct any omissions or errors (human and machine). If errors or omissions are indicated on your audit sheet, make the corrections and return the sheet to the Student Records Office immediately.

8. If you have any questions please call 973-3546.

CLASSES END FRIDAY, AUGUST 4. GRADE ROSTERS MUST BE RETURNED TO THE STUDENT RECORDS OFFICE BY 12 NOON ON SATURDAY AUGUST 5. The Student Records office will be open on Saturday from 9 a.m. - 12 noon to accept grade sheets.

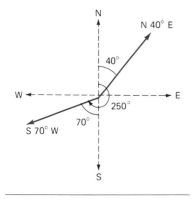

Figure 2.51

Therefore, there is a force of approximately 93.7 pounds acting down the ramp and 231.8 pounds acting against the ramp.

In certain navigation problems, the **direction** or **bearing** of a ship may be given by stating the size of an acute angle that is measured with respect to a north-south line. For example, a bearing of N40°E denotes an angle whose one side points north and whose other side points 40° east of north, as indicated in Figure 2.51. In the same figure, we have indicated a bearing of S70°W, that is, 70° west of south.

In air navigation, directions and bearing are often specified by measuring from a north-line in a **clockwise** direction. (The angles are stated as positive angles even though they are measured in a clockwise direction.) Thus, in Figure 2.51 the bearing of S70°W could also be expressed as a bearing or direction of 250°.

PROBLEM 4

If a ship sails 25 miles in the direction N40°E and then 60 miles straight east, find its distance and bearing with respect to its starting point.

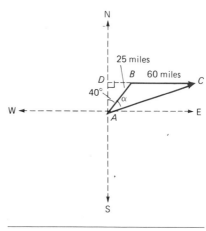

Figure 2.52

Solution   The route of the ship is indicated in Figure 2.52. The measure of $\angle ABD$ is $90° - 40° = 50°$. Therefore, the measure of $\angle ABC$ is $180° - 50° = 130°$. Then the magnitude of $\overrightarrow{AC}$ can be found by using the Law of Cosines.

$$\|\overrightarrow{AC}\|^2 = 25^2 + 60^2 - 2(25)(60)\cos 130°$$
$$= 6153.36, \text{ to the nearest hundredth}$$

Therefore,

$$\|\overrightarrow{AC}\| = 78.4 \text{ miles, to the nearest tenth of a mile.}$$

Then the measure of angle α can be found by the Law of Sines.

$$\frac{\sin \alpha}{60} = \frac{\sin 130°}{78.4}$$

$$78.4 \sin \alpha = 60 \sin 130°$$

$$\sin \alpha = \frac{60 \sin 130°}{78.4} = .5863$$

$$\alpha = 35.9°, \text{ to the nearest tenth of a degree}$$

From Figure 2.52 we see that the bearing of $\overrightarrow{AC}$ is stated in terms of $\alpha + 40°$; it is, therefore, N75.9°E.

Problem Set 2.6

For Problems 1–4, use the following vectors **u** and **v** to draw the indicated vectors.

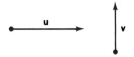

1. **u** + **v**   2. **u** − **v**   3. 2**u** + 3**v**   4. **u** − 2**v**

For Problems 5–10, use the following vectors **u** and **v** to draw the indicated vectors.

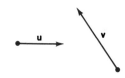

5. **u** − **v**   6. **u** + **v**   7. **v** + 2**u**   8. 2**u** + **v**

9. 2**u** − **v**   10. 2**v** − **u**

11. Suppose that an airplane flying at 250 miles per hour is headed due south but it is blown off of its course by the wind that is blowing at 40 miles per hour from the east. By what angle is the plane blown off of its intended path and what is its actual ground speed?

12. A boat that can travel 15 miles per hour in still water attempts to go directly across a river that is flowing at 4 miles per hour. By what angle is the boat pushed off of its intended path?

13. A river flows due south at 125 feet per minute. In what direction must a motorboat that can travel 500 feet per minute, be headed so that it actually travels due east?

14. A balloon is rising at 5 feet per second while a wind is blowing at 15 feet per second. Find the speed of the balloon and the angle it makes with the horizontal.

15. Two forces of 50 pounds and 75 pounds act on an object at right angles. Find the magnitude of the resultant force and the angle it makes with the larger force.

16. Two forces of 4 kilograms and 9 kilograms act on an object at right angles. Find the magnitude of the resultant force and the angle it makes with the smaller force.

17. An airplane heads in a direction of 80° at 200 kilometers per hour. A wind blowing in the direction of 170° forces the plane onto a course that is due east. Find the speed of the wind.

18. An airplane heads in a direction of 265° at 330 miles per hour. A wind blowing in the direction of 355° forces the plane onto a course that is due west. Find the speed of the wind.

19. A force of 45 pounds is acting on an object at a 20° angle to the horizontal. Find the horizontal and vertical components of the force.

20. An airplane is flying at 275 miles per hour in a direction of 235°. Find the westerly and southerly components of its velocity.

21. A boy exerts a force of 20 pounds to pull a wagon with some toys in it. The handle of the wagon makes a 35° angle with the horizontal. Find the horizontal and vertical components of the 20-pound force.

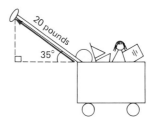

22. An automobile weighing 3200 pounds is parked on a ramp that makes a 15° angle with the horizontal. How much force is exerted parallel to the ramp and how much force is exerted perpendicular to the ramp?

23. A small car weighing 1750 pounds is parked on a ramp that makes a 27° angle with the horizontal. How much force must be applied parallel to the ramp to keep the car from rolling down the ramp?

24. A force of 30 pounds is needed to hold a barrel in place on a ramp that makes an angle of 32° with the horizontal. Find the weight of the barrel.

25. A force of 52.1 pounds is needed to keep a 75-pound lead ball from rolling down a ramp. Find the angle that the ramp makes with the horizontal.

26. A weight of 60 pounds is held by two wires as indicated in the following diagram. Find the horizontal component of force F_1.

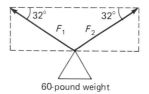

60-pound weight

27. Forces of 220 kilograms and 175 kilograms act on an object. The angle between the forces is 65°. Find the magnitude of the resultant of the two forces and also find the angle that the resultant makes with the larger force.

28. Forces of 105 pounds and 85 pounds act on an object. The angle between the forces is 42°. Find the magnitude of the resultant of the two forces and also find the angle that the resultant makes with the smaller force.

29. If a ship sails 35 miles in the direction N32°W and then 70 miles straight west, find its distance and bearing with respect to its starting point.

30. If a ship sails 50 miles in the direction S47°E and then 80 miles straight east, find its distance and bearing with respect to its starting point.

31. Suppose that a boat travels 70 miles in the direction N50°E and then travels 85 miles in the direction S65°E. Find its distance and bearing with respect to its starting point.

32. Suppose that a boat travels 40 miles in the direction S27°W and then 90 miles straight east. Find its distance and bearing with respect to its starting point.

Miscellaneous Problems

33. Give a geometric argument that vector addition is a commutative operation.

34. Give a geometric argument that vector addition is an associative operation.

35. Give a geometric argument that $k(\mathbf{u} + \mathbf{v}) = k\mathbf{u} + k\mathbf{v}$ for any vectors $\mathbf{u}$ and $\mathbf{v}$, and any real number k.

Chapter Summary

If θ is an angle in standard position and $P(x, y)$ is *any point* on the terminal side of θ, then

$$\sin \theta = \frac{y}{r}, \qquad \csc \theta = \frac{r}{y}$$

$$\cos \theta = \frac{x}{r}, \qquad \sec \theta = \frac{r}{x}$$

$$\tan \theta = \frac{y}{x}, \qquad \cot \theta = \frac{x}{y}.$$

If $r = 1$, then $\sin \theta = y$ and $\cos \theta = x$.

If θ is any angle in standard position with its terminal side in one of the four quadrants, then the **reference angle** of θ (we have called it θ') is the acute angle formed by the terminal side of θ and the x-axis.

The trigonometric functions of any angle θ are equal to those of the reference angle associated with θ, except possibly for the sign. The sign can be determined by considering the quadrant in which the terminal side of θ lies.

You should be able to find, without a calculator or a table, the exact trigonometric functional values of any angle having a reference angle of 30°, 45°, or 60°.

Be sure that you have the following relationships at your fingertips:

$$\csc \theta = \frac{1}{\sin \theta}, \qquad \sec \theta = \frac{1}{\cos \theta}, \qquad \cot \theta = \frac{1}{\tan \theta}$$

$$\tan \theta = \frac{\sin \theta}{\cos \theta}, \qquad \cot \theta = \frac{\cos \theta}{\sin \theta}, \qquad \sin^2 \theta + \cos^2 \theta = 1,$$

$$\sin(-\theta) = -\sin \theta, \qquad \cos(-\theta) = \cos \theta, \qquad \tan(-\theta) = -\tan \theta$$

Problem solving is the central theme of Sections 2.3–2.6.

Solving a right triangle can be analyzed as follows.

1. If an acute angle and the length of one side are known, then the other acute angle is the complement of the one given. The lengths of the other two sides can be found by solving an equation using the appropriate trigonometric functions.

2. If the lengths of two sides are given, the third side can be found by using the Pythagorean Theorem. The two acute angles can be determined from equations using the appropriate trigonometric functions.

The following relationships are referred to as the **Law of Cosines**.

$$a^2 = b^2 + c^2 - 2bc \cos A$$
$$b^2 = a^2 + c^2 - 2ac \cos B$$
$$c^2 = a^2 + b^2 - 2ab \cos C$$

If two sides and the included angle (SAS) of a triangle are known or if three sides (SSS) are known, then the Law of Cosines can be used to solve the triangle.

The following relationships are referred to as the **Law of Sines**.

$$\frac{a}{\sin A} = \frac{b}{\sin B} = \frac{c}{\sin C}$$

If two angles and a side of a triangle are known, then the Law of Sines can be used to solve the triangle.

If in a triangle two sides and an angle opposite one of them are known, then two, one, or no triangles may be determined.

Vectors—as directed line segments—play an important role in many applications because they can be used to represent quantities that have both magnitude and direction. We call two vectors equal if they have the same magnitude and same direction. Vectors can be added and multiplied by real numbers called scalars. Problems 1–4 of Section 2.6 provide a good review of some of the applications of vectors.

Chapter 2 Review Problem Set

For Problems 1–4, point P is a point on the terminal side of θ and θ is in standard position. Find $\sin \theta$, $\cos \theta$, and $\tan \theta$.

1. $P(-6, 8)$ 2. $P(1, -3)$ 3. $P(-2, -4)$ 4. $P(-5, -12)$

For Problems 5–16, find exact values without using a calculator or a table.

5. $\cos(-150°)$ 6. $\sin(-330°)$ 7. $\csc 45°$ 8. $\sec 120°$

9. $\tan 675°$ 10. $\cot 480°$ 11. $\sin \dfrac{7\pi}{3}$ 12. $\cos \dfrac{13\pi}{3}$

13. $\sec\left(-\dfrac{3\pi}{2}\right)$ 14. $\csc(-\pi)$ 15. $\cot\dfrac{7\pi}{4}$ 16. $\tan\dfrac{5\pi}{4}$

17. If $\sin\theta > 0$ and $\tan\theta < 0$, what quadrant contains the terminal side of θ?

18. If $\sec\theta > 0$ and $\csc\theta < 0$, what quadrant contains the terminal side of θ?

19. Find $\tan\theta$ if the terminal side of θ lies on the line $y = -5x$ in the fourth quadrant.

20. If $\sin\theta = -\frac{5}{13}$ and θ is a fourth-quadrant angle, find $\cos\theta$.

21. If $\sin\theta = .7986$, find $\sin(-\theta)$.

22. If $\cos\theta = -.4067$, find $\cos(-\theta)$.

23. If $\sin\theta = .7615$, find θ, to the nearest tenth of a degree, such that $0° < \theta < 90°$.

24. If $\cos\theta = .3697$, find θ, to the nearest tenth of a degree, such that $0° < \theta < 90°$.

25. If $\tan\theta = 13.6174$, find θ, to the nearest tenth of a degree, such that $0° < \theta < 90°$.

26. The sides of a right triangle are of length 8 feet, 15 feet, and 17 feet. Find the size of the angle, to the nearest tenth of a degree, opposite the 15-foot side.

27. From a point 80 feet from the base of a tree, the angle of elevation to the top of the tree is $28°$. Find the height of the tree to the nearest tenth of a foot.

28. A 50-foot antenna stands on top of a building. From a point on the ground away from the base of the building the angle of elevation to the top of the antenna is $51°$ and the angle of elevation to the bottom of the antenna is $42°$. Find the height of the building to the nearest foot.

29. In triangle ABC, if $A = 74°$, $c = 19$ miles, and $b = 27$ miles, find a to the nearest tenth of a mile.

30. In triangle ABC, $A = 118.2°$, $C = 17.3°$, and $c = 56$ yards. Find b to the nearest tenth of a yard.

31. If $A = 29°$, $a = 19$ meters, and $c = 37$ meters, how many triangles are determined?

32. In the accompanying figure, find x to the nearest tenth of a foot.

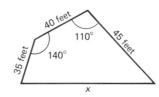

33. A triangular lot measures 52 yards by 47 yards by 85 yards. Find, to the nearest tenth of a degree, the size of the angle opposite the longest side.

34. A plane travels 350 miles due east after takeoff, then adjusts its course $20°$ northward and flies another 175 miles. How far is it from its point of departure? Express your answer to the nearest mile.

35. Two points, A and B, are located on opposite sides of a river. Point C is located 350 yards from B on the same side of the river as B. In $\triangle ABC$, $B = 69°$, and $C = 43°$. Find the distance between A and B to the nearest yard.

36. Two people 60 feet apart are in line with the base of a tower. The angle of elevation to the top of the tower from one person is $39.6°$ and from the other person is $27.4°$. Find the height of the tower to the nearest tenth of a foot.

Use vectors to help solve Problems 37–40.

37. A lawn mower is being pushed by applying a force of 45 pounds. The handle of the mower makes an angle of 40° with the horizontal. How much force is being applied horizontally and how much force is being applied downward?

38. A car weighing 2750 pounds is parked on a ramp that makes a 32° angle with the horizontal. How much force must be applied to keep the car from rolling down the ramp?

39. Forces of 130 pounds and 165 pounds act on an object. The angle between the forces is 28°. Find the magnitude of the resultant of the two forces and also find the angle that the resultant makes with the larger force.

40. A ship sails 70 miles straight west and then 110 miles in the direction N20°W. Find its distance and bearing with respect to its starting point.

3 Graphing Trigonometric Functions

Courtesy, TRW Inc.

What are the graphs of the basic trigonometric functions? Can we apply our knowledge of the graphing variations of basic curves to the graphing of trigonometric functions? Do the basic trigonometric functions have inverses? The focus of this chapter is to provide meaningful answers to these questions.

3.1

Unit Circle: Sine and Cosine Curves

Consider the circle $x^2 + y^2 = 1$, which has its center at the origin and a radius of length one unit (Figure 3.1). This is commonly referred to as the **unit circle**. The terminal side of any angle in standard position intersects the unit circle at a point $P(x, y)$ such that $r = 1$. (Note that this is equivalent to choosing a point on the terminal side such that $r = 1$, as we did in Chapter 2.) Again it should be observed that $\sin \theta = y$ and $\cos \theta = x$; so the coordinates of point P are $(\cos \theta, \sin \theta)$. In other words, the trigonometric functions $\sin \theta$ and $\cos \theta$ can be thought of as the **ordinate** and **abscissa** values, respectively, of the point of intersection of the terminal side of θ and the unit circle.

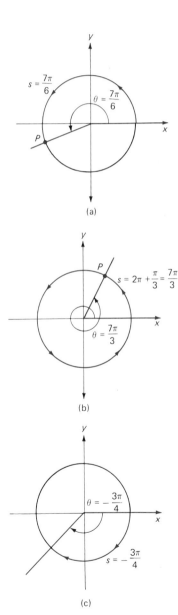

(a)

(b)

(c)

Figure 3.3

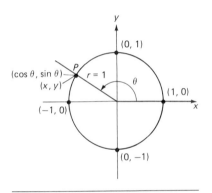

Figure 3.1

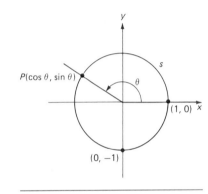

Figure 3.2

The formula for arc length, $s = r\theta$, developed in Section 1.4 takes on special significance when applied to the unit circle. In Figure 3.2, consider a central angle θ in standard position intercepting an arc s on the unit circle. Since $r = 1$, the formula $s = r\theta$ becomes $s = \theta$. That is to say, numerically the length of the arc from $(1, 0)$ to P equals the measure of angle θ in radians. For example, if

$$\theta = \frac{7\pi}{6}, \quad \text{then } s = \frac{7\pi}{6} \text{ units (Figure 3.3(a))};$$

$$\theta = \frac{7\pi}{3}, \quad \text{then } s = \frac{7\pi}{3} \text{ units (Figure 3.3(b))};$$

$$\theta = -\frac{3\pi}{4}, \quad \text{then } s = -\frac{3\pi}{4} \text{ units (Figure 3.3(c))}.$$

In general, to each real number that represents the radian measure of a central angle θ in standard position, we can associate a real number s that represents the length of the arc intercepted by the angle on the unit circle. If θ is positive, then the arc is measured from $(1, 0)$ in a counterclockwise direction to P, where P is the point of intersection of the terminal side of θ and the unit circle. If θ is negative, the arc is measured from $(1,0)$ in a clockwise direction to P. If $\theta = 0$, then $s = 0$. The point $P(x, y)$ is referred to as the **point on the unit circle that corresponds to s.**

An important consequence of the previous discussion is that we can now consider the trigonometric functions by using a domain of real numbers that are independent of any reference to angles. In such a setting, the trigonometric functions are often referred to as the **circular functions. If s is a real number and $P(x, y)$ is the point on the unit circle that corresponds to s, then**

$$\sin s = y \qquad\qquad \csc s = \frac{1}{y} \text{ (if } y \neq 0\text{)}$$

$$\cos s = x \qquad\qquad \sec s = \frac{1}{x} \text{ (if } x \neq 0\text{)}$$

$$\tan s = \frac{y}{x} \text{ (if } x \neq 0\text{)} \qquad \cot s = \frac{x}{y} \text{ (if } y \neq 0\text{)}.$$

EXAMPLE 1

Determine the circular functions $\sin s$, $\cos s$, and $\tan s$ for

(a) $s = \pi$ **(b)** $s = \dfrac{\pi}{4}$ **(c)** $s = -\dfrac{3\pi}{2}$.

Solution

(a) Since the circumference of the unit circle is 2π, the point that corresponds to $s = \pi$ is $(-1, 0)$ as indicated in Figure 3.4. Therefore,

$$\sin \pi = y = 0,$$

$$\cos \pi = x = -1,$$

and

$$\tan \pi = \frac{y}{x} = \frac{0}{-1} = 0.$$

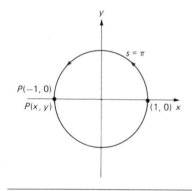

Figure 3.4

(b) Since $\pi/4 = \frac{1}{8}(2\pi)$, the point $P(x, y)$ must lie on the line $y = x$ that bisects the first quadrant (Figure 3.5). Therefore, substituting x for y in the equation $x^2 + y^2 = 1$, we obtain

$$x^2 + x^2 = 1$$

$$2x^2 = 1$$

$$x^2 = \frac{1}{2}$$

$$x = \sqrt{\frac{1}{2}} = \frac{\sqrt{2}}{2}. \qquad \text{x and y are positive because P is in the first quadrant.}$$

The coordinates of P are $(\sqrt{2}/2, \sqrt{2}/2)$ and the circular functions are determined.

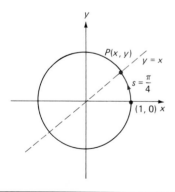

Figure 3.5

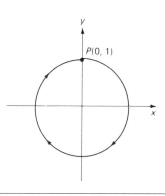

Figure 3.6

$$\sin\frac{\pi}{4} = y = \frac{\sqrt{2}}{2}$$

$$\cos\frac{\pi}{4} = x = \frac{\sqrt{2}}{2}$$

$$\tan\frac{\pi}{4} = \frac{y}{x} = \frac{\dfrac{\sqrt{2}}{2}}{\dfrac{\sqrt{2}}{2}} = 1$$

(c) Since $-\dfrac{3\pi}{2} = -\dfrac{3}{4}(2\pi)$, the point $P(x, y)$ that corresponds to $s = -\dfrac{3\pi}{2}$ is $(0, 1)$ as indicated in Figure 3.6. Therefore,

$$\sin\left(-\frac{3\pi}{2}\right) = y = 1$$

$$\cos\left(-\frac{3\pi}{2}\right) = x = 0$$

$$\tan\left(-\frac{3\pi}{2}\right) = \frac{y}{x} = \frac{1}{0}. \qquad \text{(undefined)}$$

In Example 1 we determined the values of some trigonometric (circular) functions without any reference to angles, but don't get the wrong idea. We are not trying to completely eliminate the relationships between angles and trigonometric functions. Quite the contrary, there are many applications of trigonometry that are based on an angle definition of the trigonometric functions. However, there are also numerous applications—especially in the calculus—that use the trigonometric (circular) functions without any reference to angles. So we need to be able to work with these functions in different settings.

The Sine Curve

As we graph the trigonometric functions using an xy-coordinate system, equations of the type $y = \sin x$, $y = \cos x$, $y = \tan x$, and so on, will be used. Be careful not to confuse the use of the variable x in these equations with the x employed earlier in the unit circle approach (where x denoted the x-coordinate of the point P on the unit circle). What is the graph of the function $y = \sin x$? We could begin by making a fairly extensive table of values and then plot the corresponding points. However, in this case our previous work with the sine function gives us some initial guidance. For example, we know that the sine function behaves as follows.

AS x INCREASES	THE SINE FUNCTION
from 0 to $\pi/2$	increases from 0 to 1
from $\pi/2$ to π	decreases from 1 to 0
from π to $3\pi/2$	decreases from 0 to -1
from $3\pi/2$ to 2π	increases from -1 to 0

Furthermore, we know from our work with coterminal angles (or the unit circle) that the sine function will repeat itself every 2π radians. More formally, we say that the sine function has a **period** of 2π. (A precise definition of the term "period" is given in the next section.) Now with these ideas in mind and our knowledge of some specific angles, let's set up a table of values allowing x to vary from 0 to 2π at intervals of $\pi/6$. (For graphing purposes we will use decimal approximations of the radical expressions such as $\sqrt{3}/2$.) Plotting the points, (x, y), determined by the table and connecting them with a smooth curve produces Figure 3.7.

x	$y = \sin x$
0	0
$\pi/6$	$\frac{1}{2}$
$\pi/3$	$\frac{\sqrt{3}}{2} \approx 0.87$
$\pi/2$	1
$2\pi/3$	$\frac{\sqrt{3}}{2} \approx 0.87$
$5\pi/6$	$\frac{1}{2}$
π	0
$7\pi/6$	$-\frac{1}{2}$
$4\pi/3$	$-\frac{\sqrt{3}}{2} \approx -0.87$
$3\pi/2$	-1
$5\pi/3$	$-\frac{\sqrt{3}}{2} \approx -0.87$
$11\pi/6$	$-\frac{1}{2}$
2π	0

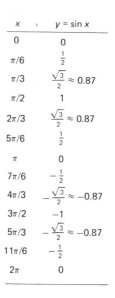

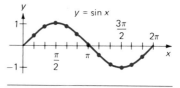

Figure 3.7

Now we can use the fact that the sine function has a period of 2π and sketch the curve from -4π to 3π as in Figure 3.8. The curve continues to repeat itself every 2π units in both directions indefinitely.

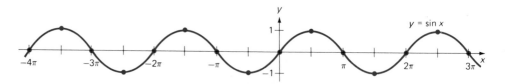

Figure 3.8

The Cosine Curve

As you might expect from the definitions of the sine and cosine functions, their graphs are very similar. The cosine function also has a period of 2π; that is to say, the graph repeats itself every 2π units. In Figure 3.9 we have graphed $y = \cos x$, for x between 0 and 2π, by plotting the points determined by the following table. Figure 3.10 depicts the cosine curve for $-2\pi \le x \le 4\pi$. Note that the cosine curve is identical to the sine curve, but moved $\pi/2$ units to the left. In other words,

$$\sin\left(x + \frac{\pi}{2}\right) = \cos x.$$

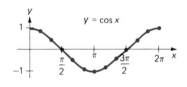

We will verify this relationship in the next chapter.

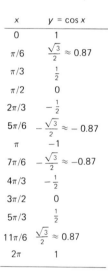

x	$y = \cos x$
0	1
$\pi/6$	$\frac{\sqrt{3}}{2} \approx 0.87$
$\pi/3$	$\frac{1}{2}$
$\pi/2$	0
$2\pi/3$	$-\frac{1}{2}$
$5\pi/6$	$-\frac{\sqrt{3}}{2} \approx -0.87$
π	-1
$7\pi/6$	$-\frac{\sqrt{3}}{2} \approx -0.87$
$4\pi/3$	$-\frac{1}{2}$
$3\pi/2$	0
$5\pi/3$	$\frac{1}{2}$
$11\pi/6$	$\frac{\sqrt{3}}{2} \approx 0.87$
2π	1

Figure 3.9

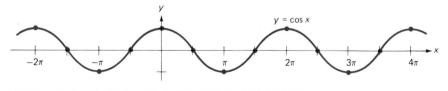

Figure 3.10

Variations of Sine and Cosine Curves

Let's briefly review a few ideas from our previous graphing experiences. Recall the following ideas from graphing parabolas.

1. The graph of $y = x^2 + 2$ is the graph of $y = x^2$ *moved up 2 units.* ($y = x^2 + 2$ could be written as $y = 2 + x^2$.)

2. The graph of $y = (x - 3)^2$ is the graph of $y = x^2$ *moved 3 units to the right.*

3. The graph of $y = -x^2$ is the graph of $y = x^2$ *reflected across the x-axis.*

Now let's apply these same ideas to the sine and cosine graphs. Furthermore, since we now know the general shapes of the sine and cosine curves, it is no longer necessary to plot so many points.

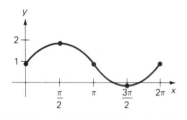

Figure 3.11

EXAMPLE 2

Graph $y = 1 + \sin x$ for $0 \le x \le 2\pi$.

Solution The graph of $y = 1 + \sin x$ is the graph of $y = \sin x$ *moved up 1 unit* (Figure 3.11).

EXAMPLE 3

Graph $y = \sin[x - (\pi/2)]$ for $\pi/2 \le x \le 5\pi/2$.

Solution The graph of $y = \sin[x - (\pi/2)]$ is the graph of $y = \sin x$ *moved $\pi/2$ units to the right.* Thus, in the interval $\pi/2 \le x \le 5\pi/2$ we get Figure 3.12.

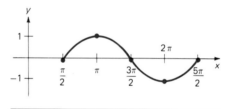

Figure 3.12

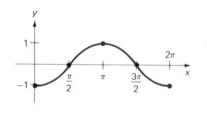

Figure 3.13

EXAMPLE 4

Graph $y = -\cos x$ for $0 \le x \le 2\pi$.

Solution The graph of $y = -\cos x$ is the graph of $y = \cos x$ *reflected across the x-axis* (Figure 3.13).

Recall that in Chapter 2 we established that $\sin(-x) = -\sin x$. Therefore, the graph of $y = \sin(-x)$ is the same as the graph of $y = -\sin x$. Also, since $\cos(-x) = \cos x$, the graph of $y = \cos(-x)$ is the same as the graph of $y = \cos x$.

Problem Set 3.1

For Problems 1–12, sketch the unit circle and indicate the given arc s along with its initial and terminal points. Then determine the exact values of the six circular functions of s without using a calculator or a table.

1. $s = \dfrac{3\pi}{4}$ 2. $s = \dfrac{7\pi}{4}$ 3. $s = \dfrac{3\pi}{2}$ 4. $s = 2\pi$

5. $s = -\dfrac{\pi}{4}$ 6. $s = -\pi$ 7. $s = \dfrac{4\pi}{3}$ 8. $s = \dfrac{7\pi}{6}$

9. $s = \dfrac{5\pi}{2}$ 10. $s = \dfrac{7\pi}{3}$ 11. $s = -\dfrac{11\pi}{4}$ 12. $s = -\dfrac{15\pi}{4}$

For Problems 13–34, graph each of the functions in the indicated interval.

13. $y = 2 + \sin x$, $-2\pi \le x \le 2\pi$ 14. $y = -1 + \sin x$, $-2\pi \le x \le 2\pi$

15. $y = 3 + \cos x$, $-2\pi \le x \le 2\pi$ 16. $y = -2 + \cos x$, $-2\pi \le x \le 2\pi$

17. $y = -\sin x$, $0 \le x \le 2\pi$ 18. $y = 1 - \sin x$, $0 \le x \le 2\pi$

19. $y = 1 - \cos x$, $0 \le x \le 2\pi$ 20. $y = -2 - \cos x$, $0 \le x \le 2\pi$

21. $y = \sin\left(x + \dfrac{\pi}{2}\right)$, $-\dfrac{\pi}{2} \le x \le \dfrac{3\pi}{2}$ 22. $y = \sin(x - \pi)$, $\pi \le x \le 3\pi$

23. $y = \cos\left(x - \dfrac{\pi}{2}\right)$, $\dfrac{\pi}{2} \le x \le \dfrac{5\pi}{2}$ 24. $y = \cos(x + \pi)$, $-\pi \le x \le \pi$

25. $y = 1 + \sin(x - \pi)$, $\pi \le x \le 3\pi$

26. $y = -2 + \sin\left(x - \dfrac{\pi}{2}\right)$, $\dfrac{\pi}{2} \le x \le \dfrac{5\pi}{2}$

27. $y = -1 + \cos(x + \pi)$, $-\pi \le x \le \pi$

28. $y = 1 + \cos\left(x - \dfrac{\pi}{2}\right)$, $\dfrac{\pi}{2} \le x \le \dfrac{5\pi}{2}$

29. $y = -\sin(x + \pi)$, $-\pi \le x \le \pi$

30. $y = -\cos\left(x + \dfrac{\pi}{2}\right)$, $-\dfrac{\pi}{2} \le x \le \dfrac{3\pi}{2}$

31. $y = 1 - \sin(x - \pi)$, $\pi \le x \le 3\pi$

32. $y = 2 - \cos(x + \pi)$, $-\pi \le x \le \pi$

33. $y = \sin(-x)$, $-2\pi \le x \le 2\pi$ (*Don't forget that* $\sin(-x) = -\sin x$.)

34. $y = \cos(-x)$, $-2\pi \le x \le 2\pi$ (*Don't forget that* $\cos(-x) = \cos x$.)

Miscellaneous Problems

For Problems 35–38, graph each of the functions in the interval $0 \le x \le 2\pi$.

35. $y = 2\sin x$ 36. $y = 3\sin x$ 37. $y = \tfrac{1}{2}\sin x$ 38. $y = -2\sin x$

39. Graph $y = \sin 2x$ in the interval $0 \le x \le 2\pi$.

40. Graph $y = \sin 4x$ in the interval $0 \le x \le 2\pi$.

41. Graph $y = \sin \tfrac{1}{2}x$ in the interval $0 \le x \le 4\pi$.

3.2

Period, Amplitude, and Phase Shift

Let's begin this section by summarizing some of the ideas from the previous section that pertain to the graphing of variations of the sine and cosine curves. We found that both the sine and cosine functions have a period of 2π. This means that their graphs repeat themselves every 2π units. The following general definition is helpful.

DEFINITION 3.1

A function f is called **periodic** if there exists a positive real number p such that

$$f(x + p) = f(x)$$

for all x in the domain of f. The smallest value for p is called the **period** of the function.

From our work with the sine and cosine functions, it is evident that 2π is the smallest positive number such that $\sin(x + 2\pi) = \sin x$ and $\cos(x + 2\pi) = \cos x$. Thus, as before, we conclude that both the sine and cosine functions are periodic with a period of 2π.

Now suppose we consider the graph of a function having an equation of the form $y = \sin bx$, where $b > 0$. One cycle of the graph is completed as bx increases from 0 to 2π. When $bx = 0$, $x = 0$ and when $bx = 2\pi$, $x = 2\pi/b$. A similar line of reasoning holds for $y = \cos bx$; therefore we can state: **the period of $y = \sin bx$ and also of $y = \cos bx$, where $b > 0$, is $2\pi/b$.** If $b < 0$, then we can first apply the appropriate property, $\sin(-x) = -\sin x$ or $\cos(-x) = \cos x$. For example, to graph $y = \sin(-3x)$ we can first change to $y = -\sin 3x$ and then proceed.

EXAMPLE 1

Find the period of $y = \cos 3x$ and sketch the graph for one period beginning at $x = 0$.

Solution The period of $y = \cos 3x$ is $2\pi/3$. Therefore, let's divide the interval from 0 to $2\pi/3$ on the x-axis into 4 equal subintervals. Each subinterval will be of length

$$\frac{\dfrac{2\pi}{3} - 0}{4} = \frac{\dfrac{2\pi}{3}}{4} = \frac{2\pi}{12} = \frac{\pi}{6}.$$

Then we can plot points determined by the endpoint values of the subintervals and sketch the curve as in Figure 3.14.

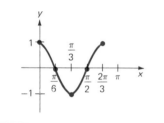

Figure 3.14

EXAMPLE 2

Find the period of $y = \sin \pi x$ and sketch the graph for one period beginning at $x = 0$.

Solution The period of $y = \sin \pi x$ is $2\pi/\pi = 2$. Again, by dividing the interval from 0 to 2 into 4 equal subintervals, we can plot points determined by the endpoint values of the subintervals and sketch the curve as in Figure 3.15.

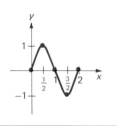

Figure 3.15

Amplitude

Consider the graph of $y = 2 \sin x$. It should be evident that for each x-coordinate, the y-coordinate is twice that of the corresponding y-coordinate on the basic sine

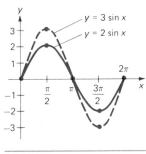

Figure 3.16

graph. Likewise, the corresponding y-coordinates of $y = 3 \sin x$ are three times those of the basic sine graph. Thus, in Figure 3.16 we have sketched the graphs of $y = 2 \sin x$ and $y = 3 \sin x$ in the interval from 0 to 2π.

The maximum functional value attained by $y = 2 \sin x$ is 2 and by $y = 3 \sin x$ is 3. Each of these maximum functional values is called the **amplitude** of the graph. In general, we can state that **the amplitude of the graph of $y = a \sin x$ or $y = a \cos x$ is |a|**. For example, the amplitude of $y = \frac{1}{2} \sin x$ is $|\frac{1}{2}| = \frac{1}{2}$ and the amplitude of $y = -2 \cos x$ is $|-2| = 2$. Together the concepts of period and amplitude help us graph functions of the form $y = a \sin bx$ or $y = a \cos bx$.

EXAMPLE 3

Find the period and amplitude of $y = -3 \sin \frac{1}{2}x$, and sketch the curve for one period beginning at $x = 0$.

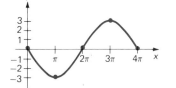

Figure 3.17

Solution The period is $2\pi/\frac{1}{2} = 4\pi$ and the amplitude is $|-3| = 3$. The curve $y = -3 \sin \frac{1}{2}x$ is an x-axis reflection of $y = 3 \sin \frac{1}{2}x$. Thus, we obtain Figure 3.17.

Phase Shift

In the previous section we found that the graph of $y = \sin[x - (\pi/2)]$ is the basic sine curve shifted $\pi/2$ units to the right. Likewise, the graph of $y = \cos(x + \pi)$ is the basic cosine curve shifted π units to the left. Each of the numbers, $\pi/2$ and π, which represent the amount of shift, is called the **phase shift** of the graph. In general, **the phase shift of $y = \sin(x - c)$ or $y = \cos(x - c)$ is |c|. If c is positive, the shift is to the ~~right~~ right and if c is negative the shift is to the ~~left~~ Left.**

Now let's pull together the concepts of period, amplitude, and phase shift in one very useful property.

PROPERTY 3.1

Consider the functions $y = a \sin b(x - c)$ and $y = a \cos b(x - c)$, where $b > 0$.

1. The period of both curves is $2\pi/b$.
2. The amplitude of both curves is $|a|$.
3. The phase shift of both curves is $|c|$. The shift is to the ~~right~~ *left* if c is positive and to the ~~left~~ *right* if c is negative.

EXAMPLE 4

Find the period, amplitude, and phase shift of $y = 2\sin(x - \pi)$, and sketch the curve.

Solution By applying Property 3.1 we can determine the following information directly from the equation.

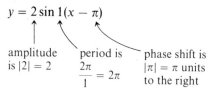

$$y = 2\sin 1(x - \pi)$$

amplitude is $|2| = 2$

period is $\dfrac{2\pi}{1} = 2\pi$

phase shift is $|\pi| = \pi$ units to the right

Thus, one complete cycle of the sine curve having an amplitude of 2 is contained in the interval from $x = \pi$ to $x = \pi + 2\pi = 3\pi$. The curve repeats itself every 2π units in both directions (Figure 3.18).

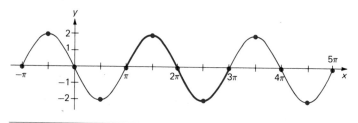

Figure 3.18

EXAMPLE 5

Find the period, amplitude, and phase shift of $y = \tfrac{1}{2}\sin(2x + \pi)$ and sketch the curve.

Solution First, let's change the form of the equation so that Property 3.1 can be applied.

$$y = \frac{1}{2}\sin(2x + \pi)$$

$$= \frac{1}{2}\sin 2\left(x + \frac{\pi}{2}\right)$$

$$= \frac{1}{2}\sin 2\left(x - \left(-\frac{\pi}{2}\right)\right)$$

Now we can obtain some information about the graph as follows.

$$y = \frac{1}{2}\sin 2\left(x - \left(-\frac{\pi}{2}\right)\right)$$

amplitude is $\left|\dfrac{1}{2}\right| = \dfrac{1}{2}$

period is $\dfrac{2\pi}{2} = \pi$

phase shift is $\left|-\dfrac{\pi}{2}\right| = \dfrac{\pi}{2}$ units to the left

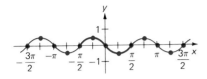

Figure 3.19

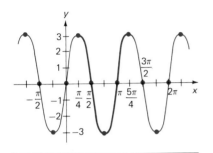

Figure 3.20

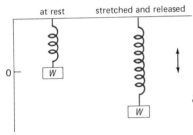

Figure 3.21

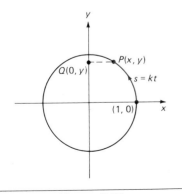

Figure 3.22

Therefore, one complete cycle of the sine curve having an amplitude of $\frac{1}{2}$ is contained in the interval from $x = -(\pi/2)$ to $x = -(\pi/2) + \pi = \pi/2$. The curve repeats itself every π units in both directions (Figure 3.19).

EXAMPLE 6

Find the period, amplitude, and phase shift of $y = 3\cos[2x - (\pi/2)]$, and sketch the curve.

Solution The given equation can be written as

$$y = 3\cos\left(2x - \frac{\pi}{2}\right)$$

$$= 3\cos 2\left(x - \frac{\pi}{4}\right).$$

From this form we can obtain the following information.

$$y = 3\cos 2\left(x - \frac{\pi}{4}\right)$$

amplitude   period is   phase shift is
is $|3| = 3$ $\dfrac{2\pi}{2} = \pi$ $\left|\dfrac{\pi}{4}\right| = \dfrac{\pi}{4}$ units to the right

Therefore, one complete cycle of the cosine curve having an amplitude of 3 is contained in the interval from $x = \pi/4$ to $x = (\pi/4) + \pi = 5\pi/4$. The curve repeats itself every π units in both directions (Figure 3.20).

Simple Harmonic Motion

Many phenomena in our world behave in a cyclic or rhythmic manner. Often such behavior can be mathematically described using variations of the sine and cosine curves. Such phenomena occur in a variety of applications ranging from alternating current in electricity to the sound waves generated by a vibrating tuning fork. There is a large class of problems known as simple harmonic motion problems. An example of simple harmonic motion is illustrated in Figure 3.21. Suppose that a weight, W, is attached to a spring. If the weight is pulled down and then released, it will oscillate up and down about the rest or equilibrium point marked 0 in Figure 3.21. Vibratory motion of this type is called simple harmonic motion.

To obtain a mathematical model of simple harmonic motion, let's use the unit circle in Figure 3.22. Consider a point $P(x, y)$ moving at a constant rate of k radians per unit of time in a counterclockwise direction around the circle. Using $(1, 0)$ as the initial position of P (when $t = 0$), then the arc distance s is given by $s = kt$ after t units of time. So, $y = \sin s = \sin kt$. Now consider point Q, which is the projection of P on the vertical axis. As P moves around the circle, point Q oscillates up and down (simple harmonic motion). At any time t, the distance of point Q from the center of the circle is given by $y = \sin kt$.

Problem Set 3.2

For Problems 1–18, graph the given function in the indicated interval.

1. $y = \cos 2x, \quad 0 \le x \le 2\pi$

2. $y = \cos \pi x, \quad 0 \le x \le 2$

3. $y = \sin 3x, \quad -\dfrac{2\pi}{3} \le x \le \dfrac{2\pi}{3}$

4. $y = \sin 2x, \quad -2\pi \le x \le 2\pi$

5. $y = 2\cos x, \quad 0 \le x \le 2\pi$

6. $y = -3\cos x, \quad 0 \le x \le 2\pi$

7. $y = \dfrac{1}{2}\sin x, \quad -2\pi \le x \le 2\pi$

8. $y = \dfrac{1}{2}\cos x, \quad -2\pi \le x \le 2\pi$

9. $y = 2\sin\dfrac{1}{2}x, \quad 0 \le x \le 4\pi$

10. $y = 3\cos\dfrac{1}{2}x, \quad 0 \le x \le 4\pi$

11. $y = \cos\left(x + \dfrac{\pi}{2}\right), \quad -\dfrac{\pi}{2} \le x \le \dfrac{3\pi}{2}$

12. $y = \sin(x - \pi), \quad \pi \le x \le 3\pi$

13. $y = -\sin\left(x - \dfrac{\pi}{2}\right), \quad \dfrac{\pi}{2} \le x \le \dfrac{5\pi}{2}$

14. $y = -\cos(x - \pi), \quad \pi \le x \le 3\pi$

15. $y = \sin(-2x), \quad 0 \le x \le 2\pi$

16. $y = \sin(-\pi x), \quad 0 \le x \le 4$

17. $y = \cos(-\pi x), \quad 0 \le x \le 4$

18. $y = \cos(-2x), \quad 0 \le x \le 2\pi$

For Problems 19–40, find the period, amplitude, and phase shift of the given function and draw the graph of the function.

19. $y = 3\sin\left(x + \dfrac{\pi}{2}\right)$

20. $y = \dfrac{1}{2}\sin\left(x - \dfrac{\pi}{2}\right)$

21. $y = 2\cos(x - \pi)$

22. $y = 3\cos\left(x + \dfrac{\pi}{2}\right)$

23. $y = \dfrac{1}{2}\cos\left(x + \dfrac{\pi}{4}\right)$

24. $y = 2\sin\left(x - \dfrac{\pi}{3}\right)$

25. $y = -2\sin(x + \pi)$

26. $y = -3\cos(x + \pi)$

27. $y = 2\sin 2(x - \pi)$

28. $y = 3\cos 2\left(x - \dfrac{\pi}{2}\right)$

29. $y = \dfrac{1}{2}\cos 3(x + \pi)$

30. $y = 2\sin 3(x - \pi)$

31. $y = \sin(2x - \pi)$

32. $y = \cos\left(2x + \dfrac{\pi}{2}\right)$

33. $y = 2\cos(3x - \pi)$

34. $y = 4\sin(2x + \pi)$

35. $y = \dfrac{1}{2}\sin(3x - \pi)$

36. $y = 2\sin\left(\dfrac{1}{2}x - \dfrac{\pi}{2}\right)$

37. $y = 2\cos\left(\dfrac{1}{2}x + \dfrac{\pi}{2}\right)$

38. $y = -2\sin(2x - \pi)$

39. $y = -3\sin(2x + \pi)$

40. $y = -\dfrac{1}{2}\cos\left(2x - \dfrac{\pi}{2}\right)$

Miscellaneous Problems

Graph each of the following functions.

41. $y = 1 + 2\sin\left(x - \dfrac{\pi}{2}\right)$

42. $y = 2 - 3\cos\left(x + \dfrac{\pi}{2}\right)$

43. $y = -1 + 2\cos(2x + \pi)$

44. $y = -2 + 2\sin\left(2x - \dfrac{\pi}{2}\right)$

3.3

Graphing the Other Basic Trigonometric Functions

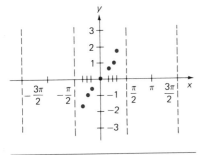

x	$y = \tan x$
$-\pi/3$	$-\sqrt{3} \approx -1.73$
$-\pi/4$	-1
$-\pi/6$	$-\frac{\sqrt{3}}{3} \approx -0.58$
0	0
$\pi/6$	$\frac{\sqrt{3}}{3} \approx 0.58$
$\pi/4$	1
$\pi/3$	$\sqrt{3} \approx 1.73$

Recall that $\tan x = (\sin x)/(\cos x)$ and therefore it is not defined at those values of x for which $\cos x = 0$ (namely, at $x = \pm(\pi/2)$, $\pm(3\pi/2)$, $\pm(5\pi/2)$, etc.). Thus, to graph $y = \tan x$, we will first use vertical dashed lines to indicate those values of x where the tangent function is undefined (Figure 3.23).

Next, let's consider some functional values in the interval between $x = -(\pi/2)$ and $x = \pi/2$ as indicated in the table. The points determined by these values are plotted in Figure 3.23. Notice that $\tan x$ is getting larger as x increases from $\pi/6$ to $\pi/4$ to $\pi/3$. By choosing more values of x between $\pi/3$ and $\pi/2$, and using a calculator, we obtain the following results.

$$\tan 1.1 = 2.0 \qquad \tan 1.2 = 2.6$$
$$\tan 1.3 = 3.6 \qquad \tan 1.4 = 5.8$$
$$\tan 1.5 = 14.1 \qquad \tan 1.55 = 48.1$$

(Perhaps you should check these values with your calculator. Don't forget to put the calculator in the radian mode.) As x approaches $\pi/2$ from the left side, $\tan x$ is getting larger and larger. The line $x = \pi/2$ is an asymptote. Likewise, the line $x = -(\pi/2)$ is an asymptote and the graph of $y = \tan x$ for $-(\pi/2) < x < (\pi/2)$ is shown in Figure 3.24.

Using a similar approach we can graph $y = \tan x$ in the intervals $\pi/2 < x < (3\pi/2)$ and $-(3\pi/2) < x < -(\pi/2)$ to obtain Figure 3.25. Note that the tangent function has a period of π.

Figure 3.23

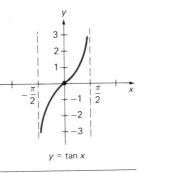

$y = \tan x$

Figure 3.24

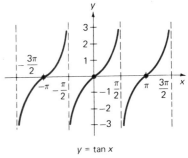

$y = \tan x$

Figure 3.25

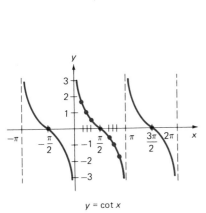

$y = \cot x$

Figure 3.26

We can graph the cotangent function in much the same manner that we graphed the tangent function. First, because $\cot x = \cos x/\sin x$, where $\sin x \neq 0$, we know that the cotangent function is not defined at $x = 0$, $\pm\pi$, $\pm2\pi$, etc. These values locate the vertical asymptotes as indicated in Figure 3.26. As the reciprocal of the tangent function, the cotangent function also has a period of π.

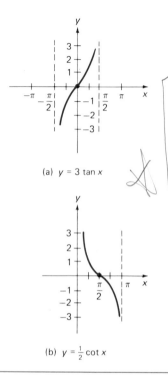

(a) $y = 3 \tan x$

(b) $y = \frac{1}{2} \cot x$

Figure 3.27

Finally, by plotting a few points $(\pi/6, \sqrt{3})$, $(\pi/4, 1)$, $(\pi/3, \sqrt{3}/3)$, $(2\pi/3, -(\sqrt{3}/3))$, $(3\pi/4, -1)$, and $(5\pi/6, -\sqrt{3})$ in the interval $0 < x < \pi$, we can sketch the cotangent curve as in Figure 3.26.

Variations of Tangent and Cotangent Curves

We have seen that both the tangent and cotangent functions have a period of π; that is, they repeat themselves every π units. This also means that their asymptotes are π units apart. Furthermore, because the tangent and cotangent functions increase and decrease without bound, the concept of amplitude has no meaning. However, to see the effect of a number a on the graph of a function of the form $y = a \tan x$ or $y = a \cot x$, the graphs of one period of $y = 3 \tan x$ and $y = \frac{1}{2} \cot x$ are shown in Figure 3.27. Notice how the 3 in $y = 3 \tan x$ sort of "stretches" the tangent curve and the $\frac{1}{2}$ in $y = \frac{1}{2} \cot x$ sort of "compresses" the cotangent curve. These facts need to be kept in mind as we use the general equations $y = a \tan b(x - c)$ and $y = a \cot b(x - c)$, where $b > 0$, to study variations of the basic tangent and cotangent curves. The number a affects ordinate values but has no significance in terms of amplitude. In each case, **the period is determined by π/b and the phase shift is again $|c|$.** Let's consider some examples.

EXAMPLE 1

Find the period and phase shift of $y = 2 \tan(x - (\pi/4))$, and sketch the curve.

Solution From the equation we can determine the period and the phase shift.

$$y = 2 \tan 1\left(x - \frac{\pi}{4}\right)$$

Amplitude Hieght or stretch

period is $\dfrac{\pi}{1} = \pi$

phase shift is $\dfrac{\pi}{4}$ units to the right

Let's shift the asymptotes $x = -(\pi/2)$ and $x = \pi/2$ of $y = \tan x$ to the right $\pi/4$ units. So we have asymptotes at $x = -(\pi/4)$ and at $x = 3\pi/4$ as shown in Figure 3.28. The curve crosses the x-axis at $x = \pi/4$ since $y = 0$ at $x = \pi/4$. The points $(0, -2)$ and $(\pi/2, 2)$ help determine the shape of the curve.

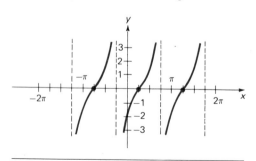

Figure 3.28

Figure 3.28 shows the graph through three periods.

Before considering the next example, let's make an important observation. Remember that two asymptotes of the curve $y = \cot x$ are located at $x = 0$ and at $x = \pi$. Now consider the function $y = \cot 2x$. If $2x = 0$, then $x = 0$ and if $2x = \pi$, then $x = \pi/2$. So, two asymptotes of the curve $y = \cot 2x$ are located at $x = 0$ and $x = \pi/2$. Keep this in mind as we consider the next example.

EXAMPLE 2

Find the period and phase shift of $y = \frac{1}{2}\cot(2x + \pi)$, and sketch the curve.

Solution First, let's change the form of the equation so that it yields some useful information.

$$y = \frac{1}{2}\cot(2x + \pi)$$

$$= \frac{1}{2}\cot 2\left(x + \frac{\pi}{2}\right)$$

period is   phase shift is $\dfrac{\pi}{2}$

$\dfrac{\pi}{2}$ units to the left

Let's shift the asymptotes, $x = 0$ and $x = \pi/2$ of $y = \cot 2x$, to the left $\pi/2$ units. So we have asymptotes at $x = -(\pi/2)$ and at $x = 0$, as shown in Figure 3.29. The curve crosses the x-axis at $x = -(\pi/4)$ since $f(-(\pi/4)) = 0$. The points $(-(\pi/8), -(1/2))$ and $(-(3\pi/8), 1/2)$ help determine the shape of the curve. Figure 3.29 shows the graph through three periods.

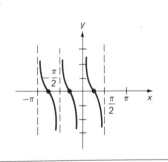

Figure 3.29

Cosecant and Secant Curves

The graphs of $y = \csc x$ and $y = \sec x$ can be sketched rather easily by using the reciprocal relationships $\csc x = 1/\sin x$ and $\sec x = 1/\cos x$. In Figures 3.30 and 3.31 we first drew the sine and cosine curves (dashed curves) and then used those

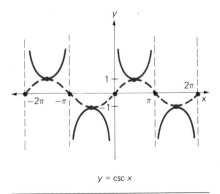

$y = \csc x$

Figure 3.30

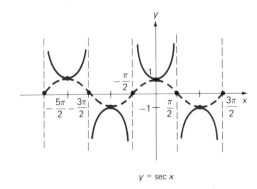

$y = \sec x$

Figure 3.31

curves to help sketch the cosecant and secant curves, respectively. The following general properties should be noted.

1. Because all functional values of $\sin x$ and $\cos x$ are between ± 1, inclusive, we know that all functional values for $\csc x$ and $\sec x$ are greater than or equal to 1, or less than or equal to -1.

2. For graphing $y = \csc x$, vertical asymptotes exist at $x = 0, \pm\pi, \pm 2\pi$, etc.

3. For graphing $y = \sec x$, vertical asymptotes exist at $x = \pm(\pi/2)$, $\pm(3\pi/2)$, $\pm(5\pi/2)$, etc.

4. Both the cosecant function and the secant function have a period of 2π.

The two equations $y = a \csc b(x - c)$ and $y = a \sec b(x - c)$ are used to express variations of the cosecant and secant curves, respectively. As with tangent and cotangent curves, the concept of amplitude has no meaning with cosecant and secant curves. Thus, the number a simply affects ordinate values but has no significance relative to amplitude. **For both $y = a \csc b(x - c)$ and $y = a \sec b(x - c)$, where $b > 0$, the period is determined by $2\pi/b$ and the phase shift is again $|c|$.**

In Figure 3.30, notice that three asymptotes of the curve $y = \csc x$ are located at $x = 0$, $x = \pi$, and $x = 2\pi$. Now consider the function $y = \csc 2x$.

If $2x = 0$, then $x = 0$.

If $2x = \pi$, then $x = \pi/2$.

If $2x = 2\pi$, then $x = \pi$.

So, the curve $y = \csc 2x$ has three asymptotes at $x = 0$, $x = \pi/2$, and $x = \pi$.

EXAMPLE 3

Find the period and phase shift of $y = \csc 2(x - (\pi/4))$, and sketch the curve.

Solution From the equation we can determine the period and phase shift.

$$y = \csc 2\left(x - \frac{\pi}{4}\right)$$

period is $\dfrac{2\pi}{2} = \pi$

phase shift is $\dfrac{\pi}{4}$ units to the right

Let's shift the asymptotes, $x = 0$, $x = \pi/2$, and $x = \pi$ of $y = \csc 2x$, to the right $\pi/4$ units. So, we have asymptotes at $x = \pi/4$, $x = 3\pi/4$, and $x = 5\pi/4$ as indicated in Figure 3.32. The points $(\pi/2, 1)$ and $(\pi, -1)$ determine the turning points. One period of the graph is shown in Figure 3.32.

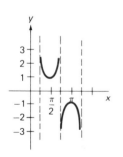

Figure 3.32

Problem Set 3.3

For Problems 1–22, graph each of the functions in the indicated interval.

1. $y = \tan x$, $\quad 0 \le x \le \pi$

2. $y = -\tan x$, $\quad -\dfrac{\pi}{2} \le x \le \dfrac{\pi}{2}$

DO every other odd problem

3. $y = -2 \tan x, \quad -\dfrac{\pi}{2} \le x \le \dfrac{\pi}{2}$

4. $y = \dfrac{1}{2} \tan x, \quad -\dfrac{\pi}{2} \le x \le \dfrac{\pi}{2}$

5. $y = 2 + \tan x, \quad -\dfrac{\pi}{2} \le x \le \dfrac{\pi}{2}$

6. $y = -1 + \tan x, \quad -\dfrac{\pi}{2} \le x \le \dfrac{\pi}{2}$

7. $y = \tan(-x), \quad -\dfrac{\pi}{2} \le x \le \dfrac{\pi}{2}$

8. $y = \tan 2x, \quad -\dfrac{\pi}{4} \le x \le \dfrac{\pi}{4}$

9. $y = 1 + \cot x, \quad 0 \le x \le 2\pi$

10. $y = -2 + \cot x, \quad 0 \le x \le 2\pi$

11. $y = \cot(-x), \quad 0 \le x \le \pi$

12. $y = -\cot x, \quad 0 \le x \le \pi$

13. $y = \cot 3x, \quad 0 \le x \le \pi$

14. $y = \cot \pi x, \quad -1 \le x \le 1$

15. $y = \csc 2x, \quad 0 \le x \le \pi$

16. $y = -\csc x, \quad 0 \le x \le 2\pi$

17. $y = 3 \csc \pi x, \quad 0 \le x \le 2$

18. $y = 1 + \csc x, \quad 0 \le x \le 2\pi$

19. $y = -\sec x, \quad -\dfrac{\pi}{2} \le x \le \dfrac{3\pi}{2}$

20. $y = \sec 3x, \quad -\dfrac{\pi}{6} \le x \le \dfrac{\pi}{2}$

21. $y = \sec(-x), \quad -\dfrac{\pi}{2} \le x \le \dfrac{3\pi}{2}$

22. $y = 1 + \sec x, \quad -\dfrac{\pi}{2} \le x \le \dfrac{3\pi}{2}$

For Problems 23–36, find the period and phase shift of the given function, and sketch the graph through two periods.

23. $y = \tan\left(x + \dfrac{\pi}{2}\right)$

24. $y = \tan(x - \pi)$

25. $y = 2\tan\left(x - \dfrac{\pi}{4}\right)$

26. $y = 3\tan\left(x + \dfrac{\pi}{4}\right)$

27. $y = \tan(2x + \pi)$

28. $y = \tan(3x - \pi)$

29. $y = -2\tan\left(x - \dfrac{\pi}{2}\right)$

30. $y = -\tan(x + \pi)$

31. $y = \tan(3x - 2\pi)$

32. $y = \tan(2x + 3\pi)$

33. $y = \dfrac{1}{2}\cot\left(x - \dfrac{\pi}{4}\right)$

34. $y = -\dfrac{1}{2}\cot\left(x + \dfrac{\pi}{2}\right)$

35. $y = \cot\dfrac{1}{2}\left(x + \dfrac{\pi}{2}\right)$

36. $y = \cot(2x - 2\pi)$

For Problems 37–46, find the period and phase shift of the given function, and sketch the graph through one period.

37. $y = \csc\left(x - \dfrac{\pi}{2}\right)$

38. $y = \csc(x + \pi)$

39. $y = -\csc\left(x + \dfrac{\pi}{4}\right)$

40. $y = -\csc\left(x - \dfrac{\pi}{2}\right)$

41. $y = \csc(2x + \pi)$

42. $y = \csc\dfrac{1}{2}\left(x + \dfrac{\pi}{2}\right)$

43. $y = \sec(x + \pi)$

44. $y = 2\sec\left(x - \dfrac{\pi}{2}\right)$

45. $y = -2\sec\left(x - \dfrac{\pi}{4}\right)$

46. $y = \sec(3x - 3\pi)$

3.4

More on Graphing and Inverse Functions

Suppose that we want to graph the function $y = \sin x + \cos x$. It would be possible to make an extensive table of values, plot the points, and sketch the curve. However, another approach is to consider $y = \sin x + \cos x$ as the sum of the sine function and the cosine function. The addition of ordinate values can be done with a ruler or a compass. Let's illustrate this approach, which is called **graphing by the addition of ordinates**.

EXAMPLE 1

Graph $y = \sin x + \cos x$ for $0 \le x \le 2\pi$.

Solution First, let's sketch the sine and cosine curves for $0 \le x \le 2\pi$ on the same set of axes, as indicated by the dashed curves in Figure 3.33. Then for various values of x we can add y-coordinates geometrically. For example, at $x = \pi/4$ we can use a compass or a ruler to add the y-coordinate of the cosine curve to the y-coordinate of the sine curve. Plotting a sufficient number of points in this manner produces the graph of $y = \sin x + \cos x$, as indicated by the solid curve in Figure 3.33.

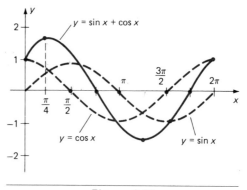

Figure 3.33

EXAMPLE 2

Graph $y = \cos x - \sin x$ for $0 \le x \le 2\pi$.

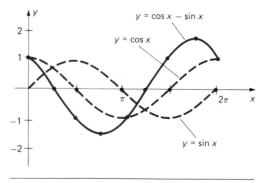

Figure 3.34

Solution Again let's begin by sketching the sine and cosine curves on the same set of axes, as shown in Figure 3.34. Now using a compass or ruler we can subtract y-coordinates of the sine curve from corresponding y-coordinates of the cosine curve. The solid curve in Figure 3.34 is the graph of $y = \cos x - \sin x$.

Another approach to Example 2 would be to first change $y = \cos x - \sin x$ to $y = \cos x + (-\sin x)$. Then the curves $y = \cos x$ and $y = -\sin x$ could be sketched, and corresponding y-coordinates added to produce the final graph.

Inverse Functions

We reviewed the concept of function in Chapter 1 as a prerequisite for our work with trigonometric functions in subsequent chapters. Now it is time to review the concept of inverse function so that we can study the inverse trigonometric functions in the next section.

Graphically, the distinction between a relation and a function can be easily recognized. In Figure 3.35 we have sketched four graphs. Which of these are graphs of functions and which are graphs of relations that are not functions? Think in terms of "to each member of the domain there is assigned one and only one member of the range;" this is the basis for what is known as the **vertical line test for functions**. Because each value of x produces only one value of y, any vertical line drawn through a graph of a function must not intersect the graph in more than one point. Therefore, parts (a) and (c) of Figure 3.35 are graphs of functions and parts (b) and (d) are graphs of relations that are not functions.

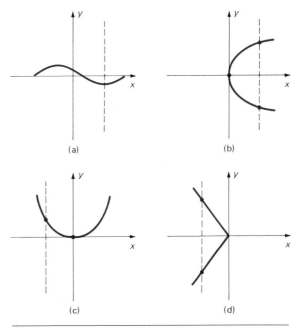

(a)

(b)

(c)

(d)

Figure 3.35

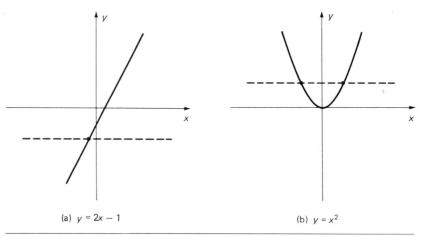

(a) $y = 2x - 1$ (b) $y = x^2$

Figure 3.36

There is also a useful distinction made between two basic types of functions. Consider the graphs of the two functions $y = 2x - 1$ and $y = x^2$ in Figure 3.36. In part (a) any **horizontal line** will intersect the graph in no more than one point. Therefore, any value of y has only one value of x associated with it. Any function that has the additional property of having only one value of x associated with each value of y is called a **one-to-one function**. The function $y = x^2$ is not a one-to-one function because the horizontal line in part (b) of Figure 3.36 intersects the parabola in two points.

In terms of ordered pairs, a one-to-one function does not contain any ordered pairs having the same second component. For example,

$$f = \{(1, 3), (2, 6), (4, 12)\}$$

is a one-to-one function, but

$$g = \{(1, 2), (2, 5), (-2, 5)\}$$

is not a one-to-one function.

If the components of each ordered pair of a given one-to-one function are interchanged, the resulting function and the given function are called **inverses** of each other. Thus,

$$\{(1, 3), (2, 6), (4, 12)\} \quad \text{and} \quad \{(3, 1), (6, 2), (12, 4)\}$$

are **inverse functions**. The inverse of a function f is denoted by f^{-1} (read "f inverse" or "the inverse of f"). If (a, b) is an ordered pair of f, then (b, a) is an ordered pair of f^{-1}. The domain and range of f^{-1} are the range and domain, respectively, of f.

> **Remark:** Do not confuse the -1 in f^{-1} with a negative exponent. The symbol f^{-1} does not mean $1/f^1$, but refers to the inverse function of function f.

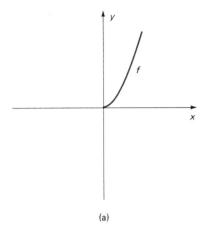

(a)

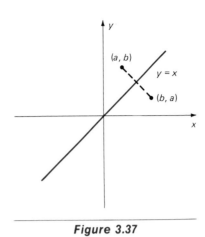

Figure 3.37

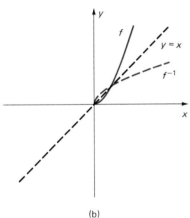

(b)

Figure 3.38

Graphically, two functions that are inverses of each other are mirror images with reference to the line $y = x$. This is due to the fact that ordered pairs (a, b) and (b, a) are mirror images with respect to the line $y = x$ as illustrated in Figure 3.37. Therefore, if the graph of a function f is known, as in Figure 3.38(a), then the graph of f^{-1} can be determined by reflecting f across the line $y = x$ (Figure 3.38(b)).

Another useful way of viewing inverse functions is in terms of composition. Basically, inverse functions *undo* each other and this can be more formally stated as follows. If f and g are inverses of each other, then

1. $(f \circ g)(x) = f(g(x)) = x$ for all x in domain of g; and
2. $(g \circ f)(x) = g(f(x)) = x$ for all x in domain of f.

As we will see in a moment, this relationship of inverse functions can be used to verify whether two functions are indeed inverses of each other.

Finding Inverse Functions

The idea of inverse functions *undoing each other* provides the basis for an informal approach to finding the inverse of a function. Consider the function

$$f(x) = 2x + 1.$$

To each x this function assigns *twice x plus 1*. To *undo* this function, we could *subtract 1 and divide by 2*. So, the inverse should be

$$f^{-1}(x) = \frac{x - 1}{2}.$$

Now let's verify that f and f^{-1} are inverses of each other.

$$(f \circ f^{-1})(x) = f(f^{-1}(x)) = f\left(\frac{x - 1}{2}\right) = 2\left(\frac{x - 1}{2}\right) + 1 = x$$

and

$$(f^{-1} \circ f)(x) = f^{-1}(f(x)) = f^{-1}(2x + 1) = \frac{\cdot 2x + 1 - 1}{2} = x.$$

Thus, the inverse of f is given by

$$f^{-1}(x) = \frac{x - 1}{2}$$

Let's consider another example of finding an inverse function by the *undoing* process.

EXAMPLE 3

Find the inverse of $f(x) = 3x - 5$.

Solution To each x, the function f assigns *three times x minus 5*. To *undo* this we can *add 5 and then divide by 3*. So, the inverse should be

$$f^{-1}(x) = \frac{x + 5}{3}$$

To verify that f and f^{-1} are inverses we can show that

$$(f \circ f^{-1})(x) = f(f^{-1}(x)) = f\left(\frac{x + 5}{3}\right) = 3\left(\frac{x + 5}{3}\right) - 5 = x$$

and

$$(f^{-1} \circ f)(x) = f^{-1}(f(x)) = f^{-1}(3x - 5) = \frac{3x - 5 + 5}{3} = x.$$

Thus, f and f^{-1} are inverses and we can write

$$f^{-1}(x) = \frac{x + 5}{3}$$

The following is a more formal technique for finding the inverse of a function.

1. Let $y = f(x)$.
2. Solve for x in terms of y.
3. Interchange x and y.
4. $f^{-1}(x)$ is determined by the final equation.

The following examples illustrate this procedure.

EXAMPLE 4

Find the inverse of $f(x) = 3x - 5$.

Solution First, let $y = f(x)$, so the equation becomes

$$y = 3x - 5.$$

Second, solving this equation for x produces

$$y = 3x - 5$$
$$y + 5 = 3x$$
$$\frac{y + 5}{3} = x.$$

Third, interchanging x and y produces

$$\frac{x + 5}{3} = y.$$

Thus, the inverse of f is

$$f^{-1}(x) = \frac{x + 5}{3}$$

EXAMPLE 5

Find the inverse of $f(x) = \frac{2}{3}x + \frac{3}{5}$

Solution Let $y = f(x)$, so the equation becomes

$$y = \frac{2}{3}x + \frac{3}{5}$$

Solving for x produces

$$15y = \left(\frac{2}{3}x + \frac{3}{5}\right)15$$
$$15y = 10x + 9$$
$$15y - 9 = 10x$$
$$\frac{15y - 9}{10} = x.$$

Interchanging variables we obtain

$$\frac{15x - 9}{10} = y$$

where the inverse of f

$$f^{-1}(x) = \frac{15x - 9}{10}$$

is obtained.

To verify that f and f^{-1} are inverses, we can show that

$$(f \circ f^{-1})(x) = x \quad \text{and} \quad (f^{-1} \circ f)(x) = x. \qquad \text{You may complete this.}$$

Problem Set 3.4

For Problems 1–12, use the "addition of ordinates" approach to sketch each of the curves.

1. $y = \sin x + 2\cos x, \quad 0 \le x \le 2\pi$
2. $y = 2\sin x + \cos x, \quad 0 \le x \le 2\pi$
3. $y = \sin 2x + \cos x, \quad 0 \le x \le 2\pi$
4. $y = \sin x + \cos 2x, \quad 0 \le x \le 2\pi$
5. $y = \sin x + \cos \frac{1}{2}x, \quad 0 \le x \le 4\pi$
6. $y = \sin \frac{1}{2}x + \cos x, \quad 0 \le x \le 4\pi$
7. $y = \sin x - \cos x, \quad 0 \le x \le 2\pi$
8. $y = 3\cos x - \sin x, \quad 0 \le x \le 2\pi$
9. $y = 2\cos x - \frac{1}{2}\sin 2x, \quad 0 \le x \le 2\pi$
10. $y = \cos x - \sin \frac{1}{2}x, \quad 0 \le x \le 4\pi$
11. $y = x + \sin x, \quad 0 \le x \le 2\pi$
12. $y = x - \cos x, \quad 0 \le x \le 2\pi$

For Problems 13–18, identify each of the following as (a) a graph of a function or (b) a graph of a relation that is not a function. Use the vertical line test.

13.
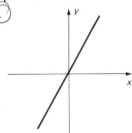

14.
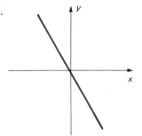

15.
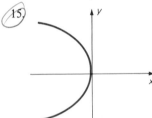

16.
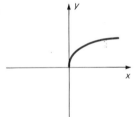

17.
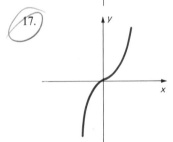

18.
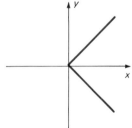

For Problems 19–24, identify each of the following as (a) the graph of a one-to-one function or (b) the graph of a function that is not one-to-one. Use the horizontal line test.

19.
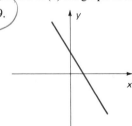

20.
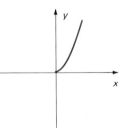

21.
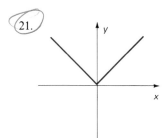

22.
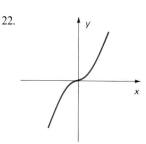

23.
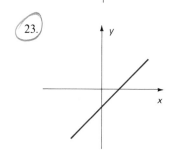

24.
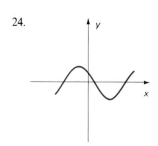

For Problems 25–28, (a) list the domain and range of the given function, (b) form the inverse function, and (c) list the domain and range of the inverse function.

25. $f = \{(1, 5), (2, 9), (5, 21)\}$

26. $f = \{(1, 1), (4, 2), (9, 3), (16, 4)\}$

27. $f = \{(0, 0), (2, 8), (-1, -1), (-2, -8)\}$

28. $f = \{(-1, 1), (-2, 4), (-3, 9), (-4, 16)\}$

For Problems 29–38, (a) find the inverse of the given function by using the "undoing" process, and (b) verify that the two functions are inverses of each other by showing that $(f \circ f^{-1})(x) = x$ and $(f^{-1} \circ f)(x) = x$.

29. $f(x) = 2x + 3$ 30. $f(x) = 4x - 5$ 31. $f(x) = \frac{1}{2}x$

32. $f(x) = -\frac{1}{3}x$ 33. $f(x) = 5x + 9$ 34. $f(x) = 3x + 7$

35. $f(x) = \frac{1}{3}x - 4$ 36. $f(x) = \frac{1}{2}x + 5$ 37. $f(x) = \frac{2}{3}x + 5$

38. $f(x) = \frac{3}{4}x - 1$

For Problems 39–48, (a) find the inverse of the given function by using the process illustrated in Examples 4 and 5, and (b) verify that the two functions are inverses of each other by showing that $(f \circ f^{-1})(x) = x$ and $(f^{-1} \circ f)(x) = x$.

39. $f(x) = 4x$ 40. $f(x) = \frac{1}{2}x$ 41. $f(x) = 2x + 9$

42. $f(x) = 3x - 11$ 43. $f(x) = -\frac{2}{3}x$ 44. $f(x) = -\frac{4}{3}x$

45. $f(x) = -3x - 4$ 46. $f(x) = -5x + 6$ 47. $f(x) = \frac{3}{4}x - \frac{5}{6}$

48. $f(x) = \frac{2}{3}x - \frac{1}{4}$

For Problems 49–56, (a) find the inverse of the given function, and (b) graph the given function and its inverse on the same set of axes.

49. $f(x) = 6x$ 50. $f(x) = \frac{3}{5}x$ 51. $f(x) = -\frac{1}{4}x$

52. $f(x) = -2x$ 53. $f(x) = 2x - 1$ 54. $f(x) = 3x + 4$

55. $f(x) = -4x + 3$ 56. $f(x) = -5x - 1$

57. Find the inverse of $f(x) = x^2$, where $x \geq 0$, and graph both f and f^{-1} on the same set of axes.

58. Find the inverse of $f(x) = x^2 + 1$, where $x \geq 0$, and graph both f and f^{-1} on the same set of axes.

59. Explain why every nonconstant linear function has an inverse.

Miscellaneous Problems

60. The composition idea can also be used to find the inverse of a function. For example, to find the inverse of $f(x) = 5x + 3$, we could proceed as follows.

$$f(f^{-1}(x)) = 5(f^{-1}(x)) + 3 \quad \text{and} \quad f(f^{-1}(x)) = x$$

Therefore, equating the two expressions for $f(f^{-1}(x))$ we obtain

$$5(f^{-1}(x)) + 3 = x$$

$$5(f^{-1}(x)) = x - 3$$

$$f^{-1}(x) = \frac{x - 3}{5}$$

Use this approach to find the inverse of each of the following functions.

(a) $f(x) = 2x + 1$ (b) $f(x) = 3x - 2$ (c) $f(x) = -4x + 5$

(d) $f(x) = -x + 1$ (e) $f(x) = 2x$ (f) $f(x) = -5x$

3.5

Inverse Trigonometric Functions

It is evident that the sine function over the domain of all real numbers is not a one-to-one function. For example, suppose that we consider the solutions for $\sin x = \frac{1}{2}$. Certainly, $\pi/6$ is a solution, but there are infinitely many more solutions, such as $5\pi/6$, $13\pi/6$, $17\pi/6$, $-(7\pi/6)$, $-(11\pi/6)$, etc., as indicated in Figure 3.39. However, we can form a one-to-one function from the sine function, and not eliminate any values from its range, by restricting the domain to the interval $(-\pi/2) \leq x \leq \pi/2$. Therefore, we have a *new* function defined by the equation $y = \sin x$ having a domain of $-(\pi/2) \leq x \leq (\pi/2)$ and a range of $-1 \leq y \leq 1$ (Figure 3.40).

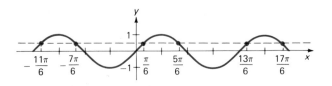

Figure 3.39

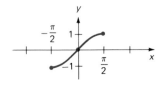

Figure 3.40

Now the inverse sine function can be defined as follows.

DEFINITION 3.2

> The **inverse sine function** is defined by
>
> $$y = \sin^{-1}x \quad \text{if and only if} \quad x = \sin y$$
>
> where $-1 \le x \le 1$ and $-(\pi/2) \le y \le \pi/2$.

In Definition 3.2 the equation $y = \sin^{-1}x$ can be read as "y is the angle whose sine is x." Therefore, $y = \sin^{-1}\frac{1}{2}$ means "y is the angle, between $-(\pi/2)$ and $\pi/2$, inclusive, whose sine is $\frac{1}{2}$;" thus, $y = \pi/6$. (The angle could also be expressed as 30°.)

EXAMPLE 1

Solve $y = \sin^{-1}(-(\sqrt{2}/2))$ for y, where $-(\pi/2) \le y \le \pi/2$.

Solution From our previous work with special angles, we should recall that the angle between $-(\pi/2)$ and $\pi/2$, inclusive, whose sine equals $-(\sqrt{2}/2)$ is $-(\pi/4)$. Therefore, $y = -(\pi/4)$.

EXAMPLE 2

Use a calculator and solve each of the following for y:
- **(a)** $y = \sin^{-1}(.7256), \quad -(\pi/2) \le y \le \pi/2$
- **(b)** $y = \sin^{-1}(-.3402), \quad -(\pi/2) \le y \le \pi/2$
- **(c)** $y = \sin^{-1}(.6378), \quad -90° \le y \le 90°$

Solution For parts **(a)** and **(b)**, be sure that your calculator is set for radian measure.
- **(a)** Enter .7256, press the INV and SIN keys in that order, and obtain $y = .812$, to three decimal places.
- **(b)** Enter $-.3402$, press the INV and SIN keys in that order, and obtain $y = -.347$, to three decimal places.
- **(c)** Set your calculator for degree measure. Then enter .6378, press the INV and SIN keys in that order, and obtain $y = 39.6°$, to the nearest tenth of a degree.

  Remark: Some calculators have a SIN⁻¹ key that will produce the inverse function directly. Note also that your calculator has been designed to yield values that agree with the range of the inverse sine function. In other words, your calculator will produce inverse sine values between $-(\pi/2)$ and $\pi/2$, inclusive, or between $-90°$ and $90°$, inclusive.

EXAMPLE 3

Evaluate $\cos(\sin^{-1}(-\frac{1}{2}))$.

Solution The expression $\cos(\sin^{-1}(-\frac{1}{2}))$ means **the cosine of the angle between** $-(\pi/2)$ **and** $\pi/2$, **inclusive, whose sine is** $-\frac{1}{2}$. From our previous work with special

x	$y = \sin^{-1}x$
-1	$-\pi/2$
$-\sqrt{3}/2$	$-\pi/3$
$-1/2$	$-\pi/6$
0	0
$1/2$	$\pi/6$
$\sqrt{3}/2$	$\pi/3$
1	$\pi/2$

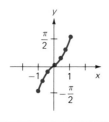

Figure 3.41

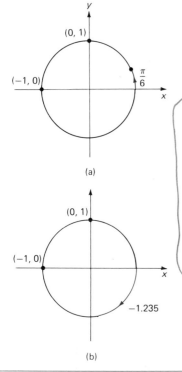

(a)

(0, 1)

(−1, 0)

(0, 1)

(−1, 0)

−1.235

(b)

Figure 3.42

angles we know that the angle between $-(\pi/2)$ and $\pi/2$, inclusive, whose sine is $-\frac{1}{2}$ is $-(\pi/6)$. Then we know that $\cos(-(\pi/6)) = \sqrt{3}/2$. Therefore,

$$\cos\left(\sin^{-1}\left(-\frac{1}{2}\right)\right) = \frac{\sqrt{3}}{2}.$$

EXAMPLE 4

Graph $y = \sin^{-1}x$.

Solution A table of values can be easily formed from our previous work with special angles. Plotting the points determined by the table and connecting them with a smooth curve produces Figure 3.41.

Remember that for two functions f and g to be inverses of each other, the following two conditions must be satisfied: (1) $f(g(x)) = x$ for all x in the domain of g, and (2) $g(f(x)) = x$ for all x in domain of f. Let's show that the two functions

$$f(x) = \sin x \quad \text{for } -\frac{\pi}{2} \le x \le \frac{\pi}{2}$$

and

$$g(x) = \sin^{-1}x \quad \text{for } -1 \le x \le 1$$

satisfy these conditions.

$$f(g(x)) = f(\sin^{-1}x) = \sin(\sin^{-1}x) = x \quad \text{for } -1 \le x \le 1$$

and

$$g(f(x)) = g(\sin x) = \sin^{-1}(\sin x) = x \quad \text{for } -\frac{\pi}{2} \le x \le \frac{\pi}{2}$$

The fact that $f(x) = \sin x$ and $g(x) = \sin^{-1}x$ are inverses could also be used for graphing purposes. That is to say, their graphs are reflections of each other through the line $y = x$. We will have you use this idea to graph $f(x) = \sin^{-1}x$ in the next set of problems.

The inverse sine function is also called the **arcsine function** and the notation **arcsin** x can be used in place of $\sin^{-1}x$. The "arc" vocabulary refers to the fact that $y = \arcsin x$ means $x = \sin y$, that is, y is a real number (that can be geometrically interpreted as arc-length on a unit circle) whose sine is x. Therefore, $\arcsin \frac{1}{2}$ can also refer to the arc indicated in Figure 3.42(a), which is $\pi/6$ units long. Likewise, $\arcsin(-.9440)$ can refer to the arc indicated in Figure 3.42(b), which is -1.235 units long.

The Inverse Cosine Function

The other trigonometric functions can also be used to introduce inverse functions. In each case, a restriction needs to be placed on the original domain to create a one-to-one function that contains the entire range of the original function. Then a corresponding inverse function can be defined. (The inverses of the cotangent, cosecant, and secant functions are given in the next problem set.)

By restricting the domain of the cosine function to real numbers between 0 and π, inclusive, a one-to-one function with a range between -1 and 1, inclusive, is obtained. Then the following definition creates the inverse cosine function.

DEFINITION 3.3

The **inverse cosine function** or **arccosine function** is defined by

$$y = \cos^{-1}x = \arccos x \quad \text{if and only if} \quad x = \cos y$$

where $-1 \le x \le 1$ and $0 \le y \le \pi$.

EXAMPLE 5

Solve $y = \cos^{-1}(-(\sqrt{3}/2))$ for y, where $0 \le y \le \pi$.

Solution The expression $\cos^{-1}(-(\sqrt{3}/2))$ can be interpreted as "the angle whose cosine is $-(\sqrt{3}/2)$." From our previous work with special angles, we know that $y = 5\pi/6$.

EXAMPLE 6

Use a calculator and solve each of the following for y:

(a) $y = \cos^{-1}(.3214), \quad 0 \le y \le \pi$
(b) $y = \arccos(.7914), \quad 0 \le y \le \pi$
(c) $y = \cos^{-1}(-.7120), \quad 0° \le y \le 180°$

Solution For parts (a) and (b), be sure that your calculator is set for radian measure.

(a) Enter .3214, press the $\boxed{\text{INV}}$ and $\boxed{\text{COS}}$ keys in that order, and obtain $y = 1.244$, to three decimal places.
(b) Enter .7914, press the $\boxed{\text{INV}}$ and $\boxed{\text{COS}}$ keys in that order, and obtain $y = .658$, to three decimal places.
(c) Set your calculator for degree measure. Then enter $-.7120$, press the $\boxed{\text{INV}}$ and $\boxed{\text{COS}}$ keys in that order, and obtain $y = 135.4°$, to the nearest tenth of a degree.

EXAMPLE 7

Evaluate $\sin(\cos^{-1}\frac{1}{2})$.

Solution The expression $\sin(\cos^{-1}\frac{1}{2})$ means **the sine of the angle, between 0 and π, inclusive, whose cosine is $\frac{1}{2}$**. We know that $\pi/3$ is the angle whose cosine is $\frac{1}{2}$ and we know that $\sin \pi/3 = \sqrt{3}/2$. Therefore, $\sin(\cos^{-1}\frac{1}{2}) = \sqrt{3}/2$.

The Inverse Tangent Function

By restricting the domain of the tangent function to real numbers between $-(\pi/2)$ and $\pi/2$, $(-(\pi/2)$ and $\pi/2$ are not included since the tangent is undefined at those values) a one-to-one function is obtained. Therefore, the inverse tangent function can be defined.

DEFINITION 3.4

> The **inverse tangent function** or **arctangent function** is defined by
>
> $$y = \tan^{-1} x = \arctan x \quad \text{if and only if} \quad x = \tan y$$
>
> where $-\infty < x < \infty$ and $-(\pi/2) < y < \pi/2$.

EXAMPLE 8

Solve $y = \tan^{-1}(-(\sqrt{3}/3))$ for y, where $-90° < y < 90°$.

Solution The expression $\tan^{-1}(-(\sqrt{3}/3))$ can be interpreted as "the angle between $-90°$ and $90°$, whose tangent is $-(\sqrt{3}/3)$." Therefore, from our previous work with special angles, we know that $y = -30°$.

EXAMPLE 9

Find an exact value for $\sin(\tan^{-1}\frac{4}{3})$ without using a calculator or a table.

Solution Let $\theta = \tan^{-1}\frac{4}{3}$; that is, θ is the angle, between $-(\pi/2)$ and $\pi/2$, whose tangent is $\frac{4}{3}$. Therefore, θ is a first-quadrant angle as indicated in Figure 3.43.

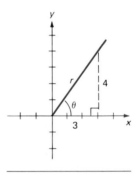

Figure 3.43

From the right triangle formed, we obtain

$$r^2 = 3^2 + 4^2 = 9 + 16 = 25$$

$$r = 5.$$

Therefore, $\sin \theta = \frac{4}{5}$ and we have $\sin(\tan^{-1}\frac{4}{3}) = \frac{4}{5}$.

Problem Set 3.5

For Problems 1–14, solve for y and express y in radian measure. Do not use a calculator or a table.

1. $y = \sin^{-1}\dfrac{\sqrt{2}}{2}$ 2. $y = \sin^{-1}\dfrac{\sqrt{3}}{2}$ 3. $y = \sin^{-1}\left(-\dfrac{\sqrt{3}}{2}\right)$

4. $y = \sin^{-1} 1$

5. $y = \cos^{-1} \dfrac{1}{2}$

6. $y = \cos^{-1}\left(-\dfrac{1}{2}\right)$

7. $y = \cos^{-1} \dfrac{\sqrt{3}}{2}$

8. $y = \cos^{-1} 0$

9. $y = \arctan 1$

10. $y = \arctan(-1)$

11. $y = \arctan \sqrt{3}$

12. $y = \arctan 0$

13. $y = \arcsin(-1)$

14. $y = \arccos(-1)$

For Problems 15–22, solve for y and express y in degree measure. Do not use a calculator or a table.

15. $y = \tan^{-1} \dfrac{\sqrt{3}}{3}$

16. $y = \tan^{-1}\left(-\dfrac{\sqrt{3}}{3}\right)$

17. $y = \cos^{-1}\left(-\dfrac{\sqrt{2}}{2}\right)$

18. $y = \cos^{-1}\left(-\dfrac{\sqrt{3}}{2}\right)$

19. $y = \sin^{-1} 0$

20. $y = \sin^{-1}\left(-\dfrac{1}{2}\right)$

21. $y = \sin^{-1}\left(-\dfrac{\sqrt{2}}{2}\right)$

22. $y = \tan^{-1}(-\sqrt{3})$

For Problems 23–34, use a calculator or the table in the back of the book to solve for y. Express answers in radians to three decimal places.

23. $y = \sin^{-1}(.3578)$

24. $y = \sin^{-1}(.8629)$

25. $y = \arcsin(-.9142)$

26. $y = \arcsin(-.1654)$

27. $y = \arccos(.5894)$

28. $y = \arccos(.0428)$

29. $y = \cos^{-1}(-.4162)$

30. $y = \cos^{-1}(-.8894)$

31. $y = \tan^{-1}(8.6214)$

32. $y = \tan^{-1}(.9145)$

33. $y = \arctan(-.1986)$

34. $y = \arctan(-56.2413)$

For Problems 35–46, use a calculator or the table in the back of the book to solve for y. Express answers to the nearest tenth of a degree.

35. $y = \sin^{-1}(.4310)$

36. $y = \sin^{-1}(.7214)$

37. $y = \sin^{-1}(-.8214)$

38. $y = \sin^{-1}(-.2318)$

39. $y = \cos^{-1}(.2644)$

40. $y = \cos^{-1}(.8419)$

41. $y = \cos^{-1}(-.1620)$

42. $y = \cos^{-1}(-.6217)$

43. $y = \tan^{-1}(14.2187)$

44. $y = \tan^{-1}(.9854)$

45. $y = \tan^{-1}(-8.2176)$

46. $y = \tan^{-1}(-21.1765)$

For Problems 47–70, evaluate each expression without using a calculator or a table.

47. $\sin\left(\cos^{-1}\left(-\dfrac{1}{2}\right)\right)$

48. $\cos\left(\sin^{-1}\dfrac{1}{2}\right)$

49. $\cos(\sin^{-1} 1)$

50. $\sin(\cos^{-1}(-1))$

51. $\tan\left(\sin^{-1}\dfrac{\sqrt{2}}{2}\right)$

52. $\tan\left(\cos^{-1}\left(-\dfrac{\sqrt{3}}{2}\right)\right)$

53. $\sin(\tan^{-1}\sqrt{3})$

54. $\cos\left(\tan^{-1}\left(-\dfrac{\sqrt{3}}{3}\right)\right)$

55. $\sin(\sin^{-1}\sqrt{2})$

56. $\cos(\cos^{-1} 0)$

57. $\cos\left(\arcsin\dfrac{4}{5}\right)$

58. $\cos\left(\arcsin\dfrac{5}{13}\right)$

59. $\sin\left(\arctan\dfrac{3}{4}\right)$

60. $\cos\left(\arctan\left(-\dfrac{4}{3}\right)\right)$

61. $\tan\left(\sin^{-1}\left(-\dfrac{4}{5}\right)\right)$

62. $\tan\left(\cos^{-1}\left(-\dfrac{5}{13}\right)\right)$ 63. $\cos\left(\sin^{-1}\dfrac{2}{3}\right)$ 64. $\sin\left(\tan^{-1}\left(-\dfrac{2}{3}\right)\right)$

65. $\tan\left(\cos^{-1}\left(-\dfrac{1}{3}\right)\right)$ 66. $\tan\left(\sin^{-1}\dfrac{2}{5}\right)$ 67. $\sec\left(\sin^{-1}\left(-\dfrac{3}{4}\right)\right)$

68. $\csc\left(\cos^{-1}\left(-\dfrac{2}{3}\right)\right)$ 69. $\cot\left(\cos^{-1}\left(-\dfrac{3}{7}\right)\right)$ 70. $\sec\left(\sin\left(-\dfrac{1}{5}\right)\right)$

71. Graph $y = \sin^{-1}x$ by reflecting $y = \sin x$, where $-\dfrac{\pi}{2} \le x \le \dfrac{\pi}{2}$, across the line $y = x$.

72. Graph $y = \cos^{-1}x$.   73. Graph $y = \tan^{-1}x$.

Miscellaneous Problems

Consider the following definitions.*

1. The inverse cotangent function is defined by

$$y = \cot^{-1}x, \quad \text{if and only if} \quad x = \cot y$$

where $-\infty < x < \infty$, $-\dfrac{\pi}{2} \le y \le \dfrac{\pi}{2}$, and $y \ne 0$.

2. The inverse cosecant function is defined by

$$y = \csc^{-1}x, \quad \text{if and only if} \quad x = \csc y$$

where $|x| \ge 1$, $-\dfrac{\pi}{2} \le y \le \dfrac{\pi}{2}$, and $y \ne 0$.

3. The inverse secant function is defined by

$$y = \sec^{-1}x, \quad \text{if and only if} \quad x = \sec y$$

where $|x| \ge 1$, $0 \le y \le \pi$, and $y \ne \dfrac{\pi}{2}$.

74. Graph $y = \cot^{-1}x$.   75. Graph $y = \csc^{-1}x$.   76. Graph $y = \sec^{-1}x$.

Chapter Summary

The basic trigonometric functions can also be interpreted as **circular functions**. Perhaps another reading of the first part of Section 3.1 would help your understanding of the circular functions.

The following ideas pertaining to the sine and cosine functions are helpful for graphing purposes.

1. Both $\sin x$ and $\cos x$ are bounded above by 1 and below by -1.

2. Both $\sin x$ and $\cos x$ have periods of 2π.

---

* These definitions are not universally agreed upon; however, they are consistent with the values given by most scientific calculators.

3. Through one period of 2π the sine and cosine functions vary as follows.

AS x INCREASES	THE SINE FUNCTION	THE COSINE FUNCTION
from 0 to $\dfrac{\pi}{2}$	increases from 0 to 1	decreases from 1 to 0
from $\dfrac{\pi}{2}$ to π	decreases from 1 to 0	decreases from 0 to -1
from π to $\dfrac{3\pi}{2}$	decreases from 0 to -1	increases from -1 to 0
from $\dfrac{3\pi}{2}$ to 2π	increases from -1 to 0	increases from 0 to 1

The following are examples of variations of the basic curves $y = \sin x$ and $y = \cos x$:

1. The graph of $y = 1 + \sin x$ is the graph of $y = \sin x$ *moved up 1 unit.*

2. The graph of $y = \sin\left(x - \dfrac{\pi}{2}\right)$ is the graph of $y = \sin x$ *moved to the right*
   $\dfrac{\pi}{2}$ *units.*

3. The graph of $y = -\cos x$ is the graph of $y = \cos x$ *reflected across the x-axis.*

A function f is called **periodic** if there exists a positive real number p such that $f(x + p) = f(x)$, for all x in the domain of f. The smallest value for p is called the **period** of the function.

The following information about equations of the form $y = a \sin b(x - c)$ or $y = a \cos b(x - c)$, where $b > 0$, is very useful for sketching their graphs.

1. The period of both curves is $2\pi/b$.
2. The amplitude of both curves is $|a|$.
3. The phase shift of both curves is $|c|$. The shift is to the ~~right~~ *left* if c is positive and to the ~~left~~ *right* if c is negative.

The tangent curve has vertical asymptotes at $x = \pm(\pi/2), \pm(3\pi/2), \pm(5\pi/2)$, and so on. Refer back to Figure 3.25 to review the shape of the tangent curve. It has a period of π.

The cotangent curve has vertical asymptotes at $x = 0, \pm\pi, \pm2\pi$, and so on. Refer back to Figure 3.26 to review the shape of the cotangent curve. It has a period of π.

The following information about equations of the form $y = a \tan b(x - c)$ or $y = a \cot b(x - c)$, where $b > 0$, is very useful for sketching their graphs.

1. The number a affects ordinate values but has no significance in terms of amplitude.

2. The period of both curves is π/b.

3. The phase shift of both curves is $|c|$.

The graphs of $y = \csc x$ and $y = \sec x$ are shown in Figure 3.30 and Figure 3.31, respectively. Both curves have a period of 2π.

The following information about equations of the form $y = a \csc b(x - c)$ or $y = a \sec b(x - c)$, where $b > 0$, is very useful for sketching their graphs.

1. The number a affects ordinate values but has no significance in terms of amplitude.

2. The period of both curves is $2\pi/b$.

3. The phase shift of both curves is $|c|$.

A one-to-one function is a function such that no two ordered pairs have the same second component.

If the components of each ordered pair of a given one-to-one function are interchanged, the resulting function and the given function are **inverses** of each other. The inverse of a function f is denoted by f^{-1}.

Graphically, two functions that are inverses of each other are mirror images with reference to the line $y = x$.

Two functions f and f^{-1} can be shown to be inverses of each other by verifying that

1. $(f^{-1} \circ f)(x) = x$ for all x in the domain of f.

2. $(f \circ f^{-1})(x) = x$ for all x in the domain of f^{-1}.

A technique for finding the inverse of a function can be described as follows.

1. Let $y = f(x)$.

2. Solve the equation for x in terms of y.

3. Interchange x and y.

4. $f^{-1}(x)$ is determined by the final equation.

Definitions 3.2–3.4 form the basis for working with the inverse sine, inverse cosine, and inverse tangent functions. Learn these definitions.

Chapter 3 Review Problem Set

For Problems 1–6, find exact values without using a calculator or a table.

1. $\sin\left(-\dfrac{5\pi}{6}\right)$

2. $\tan\left(\dfrac{9\pi}{4}\right)$

3. $\sin\left(\cos^{-1}\dfrac{\sqrt{2}}{2}\right)$

4. $\cos\left(\sin^{-1}\left(-\dfrac{12}{13}\right)\right)$

5. $\tan\left(\arcsin\dfrac{1}{2}\right)$

6. $\sin\left(\arctan\left(-\dfrac{2}{3}\right)\right)$

For Problems 7–10, solve for y and express y in radians. Do not use a calculator or a table.

7. $y = \tan^{-1}\left(-\dfrac{\sqrt{3}}{3}\right)$

8. $y = \cos^{-1}\left(-\dfrac{\sqrt{3}}{2}\right)$

9. $y = \arcsin\left(-\dfrac{1}{2}\right)$

10. $y = \arctan(-\sqrt{3})$

For Problems 11–14, solve for y and express y in degrees. Do not use a calculator or a table.

11. $y = \sin^{-1}\dfrac{\sqrt{3}}{2}$

12. $y = \cos^{-1}\left(-\dfrac{1}{2}\right)$

13. $y = \tan^{-1}(-1)$

14. $y = \sin^{-1}\left(-\dfrac{1}{2}\right)$

For Problems 15–18, use a calculator or the table in the back of the book to solve for y. Express y in radians to three decimal places.

15. $y = \cos^{-1}(-.5724)$

16. $y = \sin^{-1}(-.7219)$

17. $y = \arctan(-71.2134)$

18. $y = \arcsin(.9417)$

For Problems 19–22, use a calculator or the table in the back of the book to solve for y. Express y to the nearest tenth of a degree.

19. $y = \cos^{-1}(.2479)$

20. $y = \sin^{-1}(-.4100)$

21. $y = \tan^{-1}(-9.2147)$

22. $y = \cos^{-1}(-.5628)$

For Problems 23–36, find the period, amplitude (if it exists), and phase shift for each graph. **Do not** graph the functions.

23. $y = 4\sin\left(x + \dfrac{\pi}{4}\right)$

24. $y = -3\cos 2\left(x - \dfrac{\pi}{3}\right)$

25. $y = 2\tan 2(x + \pi)$

26. $y = \sin(3x - \pi)$

27. $y = -2\sin(\pi x + \pi)$

28. $y = 2\cos\left(\pi x - \dfrac{\pi}{2}\right)$

29. $y = -4\cos(-3x)$

30. $y = 5\sin(-2x)$

31. $y = 5\cot(3x - \pi)$

32. $y = -4\cot(4x + \pi)$

33. $y = \csc\left(4x + \dfrac{\pi}{2}\right)$

34. $y = 2\csc\left(3x - \dfrac{\pi}{4}\right)$

35. $y = 2\sec 3(x - 2)$

36. $y = 3\sec(2x - \pi)$

For Problems 37–52, graph the given function in the indicated interval.

37. $y = -\sin 2x$, $-2\pi \le x \le 2\pi$

38. $y = -1 + \cos x$, $-2\pi \le x \le 2\pi$

39. $y = \tan \pi x$, $-\dfrac{3}{2} \le x \le \dfrac{3}{2}$

40. $y = 1 + \cos(x - \pi)$, $\pi \le x \le 3\pi$

41. $y = \csc\dfrac{1}{2}x$, $0 \le x \le 4\pi$

42. $y = 1 - \sec x$, $-\dfrac{\pi}{2} \le x \le \dfrac{3\pi}{2}$

43. $y = 2\sin\left(x - \dfrac{\pi}{2}\right)$, $-\dfrac{\pi}{2} \le x \le \dfrac{5\pi}{2}$

44. $y = \cos\left(2x + \dfrac{\pi}{2}\right)$, $-\pi \le x \le \pi$

45. $y = \tan\left(x - \dfrac{\pi}{4}\right)$, $-\dfrac{\pi}{4} \le x \le \dfrac{7\pi}{4}$

46. $y = 1 + \cot 2x$, $0 \le x \le \pi$

47. $y = 2 - \csc 2x$, $0 \le x \le 2\pi$

48. $y = \sec \pi\left(x - \dfrac{1}{2}\right)$, $0 \le x \le 2$

49. $y = -\cot \pi(x + 1)$, $-2 \leq x \leq 0$

50. $y = \cot \dfrac{1}{2}\left(x - \dfrac{\pi}{4}\right)$, $\dfrac{\pi}{4} \leq x \leq \dfrac{9\pi}{4}$

51. $y = \cos x + 3 \sin x$, $0 \leq x \leq 2\pi$

52. $y = \sin 2x - \cos x$, $0 \leq x \leq 2\pi$

For Problems 53–56, determine whether f and g are inverse functions.

53. $f(x) = 7x - 2$, $g(x) = \dfrac{x + 2}{7}$

54. $f(x) = \dfrac{3}{5}x - \dfrac{1}{4}$, $g(x) = \dfrac{5}{3}x + \dfrac{5}{12}$

55. $f(x) = -\dfrac{1}{2}x + 3$, $g(x) = \dfrac{1}{2}x - 3$

56. $f(x) = -5x + 1$, $g(x) = 5x - 1$

For Problems 57–60, find the inverse of the given function.

57. $f(x) = 6x - 1$

58. $f(x) = \dfrac{2}{3}x + 7$

59. $f(x) = -\dfrac{3}{5}x - \dfrac{2}{7}$

60. $f(x) = -5x + \dfrac{2}{9}$

4 Identities and Equations

In algebra, a statement such as $3x + 4 = 7$ is called an **equation,** or sometimes more specifically, a **conditional equation.** Solving an **algebraic equation** refers to the process of finding, from a set of potential replacements, those values for the variable that will make a true statement. Likewise, in trigonometry we are confronted with **trigonometric equations** such as $2 \sin \theta = 1$. Solving a trigonometric equation also refers to the process of finding replacements for the variables that will make a true statement.

The algebraic equation $1/x + (2/x) = 3/x$ is called an **identity** because it is true for all replacements for x when both sides of the equation are defined. Similarly, in trigonometry a statement such as $\csc \theta = 1/\sin \theta$ is called a **trigonometric identity** because it is true for all values of θ for which both sides of the equation are defined. Solving trigonometric equations and verifying trigonometric identities is the major emphasis of this chapter.

4.1

Trigonometric Identities

In Chapter 2 we frequently referred to the reciprocal relationships. They are actually trigonometric identities and can be verified using the definitions of the trigonometric functions (Definition 2.1). For example,

$$\sin \theta = \frac{y}{r} \quad \text{and} \quad \csc \theta = \frac{r}{y}.$$

Therefore, because

$$\frac{r}{y} = \frac{1}{\frac{y}{r}}$$

the identity

$$\csc \theta = \frac{1}{\sin \theta}$$

is established. In a like manner, the identities

$$\sec \theta = \frac{1}{\cos \theta} \quad \text{and} \quad \cot \theta = \frac{1}{\tan \theta}$$

can be verified. Remember that an identity such as $\csc \theta = 1/\sin \theta$ is true for all values of θ when both sides of the equation are defined.

Again using the definitions of the trigonometric functions we can show that

$$\frac{\sin \theta}{\cos \theta} = \frac{\frac{y}{r}}{\frac{x}{r}} = \left(\frac{y}{r}\right)\left(\frac{r}{x}\right) = \frac{y}{x} = \tan \theta.$$

Therefore, the identity

$$\tan \theta = \frac{\sin \theta}{\cos \theta}$$

is established. Similarly, the identity

$$\cot \theta = \frac{\cos \theta}{\sin \theta}$$

can be verified.

In Chapter 2 we verified the identity

$$\sin^2\theta + \cos^2\theta = 1.$$

From this identity two additional identities can be developed as follows. Dividing both sides of $\sin^2\theta + \cos^2\theta = 1$ by $\cos^2\theta$, and simplifying produces

$$\frac{\sin^2\theta}{\cos^2\theta} + \frac{\cos^2\theta}{\cos^2\theta} = \frac{1}{\cos^2\theta}$$

$$\tan^2\theta + 1 = \sec^2\theta.$$

Likewise, by dividing both sides of $\sin^2\theta + \cos^2\theta = 1$ by $\sin^2\theta$, and simplifying yields

$$\frac{\sin^2\theta}{\sin^2\theta} + \frac{\cos^2\theta}{\sin^2\theta} = \frac{1}{\sin^2\theta}$$

$$1 + \cot^2\theta = \csc^2\theta.$$

Let's pause for a moment and list the identities we've discussed thus far.

$$\csc\theta = \frac{1}{\sin\theta}, \qquad \sec\theta = \frac{1}{\cos\theta}, \qquad \cot\theta = \frac{1}{\tan\theta},$$

$$\tan\theta = \frac{\sin\theta}{\cos\theta}, \qquad \cot\theta = \frac{\cos\theta}{\sin\theta}, \qquad \sin^2\theta + \cos^2\theta = 1,$$

$$1 + \tan^2\theta = \sec^2\theta, \qquad 1 + \cot^2\theta = \csc^2\theta$$

You should know this list; it is sometimes referred to as the **basic** or **fundamental identities** of trigonometry.

The basic identities are used throughout the remainder of this chapter to (1) determine the other functional values from a given value, (2) simplify trigonometric expressions, (3) verify additional identities, (4) derive other important formulas, and (5) aid in the solving of trigonometric equations. Let's consider examples of some of these uses at this time.

EXAMPLE 1

If $\sin\theta = \frac{3}{5}$ and $\cos\theta < 0$, find the values of the other trigonometric functions.

Solution Using $\csc\theta = 1/\sin\theta$ produces

$$\csc\theta = \frac{1}{\frac{3}{5}} = \frac{5}{3}.$$

Substituting $\frac{3}{5}$ for $\sin\theta$ in the identity $\sin^2\theta + \cos^2\theta = 1$, and solving for $\cos\theta$, we obtain

$$\sin^2\theta + \cos^2\theta = 1$$

$$\left(\frac{3}{5}\right)^2 + \cos^2\theta = 1$$

$$\cos^2\theta = 1 - \frac{9}{25}$$

$$\cos^2\theta = \frac{16}{25}$$

$$\cos\theta = -\frac{4}{5}.\qquad \text{(Remember we were given that } \cos\theta < 0.)$$

Using $\sec\theta = 1/\cos\theta$ produces

$$\sec\theta = \frac{1}{-\frac{4}{5}} = -\frac{5}{4}.$$

Finally, using $\tan\theta = \sin\theta/\cos\theta$ and $\cot\theta = 1/\tan\theta$, we obtain

$$\tan\theta = \frac{\frac{3}{5}}{-\frac{4}{5}} = \left(\frac{3}{5}\right)\left(-\frac{5}{4}\right) = -\frac{3}{4}$$

and

$$\cot\theta = \frac{1}{\tan\theta} = \frac{1}{-\frac{3}{4}} = -\frac{4}{3}.$$

You should recognize that a problem such as Example 1 can be worked in many different ways. For example, after we found that $\csc\theta = \frac{5}{3}$, the identity $1 + \cot^2\theta = \csc^2\theta$ could be used to determine the value of $\cot\theta$. Likewise, after finding that $\sec\theta = -\frac{5}{4}$, we could use $1 + \tan^2\theta = \sec^2\theta$ to determine the value of $\tan\theta$. In fact, don't forget that the entire problem could be solved by applying the definitions of the trigonometric functions as we did in Chapter 2. However, at this time we would prefer to use the basic identities as much as possible.

EXAMPLE 2

Simplify $\sin\theta\cot\theta$.

Solution Replacing $\cot\theta$ with $\cos\theta/\sin\theta$ we can proceed as follows.

$$\sin \theta \cot \theta = \sin \theta \left(\frac{\cos \theta}{\sin \theta} \right)$$

$$= \cos \theta$$

Therefore, $\sin \theta \cot \theta$ simplifies to $\cos \theta$.

EXAMPLE 3

Simplify $(\sin \theta / \csc \theta) + (\cos \theta / \sec \theta)$.

Solution

$$\frac{\sin \theta}{\csc \theta} + \frac{\cos \theta}{\sec \theta} = \frac{\sin \theta}{\dfrac{1}{\sin \theta}} + \frac{\cos \theta}{\dfrac{1}{\cos \theta}}$$

$$= \sin^2 \theta + \cos^2 \theta = 1$$

Therefore,

$$\frac{\sin \theta}{\csc \theta} + \frac{\cos \theta}{\sec \theta} = 1.$$

EXAMPLE 4

Simplify $\cos x + \cos x \tan^2 x$.

Solution By factoring and then substituting $\sec^2 x$ for $1 + \tan^2 x$, we can proceed as follows.

$$\cos x + \cos x \tan^2 x = \cos x (1 + \tan^2 x)$$

$$= \cos x \sec^2 x$$

$$= \cos x \left(\frac{1}{\cos^2 x} \right)$$

$$= \frac{1}{\cos x} = \sec x$$

Therefore, $\cos x + \cos x \tan^2 x$ simplifies to $1/\cos x$ or $\sec x$.

From Examples 2, 3, and 4 we see that the end result of simplifying a trigonometric expression may be a constant as in Example 3 or a simpler trigonometric expression as in Examples 2 and 4. In Example 4, whether we use the final result of $1/\cos x$ or $\sec x$ depends on the context of the problem being simplified.

Verifying Identities

The process of verifying trigonometric identities is much the same as simplifying trigonometric expressions except that we know the desired result in advance. Consider the following examples and be sure that you can supply reasons for all of the steps.

EXAMPLE 5

Verify the identity $\sec\theta\cot\theta = \csc\theta$.

Solution Let's simplify the left side.

$$\sec\theta\cot\theta = \left(\frac{1}{\cos\theta}\right)\left(\frac{\cos\theta}{\sin\theta}\right)$$

$$= \frac{1}{\sin\theta}$$

$$= \csc\theta$$

Therefore, we have verified that $\sec\theta\cot\theta = \csc\theta$.

EXAMPLE 6

Verify the identity $1/\sec^2 x = (1 + \sin x)(1 - \sin x)$.

Solution By finding the indicated product on the right side, we can proceed as follows.

$$(1 + \sin x)(1 - \sin x) = 1 - \sin^2 x$$

$$= \cos^2 x$$

$$= \frac{1}{\sec^2 x}$$

Therefore, we have verified that $1/\sec^2 x = (1 + \sin x)(1 - \sin x)$.

Notice that in Example 5 we transformed the left side into the right side, but in Example 6 we transformed the right side into the left side. In general, we suggest that you attempt to transform the more complicated side into the other side. In some examples, like Example 7, each side would require the same amount of work.

EXAMPLE 7

Verify the identity $\sec x - \cos x = \sin x \tan x$.

Solution A We can transform the left side into the right side as follows.

$$\sec x - \cos x = \frac{1}{\cos x} - \cos x$$

$$= \frac{1 - \cos^2 x}{\cos x}$$

$$= \frac{\sin^2 x}{\cos x}$$

$$= \sin x\left(\frac{\sin x}{\cos x}\right)$$

$$= \sin x \tan x$$

Solution B We can transform the right side into the left side as follows.

$$\sin x \tan x = \sin x \left(\frac{\sin x}{\cos x} \right)$$

$$= \frac{\sin^2 x}{\cos x}$$

$$= \frac{1 - \cos^2 x}{\cos x}$$

$$= \frac{1}{\cos x} - \cos x$$

$$= \sec x - \cos x$$

A word of caution is in order before considering additional identities. Either transforming the left side into the right side or transforming the right side into the left side is an acceptable procedure for verifying an identity. However, *do not assume* the truth of an identity at the beginning and apply properties of equality to both sides.

EXAMPLE 8

Verify the identity

$$\frac{\cos x}{1 + \sin x} + \frac{\cos x}{1 - \sin x} = 2 \sec x.$$

Solution Adding the two fractions on the left, we can transform the left side into the right side as follows.

$$\frac{\cos x}{1 + \sin x} + \frac{\cos x}{1 - \sin x} = \frac{\cos x (1 - \sin x) + \cos x (1 + \sin x)}{(1 + \sin x)(1 - \sin x)}$$

$$= \frac{\cos x - \cos x \sin x + \cos x + \cos x \sin}{(1 + \sin x)(1 - \sin x)}$$

$$= \frac{2 \cos x}{(1 + \sin x)(1 - \sin x)}$$

$$= \frac{2 \cos x}{1 - \sin^2 x}$$

$$= \frac{2 \cos x}{\cos^2 x}$$

$$= \frac{2}{\cos x}$$

$$= 2 \sec x$$

EXAMPLE 9

Verify the identity

$$\frac{\cos x + \tan x}{\sin x \cos x} = \csc x + \sec^2 x.$$

Solution Let's apply the property of fractions $(a + c)/b = a/b + (c/b)$ to the left side and then proceed as follows.

$$\frac{\cos x + \tan x}{\sin x \cos x} = \frac{\cos x}{\sin x \cos x} + \frac{\tan x}{\sin x \cos x}$$

$$= \frac{1}{\sin x} + \frac{\dfrac{\sin x}{\cos x}}{\sin x \cos x}$$

$$= \frac{1}{\sin x} + \frac{\sin x}{\sin x \cos^2 x}$$

$$= \frac{1}{\sin x} + \frac{1}{\cos^2 x}$$

$$= \csc x + \sec^2 x$$

EXAMPLE 10

Verify the identity

$$\frac{\cos x}{1 - \sin x} = \frac{1 + \sin x}{\cos x}.$$

Solution We can apply the property of fractions $a/b = ac/bc$ and multiply both the numerator and denominator of the left side by $1 + \sin x$.

$$\frac{\cos x}{1 - \sin x} = \frac{\cos x}{1 - \sin x} \cdot \frac{1 + \sin x}{1 + \sin x}$$

$$= \frac{\cos x(1 + \sin x)}{1 - \sin^2 x}$$

$$= \frac{\cos x(1 + \sin x)}{\cos^2 x}$$

$$= \frac{1 + \sin x}{\cos x}$$

We cannot outline a specific procedure that will guarantee your success at verifying identities. However, we do offer the following suggestions.

1. Know the basic identities listed at the beginning of this section. You must have these at your fingertips.

2. Attempt to transform the more complicated side into the other side.

3. Keep in mind that many properties from algebra apply to trigonometric expressions. Similar terms can be combined. Trigonometric expressions can be multiplied and factored. Trigonometric fractions can be simplified, added, subtracted, multiplied, and divided using the same properties that we use with algebraic fractions.

4. Because 1 can be substituted for $\sin^2\theta + \cos^2\theta$, it is often helpful to simplify in terms of the sine and cosine functions.

Problem Set 4.1

1. Use Definition 2.1 and prove that $\sec\theta = \dfrac{1}{\cos\theta}$ and $\cot\theta = \dfrac{1}{\tan\theta}$.

2. Use Definition 2.1 and prove that $\dfrac{\cos\theta}{\sin\theta} = \cot\theta$.

For Problems 3–12, use the basic trigonometric identities listed at the beginning of this section to help find the remaining five trigonometric functional values.

3. $\sin\theta = \frac{4}{5}$ and the terminal side of θ lies in the first quadrant.

4. $\cos\theta = -\frac{5}{13}$ and the terminal side of θ lies in the second quadrant.

5. $\tan\theta = \frac{12}{5}$ and the terminal side of θ lies in the third quadrant.

6. $\sin\theta = -\frac{8}{17}$ and the terminal side of θ lies in the fourth quadrant.

7. $\sin\theta = \frac{4}{5}$ and $\cos\theta < 0$

8. $\csc\theta = -\frac{5}{4}$ and $\sec\theta < 0$

9. $\tan\theta = -\frac{1}{2}$ and $\cos\theta > 0$

10. $\sec\theta = 3$ and $\sin\theta < 0$

11. $\csc\theta = \frac{3}{2}$ and $\sec\theta < 0$

12. $\cot\theta = \frac{1}{3}$ and $\csc\theta < 0$

For Problems 13–24, simplify the given trigonometric expression to a single trigonometric function or a constant.

13. $\cos\theta\tan\theta$

14. $\dfrac{\tan\theta}{\sin\theta}$

15. $\cos x\csc x$

16. $\sec x - \sin x\tan x$

17. $(\cos^2 x - 1)(\tan^2 x + 1)$

18. $\sin x + \sin x\cot^2 x$

19. $(\cos^2 x)(1 + \tan^2 x)$

20. $(1 - \sin^2 x)\sec^2 x$

21. $\cos\theta + \tan\theta\sin\theta$

22. $\dfrac{\sec\theta - \cos\theta}{\tan\theta}$

23. $\dfrac{\tan x\sin x}{\sec^2 x - 1}$

24. $\tan x(\sin x + \cot x\cos x)$

For Problems 25–60, verify each of the following identities.

25. $\sin\theta\sec\theta = \tan\theta$

26. $\cos\theta\tan\theta\csc\theta = 1$

27. $\sin\theta + \sin\theta\tan^2\theta = \tan\theta\sec\theta$

28. $\cos\theta + \cos\theta\cot^2\theta = \cot\theta\csc\theta$

29. $\dfrac{\sin x + \cos x}{\cos x} = 1 + \tan x$

30. $\dfrac{\sin x + \tan x}{\sin x} = 1 + \sec x$

31. $\dfrac{\tan x + \cos x}{\sin x} = \sec x + \cot x$

32. $\csc x - \sin x = \cos x\cot x$

33. $\csc x\sec x = \tan x + \cot x$

34. $\dfrac{(\tan x)(1 + \cot^2 x)}{1 + \tan^2 x} = \cot x$

35. $\tan x = \dfrac{\cot x(1 + \tan^2 x)}{1 + \cot^2 x}$

36. $\dfrac{\cot x\cos x}{\csc^2 x - 1} = \sin x$

37. $\sin x(\csc x - \sin x) = \cos^2 x$

38. $1 - 2\sin^2\theta = 2\cos^2\theta - 1$

39. $2\sec^2\theta - 1 = 1 + 2\tan^2\theta$

40. $\cos^2\theta - \sin^2\theta = 1 - 2\sin^2\theta$

41. $\cos^2\theta - \sin^2\theta = 2\cos^2\theta - 1$

42. $\dfrac{1 + \cos x}{\sin x} + \dfrac{\sin x}{1 + \cos x} = 2\csc x$

43. $\dfrac{1}{1 - \cos x} + \dfrac{1}{1 + \cos x} = 2\csc^2 x$

44. $(\sec x - \tan x)(\csc x + 1) = \cot x$

45. $(\cos x - \sin x)(\cos x + \sin x) = 1 - 2\sin^2 x$

46. $\dfrac{1 + \sec x}{\sin x + \tan x} = \csc x$

47. $\cos^2 x = \dfrac{\csc^2 x - \cot^2 x}{\sec^2 x}$

48. $\dfrac{\cos x + \tan x}{\sin x \cos x} = \csc x + \sec^2 x$

49. $\sin^4 x - \cos^4 x = 1 - 2\cos^2 x$

50. $\tan^4 x - \sec^4 x = 1 - 2\sec^2 x$

51. $(\sin x - \cos x)^2 = 1 - 2\sin x \cos x$

52. $1 - \sin x = \dfrac{\cot x - \cos x}{\cot x}$

53. $1 + \tan x = \dfrac{\sec^2 x + 2\tan x}{1 + \tan x}$

54. $\dfrac{\sin x}{\sin x + \cos x} = \dfrac{\tan x}{1 + \tan x}$

55. $\dfrac{\sin x}{1 + \cos x} = \dfrac{1 - \cos x}{\sin x}$

56. $\dfrac{\tan x}{\sec x - 1} = \dfrac{\sec x + 1}{\tan x}$

57. $\dfrac{\csc x - 1}{\cot x} = \dfrac{\cot x}{\csc x + 1}$

58. $(\tan x - \sec x)^2 = \dfrac{1 - \sin x}{1 + \sin x}$

59. $\dfrac{1}{\tan x + \cot x} = \sin x \cos x$

60. $\dfrac{\sin x + \cos x}{\sin x - \cos x} = \dfrac{\sec x + \csc x}{\sec x - \csc x}$

4.2

Trigonometric Equations

As stated in the introductory paragraph of this chapter, solving a conditional trigonometric equation such as $2\sin x = 1$ refers to the process of finding the values for the variable x that will make a true numerical statement. Because the trigonometric functions are periodic, most trigonometric equations have infinitely many solutions. However, once the solutions within one period have been found, the remainder of the solutions are easily determined. For example, $\pi/6$ and $5\pi/6$ are the solutions between 0 and 2π (remember that 2π is the period of the sine function) that satisfy the equation $\sin x = \frac{1}{2}$. Then by adding multiples of 2π to each of these, all of the solutions can be represented by $\pi/6 + 2\pi n$ and $5\pi/6 + 2\pi n$, where n is an integer. The expressions $\pi/6 + 2\pi n$ and $5\pi/6 + 2\pi n$ are referred to as the **general solutions** of the equation. Using degree measure, the general solutions could be represented by $30° + n \cdot 360°$ and $150° + n \cdot 360°$, where n is an integer.

Solving trigonometric equations requires many of the techniques used to solve algebraic equations. Let's consider examples to illustrate some of those techniques.

EXAMPLE 1

Solve $2\cos\theta + 1 = 0$, if $0° \le \theta < 360°$.

Solution

$$2\cos\theta + 1 = 0$$

$$2\cos\theta = -1 \qquad \text{Added } -1 \text{ to both sides.}$$

$$\cos\theta = -\frac{1}{2} \qquad \text{Multiplied both sides by } \frac{1}{2}.$$

From our work with special angles, we know that $\cos 120° = -\frac{1}{2}$ and $\cos 240° = -\frac{1}{2}$. The solutions are $120°$ and $240°$.

EXAMPLE 2

Solve $\sin x \cos x = 0$, if $0 \le x \le 2\pi$.

Solution By applying the property "if $ab = 0$, then $a = 0$ or $b = 0$" we can proceed as follows.

$$\sin x \cos x = 0$$

$$\sin x = 0 \quad \text{or} \quad \cos x = 0$$

We know that $\sin 0 = 0$, $\sin \pi = 0$, $\cos(\pi/2) = 0$, and $\cos(3\pi/2) = 0$. Therefore, the solutions are 0, $\pi/2$, π, and $3\pi/2$.

In Example 1, because the statement of the problem contained degree measure ($0° \le \theta < 360°$), we expressed the solutions in degrees. Likewise, in Example 2 the statement $0 \le x < 2\pi$ implies the use of real numbers or radian measure.

EXAMPLE 3

Find the general solutions for $\sin x \tan x = \sin x$.

Solution

$$\sin x \tan x = \sin x$$

$$\sin x \tan x - \sin x = 0 \qquad \text{Added } -\sin x \text{ to both sides.}$$

$$\sin x(\tan x - 1) = 0 \qquad \text{Factored left side.}$$

$$\sin x = 0 \quad \text{or} \quad \tan x - 1 = 0 \qquad \text{Applied "if } ab = 0, \text{ then } a = 0$$
$$\text{or } b = 0."$$

$$\sin x = 0 \quad \text{or} \qquad \tan x = 1$$

Since the sine function has a period of 2π, it is sufficient to find the solutions of $\sin x = 0$ for $0 \le x < 2\pi$. Those solutions are 0 and π. The general expression $n\pi$, where n is an integer, will generate all of the solutions for $\sin x = 0$. The tangent function has a period of π, so it is sufficient to find the solutions of $\tan x = 1$ for $0 \le x < \pi$. The only solution is $\pi/4$. The general expression $\pi/4 + n\pi$, where n is an integer, will generate all of the solutions for $\tan x = 1$. Therefore, the solutions for $\sin x \tan x = \sin x$ can be represented by

$$n\pi \quad \text{and} \quad \frac{\pi}{4} + n\pi, \quad \text{where } n \text{ is an integer.}$$

Note that in Example 3 we *did not* begin by dividing both sides of the original equation by $\sin x$. Doing so would cause us to lose the solutions for $\sin x = 0$. As in algebra, we want to avoid dividing both sides of an equation by a variable or an expression containing a variable.

EXAMPLE 4

Solve $2\sin^2 x + \sin x - 1 = 0$, if $0 \le x < 2\pi$.

Solution Factoring the left side, we can proceed as follows.

$$2\sin^2 x + \sin x - 1 = 0$$
$$(2\sin x - 1)(\sin x + 1) = 0$$
$$2\sin x - 1 = 0 \quad \text{or} \quad \sin x + 1 = 0$$
$$2\sin x = 1 \quad \text{or} \quad \sin x = -1$$
$$\sin x = \frac{1}{2} \quad \text{or} \quad \sin x = -1$$

If $\sin x = \frac{1}{2}$, then $x = \pi/6$ or $5\pi/6$. If $\sin x = -1$, then $x = 3\pi/2$. The solutions are $\pi/6$, $5\pi/6$, and $3\pi/2$.

Perhaps you should check these solutions by substituting them back into the original equation.

Sometimes it is necessary to make a substitution using one of the basic identities. The next example illustrates this technique.

EXAMPLE 5

Solve $\sec^2 x + \tan^2 x = 3$, if $0 \le x < 2\pi$.

Solution Using the identity $1 + \tan^2 x = \sec^2 x$, we can substitute $1 + \tan^2 x$ for $\sec^2 x$ in the given equation and proceed as follows.

$$\sec^2 x + \tan^2 x = 3$$
$$(1 + \tan^2 x) + \tan^2 x = 3$$
$$2\tan^2 x = 2$$
$$\tan^2 x = 1$$
$$\tan x = \pm 1$$

If $\tan x = 1$, then $x = \dfrac{\pi}{4}$ or $\dfrac{5\pi}{4}$. If $\tan x = -1$, then $x = \dfrac{3\pi}{4}$ or $\dfrac{7\pi}{4}$. The solutions are $\pi/4$, $3\pi/4$, $5\pi/4$, and $7\pi/4$.

Recall from algebra that squaring both sides of an equation may produce some extraneous solutions. Therefore, we have learned that potential solutions *must be checked* if the squaring property is applied.

EXAMPLE 6

Solve $\sin x + \cos x = \sqrt{2}$, if $0 \le x < 2\pi$.

Solution

$$\sin x + \cos x = \sqrt{2}$$

$$\sin x = \sqrt{2} - \cos x$$

$$\sin^2 x = 2 - 2\sqrt{2}\cos x + \cos^2 x \qquad \text{Square both sides.}$$

$$1 - \cos^2 x = 2 - 2\sqrt{2}\cos x + \cos^2 x \qquad \begin{array}{l}\text{Substitute } 1 - \cos^2 x \text{ for} \\ \sin^2 x.\end{array}$$

$$0 = 2\cos^2 x - 2\sqrt{2}\cos x + 1$$

Now we can use the quadratic formula to solve for $\cos x$.

$$\cos x = \frac{2\sqrt{2} \pm \sqrt{8 - 8}}{4}$$

$$= \frac{2\sqrt{2}}{4} = \frac{\sqrt{2}}{2}$$

If $\cos x = \sqrt{2}/2$, then $x = \pi/4$ or $7\pi/4$.

Check: $\sin x + \cos x = \sqrt{2}$ $\qquad\qquad$ $\sin x + \cos x = \sqrt{2}$

$$\sin\frac{\pi}{4} + \cos\frac{\pi}{4} \stackrel{?}{=} \sqrt{2} \qquad\qquad \sin\frac{7\pi}{4} + \cos\frac{7\pi}{4} \stackrel{?}{=} \sqrt{2}$$

$$\frac{\sqrt{2}}{2} + \frac{\sqrt{2}}{2} \stackrel{?}{=} \sqrt{2} \qquad\qquad -\frac{\sqrt{2}}{2} + \frac{\sqrt{2}}{2} \stackrel{?}{=} \sqrt{2}$$

$$\frac{2\sqrt{2}}{2} = \sqrt{2} \qquad\qquad\qquad 0 \ne \sqrt{2}$$

The only solution is $\dfrac{\pi}{4}$.

Thus far in this section we have been able to determine solutions without the use of a calculator or a table. Now let's consider two examples where approximate solutions can be obtained using a calculator or a table.

EXAMPLE 7

Approximate, to the nearest hundredth of a radian, the solutions for

$$5\sin^2 x + 7\sin x - 6 = 0, \qquad 0 \le x < 2\pi.$$

Solution Factoring the left side of the equation we can proceed as follows.

$$5\sin^2 x + 7\sin x - 6 = 0$$

$$(\sin x + 2)(5\sin x - 3) = 0$$

$$\sin x + 2 = 0 \qquad \text{or} \quad 5\sin x - 3 = 0$$

$$\sin x = -2 \quad \text{or} \qquad 5\sin x = 3$$

$$\sin x = -2 \quad \text{or} \qquad \sin x = \frac{3}{5}$$

The equation $\sin x = -2$ produces no solutions because sine values must be between -1 and 1, inclusive. If $\sin x = \frac{3}{5} = .6$, then either from Table A in the back of the book or by using a calculator, we can determine that $x = .64$, to the nearest hundredth of a radian. Because the sine function is also positive in the second quadrant, x can be $\pi - .64 = 3.14 - .64 = 2.50$, to the nearest hundredth of a radian. Therefore, the approximate solutions of the original equation are .64 and 2.50.

EXAMPLE 8

Approximate, to the nearest tenth of a degree, the solutions for

$$\tan^2 \theta + 2\tan \theta - 1 = 0, \qquad 0° \le \theta < 360°.$$

Solution Using the quadratic formula we can determine approximate values for $\tan \theta$.

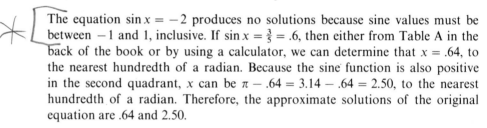

$$\tan \theta = \frac{-2 \pm \sqrt{4 + 4}}{2}$$

$$= \frac{-2 \pm \sqrt{8}}{2}$$

$$\approx \frac{-2 \pm 2.8284}{2}$$

$$\tan \theta \approx \frac{-2 + 2.8284}{2} \qquad \text{or} \quad \tan \theta \approx \frac{-2 - 2.8284}{2}$$

$$\approx .4142 \qquad\qquad\qquad\qquad \approx -2.4142$$

Now using a calculator or Table A, the approximate solutions of the original equation can be found. Tan $\theta = .4142$ implies that $\theta = 22.5°$. Since the tangent is also positive in the third quadrant, another approximate solution is $180° + 22.5° = 202.5°$. To solve $\tan \theta = -2.4142$, consider the reference angle θ' such that $\tan \theta' = 2.4142$. Solving this equation for θ' we find that $\theta' = 67.5°$. Since the tangent function is negative in the second and fourth quadrants, we obtain

$$\theta = 180° - 67.5° = 112.5°$$

or

$$\theta = 360° - 67.5° = 292.5°.$$

Therefore, the approximate solutions of the original equation are 22.5°, 112.5°, 202.5°, and 292.5°.

Problem Set 4.2

Every odd

Solve each of the following equations for θ, if $0° \le \theta < 360°$. Do not use a calculator or a table.

1. $2\sin\theta = \sqrt{3}$
2. $2\cos\theta + 1 = 0$
3. $2\cos\theta + 2 = 0$
4. $2\sin\theta + \sqrt{2} = 0$
5. $3\tan\theta + 3\sqrt{3} = 0$
6. $\tan^2\theta = 3$
7. $2\sin\theta = \sin\theta - 1$
8. $3\cos\theta + 1 = \cos\theta + 3$

Solve each of the following equations for x, if $0 \le x < 2\pi$. Do not use a calculator or a table.

9. $2\sin x + \sqrt{3} = 0$
10. $-2\cos x = \sqrt{2}$
11. $3\cos x - 2 - \cos x = 0$
12. $\sin^2 x - 1 = 0$
13. $(2\sin x + 1)(\tan x - 1) = 0$
14. $(\cos x + 1)(\sec x - 1) = 0$
15. $3\sin x + 5 = 0$
16. $\tan x \sin x = 0$

Solve each of the following equations. If the variable is θ, find all solutions such that $0° \le \theta < 360°$. If the variable is x, find all solutions such that $0 \le x < 2\pi$. Do not use a calculator or a table.

17. $2\sin^2 x = \sin x$
18. $\sin x \tan^2 x = \sin x$
19. $\tan^2\theta - \tan\theta = 0$
20. $2\tan\theta \sec\theta - \tan\theta = 0$
21. $2\cos^3\theta = \cos\theta$
22. $2\cos x = \cos x \csc x$
23. $2\cos^2 x + 3\cos x + 1 = 0$
24. $2\sin^2 x - \sin x - 1 = 0$
25. $\sec^2\theta - \sec\theta - 2 = 0$
26. $\sin\theta\cos\theta - \cos\theta + \sin\theta - 1 = 0$
27. $2\sin^2 x - \cos x - 1 = 0$
28. $2\cos^2 x - \sin x - 1 = 0$
29. $\sin^2 x + \cos x = 1$
30. $2\cos^2 x + \sin x - 2 = 0$
31. $\sin x = 1 - \cos x$
32. $\tan x + 1 = \sec x$
33. $\tan x = \cot x$
34. $\sin x - \cos x = 1$

Find *all* solutions of each of the following equations. If the variable is θ, express the solutions in degrees and if the variable is x, express the solutions in radians. Do not use a calculator or a table.

35. $2\cos\theta = \sqrt{3}$
36. $\sin\theta + 1 = 0$
37. $2\sin x + \sqrt{3} = 0$
38. $2\cos x - 1 = 0$
39. $\tan x + 1 = 0$
40. $\cot x - 1 = 0$
41. $\sec^2\theta = 4$
42. $\csc^2\theta = 4$
43. $\cot^2 x - \cot x = 0$
44. $\sec^2 x = \sec x$
45. $\csc^2 x - \csc x - 2 = 0$
46. $(\tan x - 1)(\tan x - \sqrt{3}) = 0$

Use your calculator or Table A to help find approximate solutions for θ, where $0° \le \theta < 360°$. Express the solutions to the nearest tenth of a degree.

47. $\sin \theta = -.2157$

48. $\sin \theta = .8217$

49. $\cos \theta = -.6427$

50. $\cos \theta = -.2179$

51. $\tan \theta = -3.1426$

52. $\tan \theta = 14.2789$

53. $(3 \sin \theta - 1)(2 \sin \theta + 3) = 0$

54. $(4 \sin \theta - 1)(\sin \theta + 2) = 0$

55. $6 \cos^2 \theta - 13 \cos \theta + 6 = 0$

56. $12 \cos^2 \theta - 13 \cos \theta + 3 = 0$

57. $\sin^2 \theta - 4 \sin \theta + 1 = 0$

58. $\cos^2 \theta - 3 \cos \theta + 1 = 0$

Use your calculator or Table A to help find approximate solutions for x, where $0 \le x < 2\pi$. Express the solutions to the nearest hundredth of a radian.

59. $\sin x = -.7126$

60. $\sin x = .2314$

61. $\cos x = -.8214$

62. $\cos x = -.1429$

63. $\tan x = -1.2784$

64. $\tan x = 9.1275$

65. $4 \sin^2 x + 11 \sin x - 3 = 0$

66. $12 \cos^2 x + 5 \cos x - 3 = 0$

67. $\sin^2 x - 3 \sin x - 2 = 0$

68. $\sin^2 x - \sin x - 1 = 0$

69. $\cos^2 x + \cos x - 1 = 0$

70. $2 \cos^2 x - 3 \cos x - 1 = 0$

4.3

Sum and Difference Formulas

We offer the following suggestions before you study these next two sections. First, read each section and look for the big ideas; don't be concerned about the details of proofs and worked-out examples. Then read the section again and pay attention to the details. This material is not difficult, but it is messy because of an abundance of formulas. However, you will soon see that most of the formulas are easily derived from a few key formulas.

The following identities were established in Chapter 2 and will be used at times in this section.

$$\sin(-\theta) = -\sin \theta \qquad \cos(-\theta) = \cos \theta \qquad \tan(-\theta) = -\tan \theta$$

Does $\cos(\alpha - \beta) = \cos \alpha - \cos \beta$? The following example gives an immediate response to this question.

EXAMPLE 1

Evaluate $\cos(\alpha - \beta)$ and $\cos \alpha - \cos \beta$ for $\alpha = 90°$ and $\beta = 60°$.

Solution

$$\cos(\alpha - \beta) = \cos(90° - 60°) = \cos 30° = \frac{\sqrt{3}}{2}$$

$$\cos \alpha - \cos \beta = \cos 90° - \cos 60° = 0 - \frac{1}{2} = -\frac{1}{2}$$

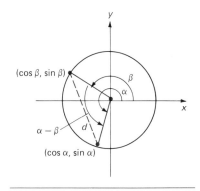

Figure 4.1

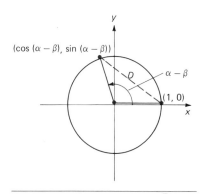

Figure 4.2

In general, $\cos(\alpha - \beta) \neq \cos \alpha - \cos \beta$. Additional examples would demonstrate that $\sin(\alpha - \beta) \neq \sin \alpha - \sin \beta$, $\cos 2\alpha \neq 2 \cos \alpha$, $\sin \frac{1}{2}\alpha \neq \frac{1}{2} \sin \alpha$, and so on. We need special formulas (identities) for these situations.

To develop a formula for $\cos(\alpha - \beta)$, let's consider two angles α and β, the angle representing $\alpha - \beta$, and the unit circle, as in Figure 4.1. The terminal side of α intersects the unit circle at the point $(\cos \alpha, \sin \alpha)$ and the terminal side of β intersects the unit circle at $(\cos \beta, \sin \beta)$. The distance, d, between the two points is given by

$$d = \sqrt{(\cos \alpha - \cos \beta)^2 + (\sin \alpha - \sin \beta)^2}.$$

Therefore,

$$\begin{aligned} d^2 &= (\cos \alpha - \cos \beta)^2 + (\sin \alpha - \sin \beta)^2 \\ &= \cos^2 \alpha - 2 \cos \alpha \cos \beta + \cos^2 \beta + \sin^2 \alpha - 2 \sin \alpha \sin \beta + \sin^2 \beta \end{aligned}$$

Since $\cos^2 \alpha + \sin^2 \alpha = 1$ and $\cos^2 \beta + \sin^2 \beta = 1$, we obtain

$$d^2 = 2 - 2 \cos \alpha \cos \beta - 2 \sin \alpha \sin \beta.$$

Now, looking back at Figure 4.1, let's construct angle $(\alpha - \beta)$ in standard position as indicated in Figure 4.2. The terminal side of $(\alpha - \beta)$ intersects the unit circle at the point $(\cos(\alpha - \beta), \sin(\alpha - \beta))$. The distance D between this point and $(1, 0)$ is given by

$$D = \sqrt{[\cos(\alpha - \beta) - 1]^2 + [\sin(\alpha - \beta) - 0]^2}.$$

Therefore,

$$\begin{aligned} D^2 &= [\cos(\alpha - \beta) - 1]^2 + [\sin(\alpha - \beta) - 0]^2 \\ &= \cos^2(\alpha - \beta) - 2 \cos(\alpha - \beta) + 1 + \sin^2(\alpha - \beta). \end{aligned}$$

Since $\cos^2(\alpha - \beta) + \sin^2(\alpha - \beta) = 1$, we obtain

$$D^2 = 2 - 2 \cos(\alpha - \beta).$$

The isosceles triangles formed in Figures 4.1 and 4.2 are congruent; thus, $d = D$ and we can equate the expressions for D^2 and d^2.

$$2 - 2 \cos(\alpha - \beta) = 2 - 2 \cos \alpha \cos \beta - 2 \sin \alpha \sin \beta$$

Subtracting 2 from both sides and then dividing both sides by -2 produces the following important identity.

$$\boxed{\cos(\alpha - \beta) = \cos \alpha \cos \beta + \sin \alpha \sin \beta}$$

The previous development assumes that α and β are positive angles with $\alpha > \beta$. However, the identity holds for all angles measured in radians or degrees; in fact, it is true for all real numbers.

EXAMPLE 2

Find an exact value for $\cos 15°$.

Find exact Value for Cos 15°

Solution Let $\alpha = 45°$ and $\beta = 30°$. Therefore,

$$\cos 15° = \cos(45° - 30°) = \cos 45° \cos 30° + \sin 45° \sin 30°$$

$$= \left(\frac{\sqrt{2}}{2}\right)\left(\frac{\sqrt{3}}{2}\right) + \left(\frac{\sqrt{2}}{2}\right)\left(\frac{1}{2}\right)$$

$$= \frac{\sqrt{6}}{4} + \frac{\sqrt{2}}{4}$$

$$= \frac{\sqrt{6} + \sqrt{2}}{4}.$$

If in the formula for $\cos(\alpha - \beta)$ we replace β with $-\beta$, we obtain

$$\cos(\alpha - (-\beta)) = \cos \alpha \cos(-\beta) + \sin \alpha \sin(-\beta).$$

Using the identities $\cos(-\beta) = \cos \beta$ and $\sin(-\beta) = -\sin \beta$, the following sum formula is produced.

$$\boxed{\cos(\alpha + \beta) = \cos \alpha \cos \beta - \sin \alpha \sin \beta}$$

EXAMPLE 3

Find an exact value for $\cos 5\pi/12$.

Solution Let $\alpha = \pi/4$ and $\beta = \pi/6$. Therefore,

$$\cos\frac{5\pi}{12} = \cos\left(\frac{\pi}{4} + \frac{\pi}{6}\right) = \cos\frac{\pi}{4}\cos\frac{\pi}{6} - \sin\frac{\pi}{4}\sin\frac{\pi}{6}$$

$$= \left(\frac{\sqrt{2}}{2}\right)\left(\frac{\sqrt{3}}{2}\right) - \left(\frac{\sqrt{2}}{2}\right)\left(\frac{1}{2}\right)$$

$$= \frac{\sqrt{6}}{4} - \frac{\sqrt{2}}{4} = \frac{\sqrt{6} - \sqrt{2}}{4}.$$

Before developing formulas for $\sin(\alpha + \beta)$ and $\sin(\alpha - \beta)$, let's consider some identities involving a 90° angle.

$$\cos(90° - \alpha) = \cos 90° \cos \alpha + \sin 90° \sin \alpha$$

$$= 0 \cdot \cos \alpha + 1 \cdot \sin \alpha$$

$$= \sin \alpha$$

Therefore,

$$\boxed{\cos(90° - \alpha) = \sin \alpha.}$$

Now substituting $90° - \alpha$ for α in the previous identity, we obtain

$$\cos(90° - (90° - \alpha)) = \sin(90° - \alpha)$$
$$\cos(90° - 90° + \alpha) = \sin(90° - \alpha)$$
$$\cos\alpha = \sin(90° - \alpha).$$

$$\sin(90° - \alpha) = \cos\alpha$$

An identity involving $\tan(90° - \alpha)$ follows directly from the sine and cosine relationships.

$$\tan(90° - \alpha) = \frac{\sin(90° - \alpha)}{\cos(90° - \alpha)} = \frac{\cos\alpha}{\sin\alpha} = \cot\alpha$$

$$\tan(90° - \alpha) = \cot\alpha$$

In Chapter 2, we verified that in a right triangle ACB, where A and B are the acute angles, any functional value of A equals the value of the cofunction of B. For example, $\sin A = \cos B$. The three previously stated identities are generalizations of this idea. For example, $\sin\alpha = \cos(90° - \alpha)$ for *any angle* α.

Now let's use the identity $\sin\alpha = \cos(90° - \alpha)$ to develop a formula for $\sin(\alpha + \beta)$.

$$\begin{aligned}
\sin(\alpha + \beta) &= \cos[90° - (\alpha + \beta)] \\
&= \cos[(90° - \alpha) - \beta] \\
&= \cos(90° - \alpha)\cos\beta + \sin(90° - \alpha)\sin\beta \\
&= \sin\alpha\cos\beta + \cos\alpha\sin\beta
\end{aligned}$$

$$\sin(\alpha + \beta) = \sin\alpha\cos\beta + \cos\alpha\sin\beta$$

Substituting $-\beta$ for β into the formula for $\sin(\alpha + \beta)$ produces

$$\begin{aligned}
\sin(\alpha + (-\beta)) &= \sin\alpha\cos(-\beta) + \cos\alpha\sin(-\beta) \\
&= \sin\alpha\cos\beta - \cos\alpha\sin\beta.
\end{aligned}$$

$$\sin(\alpha - \beta) = \sin\alpha\cos\beta - \cos\alpha\sin\beta$$

As you might expect, a formula for $\tan(\alpha + \beta)$ follows directly from the sine and cosine relationships.

$$\tan(\alpha + \beta) = \frac{\sin(\alpha + \beta)}{\cos(\alpha + \beta)} = \frac{\sin\alpha\cos\beta + \cos\alpha\sin\beta}{\cos\alpha\cos\beta - \sin\alpha\sin\beta}$$

Dividing both numerator and denominator by $\cos\alpha\cos\beta$ produces

$$\tan(\alpha+\beta)=\frac{\dfrac{\sin\alpha\cos\beta}{\cos\alpha\cos\beta}+\dfrac{\cos\alpha\sin\beta}{\cos\alpha\cos\beta}}{\dfrac{\cos\alpha\cos\beta}{\cos\alpha\cos\beta}-\dfrac{\sin\alpha\sin\beta}{\cos\alpha\cos\beta}}$$

$$=\frac{\tan\alpha+\tan\beta}{1-\tan\alpha\tan\beta}.$$

$$\tan(\alpha+\beta)=\frac{\tan\alpha+\tan\beta}{1-\tan\alpha\tan\beta}$$

If in the formula for $\tan(\alpha+\beta)$ we replace β with $-\beta$, the following result will be obtained. (We will leave the details for you as an exercise.)

$$\tan(\alpha-\beta)=\frac{\tan\alpha-\tan\beta}{1+\tan\alpha\tan\beta}$$

EXAMPLE 4

Find an exact value for $\sin\pi/12$.

Solution Let $\alpha=\pi/3$ and $\beta=\pi/4$. Therefore,

$$\sin\frac{\pi}{12}=\sin\left(\frac{\pi}{3}-\frac{\pi}{4}\right)=\sin\frac{\pi}{3}\cos\frac{\pi}{4}-\cos\frac{\pi}{3}\sin\frac{\pi}{4}$$

$$=\left(\frac{\sqrt3}{2}\right)\left(\frac{\sqrt2}{2}\right)-\left(\frac12\right)\left(\frac{\sqrt2}{2}\right)$$

$$=\frac{\sqrt6}{4}-\frac{\sqrt2}{4}=\frac{\sqrt6-\sqrt2}{4}.$$

EXAMPLE 5

Find an exact value for $\tan195°$.

Solution Let $\alpha=135°$ and $\beta=60°$. Therefore,

$$\tan195°=\tan(135°+60°)=\frac{\tan135°+\tan60°}{1-\tan135°\tan60°}$$

$$=\frac{-1+\sqrt3}{1-(-1)(\sqrt3)}$$

$$=\frac{-1+\sqrt3}{1+\sqrt3}$$

$$= \frac{-1 + \sqrt{3}}{1 + \sqrt{3}} \cdot \frac{1 - \sqrt{3}}{1 - \sqrt{3}} = \frac{-1 + 2\sqrt{3} - 3}{1 - 3}$$

$$= \frac{-4 + 2\sqrt{3}}{-2} = 2 - \sqrt{3}.$$

EXAMPLE 6

Given that $\sin \alpha = \frac{3}{5}$ with α in the first quadrant, and $\cos \beta = -\frac{7}{25}$ with β in the second quadrant, find $\cos(\alpha - \beta)$, $\sin(\alpha + \beta)$, and $\tan(\alpha + \beta)$.

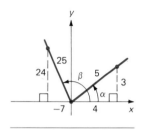

Figure 4.3

Solution If $\sin \alpha = \frac{3}{5}$ and α is in the first quadrant, then $\cos \alpha = \frac{4}{5}$ and $\tan \alpha = \frac{3}{4}$ (Figure 4.3). Likewise, if $\cos \beta = -\frac{7}{25}$ and β is in the second quadrant, then $\sin \beta = \frac{24}{25}$ and $\tan \beta = -\frac{24}{7}$.

$$\cos(\alpha - \beta) = \cos \alpha \cos \beta + \sin \alpha \sin \beta$$

$$= \left(\frac{4}{5}\right)\left(-\frac{7}{25}\right) + \left(\frac{3}{5}\right)\left(\frac{24}{25}\right) = -\frac{28}{125} + \frac{72}{125} = \frac{44}{125}$$

$$\sin(\alpha + \beta) = \sin \alpha \cos \beta + \cos \alpha \sin \beta$$

$$= \left(\frac{3}{5}\right)\left(-\frac{7}{25}\right) + \left(\frac{4}{5}\right)\left(\frac{24}{25}\right)$$

$$= -\frac{21}{125} + \frac{96}{125} = \frac{75}{125} = \frac{3}{5}$$

$$\tan(\alpha + \beta) = \frac{\tan \alpha + \tan \beta}{1 - \tan \alpha \tan \beta}$$

$$= \frac{\dfrac{3}{4} + \left(-\dfrac{24}{7}\right)}{1 - \left(\dfrac{3}{4}\right)\left(-\dfrac{24}{7}\right)}$$

$$= \frac{-\dfrac{75}{28}}{1 + \dfrac{72}{28}} = \frac{-\dfrac{75}{28}}{\dfrac{100}{28}} = -\frac{75}{100} = -\frac{3}{4}$$

EXAMPLE 7

Evaluate $\sin(\tan^{-1}\frac{1}{2} + \cos^{-1}\frac{3}{5})$.

Solution Let $\alpha = \tan^{-1}\frac{1}{2}$ and $\beta = \cos^{-1}\frac{3}{5}$. From the definitions of the inverse functions in Chapter 3, and since $\tan \alpha$ and $\cos \beta$ are positive, α and β can be considered acute angles of the following right triangles.

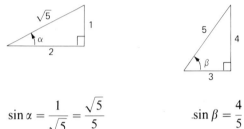

$$\sin \alpha = \frac{1}{\sqrt{5}} = \frac{\sqrt{5}}{5} \qquad\qquad \sin \beta = \frac{4}{5}$$

$$\cos \alpha = \frac{2}{\sqrt{5}} = \frac{2\sqrt{5}}{5} \qquad\qquad \cos \beta = \frac{3}{5}$$

Therefore,

$$\sin(\alpha + \beta) = \sin \alpha \cos \beta + \cos \alpha \sin \beta$$

$$= \left(\frac{\sqrt{5}}{5}\right)\left(\frac{3}{5}\right) + \left(\frac{2\sqrt{5}}{5}\right)\left(\frac{4}{5}\right)$$

$$= \frac{3\sqrt{5}}{25} + \frac{8\sqrt{5}}{25} = \frac{11\sqrt{5}}{25}.$$

Thus, $\sin(\tan^{-1}\frac{1}{2} + \cos^{-1}\frac{3}{5}) = \frac{11\sqrt{5}}{25}$.

If you are following the suggestion offered at the beginning of this section, it is now time to go back and fill in some details. Keep the following continuity pattern in mind. The first formula derived was

$$\cos(\alpha - \beta) = \cos \alpha \cos \beta + \sin \alpha \sin \beta. \tag{1}$$

Then substituting $-\beta$ for β produced

$$\cos(\alpha + \beta) = \cos \alpha \cos \beta - \sin \alpha \sin \beta. \tag{2}$$

Then applying $\cos(\alpha - \beta)$ to $\cos(90° - \alpha)$ produced

$$\cos(90° - \alpha) = \sin \alpha. \tag{3}$$

Then substituting $(\alpha + \beta)$ for α in (3) produced

$$\sin(\alpha + \beta) = \sin \alpha \cos \beta + \cos \alpha \sin \beta. \tag{4}$$

Then substituting $-\beta$ for β produced

$$\sin(\alpha - \beta) = \sin \alpha \cos \beta - \cos \alpha \sin \beta. \tag{5}$$

Then using the formulas for $\sin(\alpha + \beta)$ and $\cos(\alpha + \beta)$, we obtained

$$\tan(\alpha + \beta) = \frac{\tan \alpha + \tan \beta}{1 - \tan \alpha \tan \beta}. \tag{6}$$

Finally, substituting $-\beta$ for β produced

$$\tan(\alpha - \beta) = \frac{\tan \alpha - \tan \beta}{1 + \tan \alpha \tan \beta}. \tag{7}$$

Problem Set 4.3

1 – 23 odd

27 – 33 odd

For Problems 1–14, find the exact values without using a table or a calculator.

1. $\sin 15°$ 2. $\tan 15°$ 3. $\tan 75°$ 4. $\sin 75°$

5. $\sin 105°$ 6. $\cos 105°$ 7. $\cos 195°$ 8. $\sin 195°$

9. $\tan 255°$ 10. $\cos 345°$ 11. $\cos \dfrac{\pi}{12}$ 12. $\cos \dfrac{5\pi}{12}$

13. $\sin \dfrac{7\pi}{12}$ 14. $\sin \dfrac{11\pi}{12} = \dfrac{9\pi + 2\pi}{12} = \dfrac{9\pi}{12} + \dfrac{2\pi}{12} = \dfrac{3\pi}{4} + \dfrac{\pi}{6}$

15. Given that $\cos \alpha = \frac{3}{5}$ with α in the first quadrant, and $\sin \beta = \frac{15}{17}$ with β in the second quadrant, find $\sin(\alpha - \beta)$ and $\tan(\alpha + \beta)$.

16. If α and β are acute angles such that $\cos \alpha = \frac{4}{5}$ and $\tan \beta = \frac{8}{15}$, find $\cos(\alpha + \beta)$ and $\sin(\alpha + \beta)$.

17. If $\cos \alpha = -\frac{3}{5}$ and $\tan \beta = \frac{8}{15}$, where α is a second-quadrant angle and β is a third-quadrant angle, find $\sin(\alpha + \beta)$ and $\cos(\alpha - \beta)$.

18. If $\tan \alpha = -\frac{4}{3}$ and $\sin \beta = -\frac{3}{5}$, where α is a second-quadrant angle and β is a fourth-quadrant angle, find $\sin(\alpha - \beta)$ and $\tan(\alpha + \beta)$.

19. Given that $\tan \alpha = \frac{8}{15}$ with α in the first quadrant, and $\cos \beta = \frac{7}{25}$ with β in the fourth quadrant, find $\sin(\alpha - \beta)$ and $\cos(\alpha - \beta)$.

20. Given that $\tan \alpha = -\frac{2}{3}$ with α in the second quadrant, and $\tan \beta = \frac{3}{5}$ with β in the third quadrant, find $\tan(\alpha + \beta)$ and $\tan(\alpha - \beta)$.

For Problems 21–26, find exact values without using a calculator or a table.

21. $\sin(\tan^{-1}\frac{3}{4} + \cos^{-1}\frac{24}{25})$
22. $\cos(\tan^{-1}\frac{7}{24} - \sin^{-1}\frac{4}{5})$
23. $\tan(\arcsin\frac{15}{17} + \arccos\frac{4}{5})$
24. $\tan[\arcsin\frac{3}{5} + \arccos(-\frac{3}{5})]$
25. $\cos(\tan^{-1}\frac{1}{3} + \cos^{-1}\frac{8}{17})$
26. $\sin(\tan^{-1}\frac{1}{2} - \cos^{-1}\frac{4}{5})$

Verify each of the identities in Problems 27–38.

27. $\sin(\alpha + 90°) = \cos \alpha$
28. $\cos(\alpha + 90°) = -\sin \alpha$
29. $\sin(\alpha + \pi) = -\sin \alpha$
30. $\cos(\alpha - \pi) = -\cos \alpha$
31. $\tan(\alpha + \pi) = \tan \alpha$
32. $\tan(\alpha - \pi) = \tan \alpha$
33. $\sin(\alpha + 45°) = \dfrac{\sqrt{2}}{2}(\sin \alpha + \cos \alpha)$
34. $\cos(\alpha - 45°) = \dfrac{\sqrt{2}}{2}(\cos \alpha + \sin \alpha)$
35. $\tan\left(\alpha + \dfrac{\pi}{4}\right) = \dfrac{1 + \tan \alpha}{1 - \tan \alpha}$
36. $\tan\left(\alpha - \dfrac{\pi}{4}\right) = \dfrac{\tan \alpha - 1}{\tan \alpha + 1}$
37. $\sin(\alpha + 270°) = -\cos \alpha$
38. $\cos(\alpha - 270°) = -\sin \alpha$

39. Develop the formula for $\tan(\alpha - \beta)$ by replacing β with $-\beta$ in the formula for $\tan(\alpha + \beta)$.

40. Derive the formula $\cot(\alpha + \beta) = \dfrac{\cot \alpha \cot \beta - 1}{\cot \alpha + \cot \beta}$.

41. Derive the formula $\cot(\alpha - \beta) = \dfrac{\cot \alpha \cot \beta + 1}{\cot \beta - \cot \alpha}$.

42. If α and β are complementary angles, verify that $\sin^2 \alpha + \sin^2 \beta = 1$.

Miscellaneous Problems

43. Use the identities

$$\cos(\alpha + \beta) = \cos \alpha \cos \beta - \sin \alpha \sin \beta$$

$$\cos(\alpha - \beta) = \cos \alpha \cos \beta + \sin \alpha \sin \beta$$

to produce the following **product identities**.
   (a) $\cos \alpha \cos \beta = \frac{1}{2}[\cos(\alpha + \beta) + \cos(\alpha - \beta)]$
   (b) $\sin \alpha \sin \beta = \frac{1}{2}[\cos(\alpha - \beta) - \cos(\alpha + \beta)]$

44. Use the identities

$$\sin(\alpha + \beta) = \sin \alpha \cos \beta + \cos \alpha \sin \beta$$

$$\sin(\alpha - \beta) = \sin \alpha \cos \beta - \cos \alpha \sin \beta$$

to produce the following **product identities**.
   (a) $\sin \alpha \cos \beta = \frac{1}{2}[\sin(\alpha + \beta) + \sin(\alpha - \beta)]$
   (b) $\cos \alpha \sin \beta = \frac{1}{2}[\sin(\alpha + \beta) - \sin(\alpha - \beta)]$

45. Use the identities established in Problems 43 and 44 to express each of the following products as a sum or difference.
   (a) $\cos 3\theta \cos 2\theta$ (b) $\cos 2\theta \cos 4\theta$ (c) $\sin 3\theta \sin \theta$
   (d) $\sin 4\theta \cos 3\theta$ (e) $2 \sin 9\theta \cos 3\theta$ (f) $3 \cos \theta \sin 2\theta$

46. Use the identities from parts (a) and (b) of Problem 43 to help verify the following **sum** and **difference identities**. (*Hint*: Let $A = \alpha + \beta$ and $B = \alpha - \beta$.)
   (a) $\cos A + \cos B = 2 \cos \dfrac{A + B}{2} \cos \dfrac{A - B}{2}$

   (b) $\cos B - \cos A = 2 \sin \dfrac{A + B}{2} \sin \dfrac{A - B}{2}$

47. Use the identities from parts (a) and (b) of Problem 44 to help verify the following **sum** and **difference identities**.
   (a) $\sin A + \sin B = 2 \sin \dfrac{A + B}{2} \cos \dfrac{A - B}{2}$

   (b) $\sin A - \sin B = 2 \cos \dfrac{A + B}{2} \sin \dfrac{A - B}{2}$

48. Use the identities established in Problems 46 and 47 to help write each of the following as a product.

(a) $\cos 4\theta + \cos 2\theta$ (b) $\cos 6\theta - \cos 2\theta$ (c) $\cos \theta - \cos 5\theta$

(d) $\sin 6\theta + \sin 2\theta$ (e) $\sin 3\theta - \sin \theta$ (f) $\sin 2\theta - \sin 4\theta$

4.4

Multiple and Half-Angle Formulas

Again let's emphasize the point that $\sin 2\alpha \neq 2 \sin \alpha$. For example, $\sin 2(30°) = \sin 60° = \sqrt{3}/2$, but $2 \sin 30° = 2(\frac{1}{2}) = 1$. Thus, **multiple angle formulas** are needed.

As you might expect, the multiple angle formulas are nothing more than special cases of the sum formulas developed in the previous section. For example, in the formula for $\sin(\alpha + \beta)$, if we let $\alpha = \beta$ we obtain the following.

$$\sin(\alpha + \beta) = \sin \alpha \cos \beta + \cos \alpha \sin \beta$$

becomes

$$\sin(\alpha + \alpha) = \sin \alpha \cos \alpha + \cos \alpha \sin \alpha, \text{ or, equivalently}$$

$$\boxed{\sin 2\alpha = 2 \sin \alpha \cos \alpha.}$$

EXAMPLE 1

Find $\sin 2\alpha$ if $\sin \alpha = \frac{4}{5}$ and α is in the first quadrant.

Solution If $\sin \alpha = \frac{4}{5}$ and α is in the first quadrant, then $\cos \alpha = \frac{3}{5}$. Therefore,

$$\sin 2\alpha = 2 \sin \alpha \cos \alpha$$
$$= 2\left(\frac{4}{5}\right)\left(\frac{3}{5}\right) = \frac{24}{25}.$$

Substituting α for β in the formula for $\cos(\alpha + \beta)$ produces

$$\cos(\alpha + \beta) = \cos \alpha \cos \beta - \sin \alpha \sin \beta$$
$$\cos(\alpha + \alpha) = \cos \alpha \cos \alpha - \sin \alpha \sin \alpha$$
$$\cos 2\alpha = \cos^2 \alpha - \sin^2 \alpha.$$

Two other forms for $\cos 2\alpha$ can be obtained by using the basic identity $\sin^2 \alpha + \cos^2 \alpha = 1$. Substituting $1 - \sin^2 \alpha$ for $\cos^2 \alpha$ produces

$$\cos^2 \alpha - \sin^2 \alpha = (1 - \sin^2 \alpha) - \sin^2 \alpha$$
$$= 1 - 2 \sin^2 \alpha.$$

Similarly, substituting $1 - \cos^2 \alpha$ for $\sin^2 \alpha$ produces

$$\cos^2 \alpha - \sin^2 \alpha = \cos^2 \alpha - (1 - \cos^2 \alpha)$$
$$= 2 \cos^2 \alpha - 1.$$

$$\cos 2\alpha = \cos^2 \alpha - \sin^2 \alpha$$

or

$$\cos 2\alpha = 1 - 2\sin^2 \alpha$$

or

$$\cos 2\alpha = 2\cos^2 \alpha - 1$$

Substituting α for β in the formula for $\tan(\alpha + \beta)$ produces

$$\tan(\alpha + \beta) = \frac{\tan \alpha + \tan \beta}{1 - \tan \alpha \tan \beta}$$

$$\tan(\alpha + \alpha) = \frac{\tan \alpha + \tan \alpha}{1 - \tan \alpha \tan \alpha}$$

$$\tan 2\alpha = \frac{2\tan \alpha}{1 - \tan^2 \alpha}.$$

EXAMPLE 2

Find $\sin 2\theta$, $\cos 2\theta$, and $\tan 2\theta$ if $\sin \theta = -\frac{5}{13}$ and θ is a fourth-quadrant angle.

Solution The following figure depicts the situation.

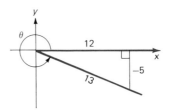

Therefore, $\cos \theta = \frac{12}{13}$ and $\tan \theta = -\frac{5}{12}$. Now we can use the double-angle identities.

$$\sin 2\theta = 2\sin \theta \cos \theta$$

$$= 2\left(-\frac{5}{13}\right)\left(\frac{12}{13}\right) = -\frac{120}{169}$$

$$\cos 2\theta = \cos^2 \theta - \sin^2 \theta$$

$$= \left(\frac{12}{13}\right)^2 - \left(-\frac{5}{13}\right)^2$$

$$= \frac{144}{169} - \frac{25}{169} = \frac{119}{169}$$

$$\tan 2\theta = \frac{2\tan\theta}{1 - \tan^2\theta}$$

$$= \frac{2\left(-\dfrac{5}{12}\right)}{1 - \left(-\dfrac{5}{12}\right)^2}$$

$$= \frac{-\dfrac{10}{12}}{1 - \dfrac{25}{144}}$$

$$= \frac{-\dfrac{5}{6}}{\dfrac{119}{144}} = \left(-\frac{5}{6}\right)\left(\frac{144}{119}\right) = -\frac{120}{119}$$

Remark: In Example 2, after finding the values for $\sin 2\theta$ and $\cos 2\theta$, the identity $\tan 2\theta = \sin 2\theta / \cos 2\theta$ could also be used to find the value of $\tan 2\theta$.

The double-angle identities provide a broader base for proving identities and solving trigonometric equations, as the next examples illustrate.

EXAMPLE 3

Verify the identity $\cos 2\alpha = \cos^4\alpha - \sin^4\alpha$.

Solution The right side can be factored as the difference of two squares.

$$\cos^4\alpha - \sin^4\alpha = (\cos^2\alpha + \sin^2\alpha)(\cos^2\alpha - \sin^2\alpha)$$

We know that $\cos^2\alpha + \sin^2\alpha = 1$ and $\cos^2\alpha - \sin^2\alpha = \cos 2\alpha$. Therefore,

$$\cos^4\alpha - \sin^4\alpha = 1 \cdot \cos 2\alpha = \cos 2\alpha.$$

EXAMPLE 4

Solve the equation $\cos 2\theta - \cos\theta = 0$, where $0° \le \theta < 360°$.

Solution We can substitute $2\cos^2\theta - 1$ for $\cos 2\theta$ and proceed as follows.

$$\cos 2\theta - \cos\theta = 0$$

$$(2\cos^2\theta - 1) - \cos\theta = 0$$

$$2\cos^2\theta - \cos\theta - 1 = 0$$

$$(2\cos\theta + 1)(\cos\theta - 1) = 0$$

$$2\cos\theta + 1 = 0 \quad \text{or} \quad \cos\theta - 1 = 0$$

$$2\cos\theta = -1 \quad \text{or} \quad \cos\theta = 1$$

$$\cos\theta = -\frac{1}{2} \quad \text{or} \quad \cos\theta = 1$$

If $\cos\theta = -\frac{1}{2}$, then $\theta = 120°$ or $240°$. If $\cos\theta = 1$, then $\theta = 0°$. The solutions are $0°$, $120°$, and $240°$.

EXAMPLE 5

Express $\sin 3\theta$ in terms of $\sin\theta$.

Solution

$$
\begin{aligned}
\sin 3\theta &= \sin(2\theta + \theta) \\
&= \sin 2\theta \cos\theta + \cos 2\theta \sin\theta \\
&= (2\sin\theta\cos\theta)\cos\theta + (1 - 2\sin^2\theta)\sin\theta \\
&= 2\sin\theta\cos^2\theta + \sin\theta - 2\sin^3\theta \\
&= 2\sin\theta(1 - \sin^2\theta) + \sin\theta - 2\sin^3\theta \\
&= 2\sin\theta - 2\sin^3\theta + \sin\theta - 2\sin^3\theta \\
&= -4\sin^3\theta + 3\sin\theta
\end{aligned}
$$

Half-Angle Formulas

Half-angle formulas for $\sin\alpha/2$ and $\cos\alpha/2$ are a direct consequence of the identities for $\cos 2\alpha$. First, let's solve the equation $\cos 2\alpha = 1 - 2\sin^2\alpha$ for $\sin\alpha$.

$$
\begin{aligned}
1 - 2\sin^2\alpha &= \cos 2\alpha \\
-2\sin^2\alpha &= \cos 2\alpha - 1 \\
\sin^2\alpha &= \frac{1 - \cos 2\alpha}{2} \\
\sin\alpha &= \pm\sqrt{\frac{1 - \cos 2\alpha}{2}}
\end{aligned}
$$

Now substituting $\alpha/2$ for α produces

$$
\sin\frac{\alpha}{2} = \pm\sqrt{\frac{1 - \cos\alpha}{2}}.
$$

A formula for $\cos\alpha/2$ can be obtained by solving $\cos 2\alpha = 2\cos^2\alpha - 1$ for $\cos\alpha$.

$$
\begin{aligned}
2\cos^2\alpha - 1 &= \cos 2\alpha \\
2\cos^2\alpha &= 1 + \cos 2\alpha \\
\cos^2\alpha &= \frac{1 + \cos 2\alpha}{2} \\
\cos\alpha &= \pm\sqrt{\frac{1 + \cos 2\alpha}{2}}
\end{aligned}
$$

Again substituting $\alpha/2$ for α produces

$$\cos\frac{\alpha}{2} = \pm\sqrt{\frac{1+\cos\alpha}{2}}.$$

In the formulas for $\sin\alpha/2$ and $\cos\alpha/2$, the choice of the plus or minus sign is determined by the quadrant in which $\alpha/2$ lies. For example, if $\alpha/2$ is in the first or second quadrant, then we would use

$$\sin\frac{\alpha}{2} = \sqrt{\frac{1-\cos\alpha}{2}}.$$

However, if $\alpha/2$ is a third- or fourth-quadrant angle, then we would use

$$\sin\frac{\alpha}{2} = -\sqrt{\frac{1-\cos\alpha}{2}}.$$

To obtain a formula for $\tan\alpha/2$ we can proceed as follows.

$$\tan\frac{\alpha}{2} = \frac{\sin\frac{\alpha}{2}}{\cos\frac{\alpha}{2}} = \frac{\pm\sqrt{\frac{1-\cos\alpha}{2}}}{\pm\sqrt{\frac{1+\cos\alpha}{2}}} = \pm\sqrt{\frac{1-\cos\alpha}{1+\cos\alpha}}$$

An alternative form for the $\tan\alpha/2$ formula can be obtained by multiplying the radicand by a form of one, namely, $(1-\cos\alpha)/(1-\cos\alpha)$.

$$\tan\frac{\alpha}{2} = \pm\sqrt{\frac{1-\cos\alpha}{1+\cos\alpha}\cdot\frac{1-\cos\alpha}{1-\cos\alpha}}$$

$$= \pm\sqrt{\frac{(1-\cos\alpha)^2}{1-\cos^2\alpha}}$$

$$= \pm\sqrt{\frac{(1-\cos\alpha)^2}{\sin^2\alpha}}$$

$$= \frac{1-\cos\alpha}{\sin\alpha}$$

We no longer need the $\pm$ sign because $1-\cos\alpha$ is never negative, and $\sin\alpha$ and $\tan\alpha/2$ will always agree in sign. For example, if $0 < \alpha < \pi$, then $0 < \alpha/2 < \pi/2$ and therefore both $\sin\alpha$ and $\tan\alpha/2$ are positive. If $\pi < \alpha < 2\pi$, then $\pi/2 < \alpha/2 < \pi$ and both $\sin\alpha$ and $\tan\alpha/2$ are negative. Furthermore, the form of $(1-\cos\alpha)/\sin\alpha$ can be changed as follows.

$$\frac{1-\cos\alpha}{\sin\alpha}\cdot\frac{1+\cos\alpha}{1+\cos\alpha} = \frac{1-\cos^2\alpha}{\sin\alpha(1+\cos\alpha)}$$

$$= \frac{\sin^2\alpha}{\sin\alpha(1+\cos\alpha)}$$

$$= \frac{\sin\alpha}{1+\cos\alpha}$$

Therefore, the following general formulas can be stated.

$$\tan\frac{\alpha}{2} = \pm\sqrt{\frac{1-\cos\alpha}{1+\cos\alpha}}$$

or

$$\tan\frac{\alpha}{2} = \frac{1-\cos\alpha}{\sin\alpha}$$

or

$$\tan\frac{\alpha}{2} = \frac{\sin\alpha}{1+\cos\alpha}$$

EXAMPLE 6

If $\cos\alpha = -\frac{4}{5}$ and α is in the third quadrant, find $\sin\alpha/2$, $\cos\alpha/2$, and $\tan\alpha/2$.

Solution Because α is in the third quadrant, $\alpha/2$ is in the second quadrant; thus, $\sin\alpha/2$ is positive and $\cos\alpha/2$ is negative.

$$\sin\frac{\alpha}{2} = \sqrt{\frac{1-\cos\alpha}{2}} = \sqrt{\frac{1-\left(-\frac{4}{5}\right)}{2}} = \sqrt{\frac{1+\frac{4}{5}}{2}}$$

$$= \sqrt{\frac{\frac{9}{5}}{2}} = \sqrt{\frac{9}{10}} = \frac{3\sqrt{10}}{10}$$

$$\cos\frac{\alpha}{2} = -\sqrt{\frac{1+\cos\alpha}{2}} = -\sqrt{\frac{1+\left(-\frac{4}{5}\right)}{2}} = -\sqrt{\frac{\frac{1}{5}}{2}}$$

$$= -\sqrt{\frac{1}{10}} = -\frac{\sqrt{10}}{10}$$

If $\cos\alpha = -\frac{4}{5}$ and α is in the third quadrant, then $\sin\alpha = -\frac{3}{5}$. Therefore,

$$\tan\frac{\alpha}{2} = \frac{1-\cos\alpha}{\sin\alpha} = \frac{1-\left(-\frac{4}{5}\right)}{-\frac{3}{5}} = \frac{\frac{9}{5}}{-\frac{3}{5}} = -3.$$

EXAMPLE 7

Find an exact value for $\tan 67.5°$.

Solution

$$\tan 67.5° = \tan\frac{1}{2}(135°)$$

$$= \frac{1-\cos 135°}{\sin 135°}$$

$$= \frac{1 - \left(-\dfrac{\sqrt{2}}{2}\right)}{\dfrac{\sqrt{2}}{2}} = \frac{1 + \dfrac{\sqrt{2}}{2}}{\dfrac{\sqrt{2}}{2}}$$

$$= \frac{\dfrac{2 + \sqrt{2}}{2}}{\dfrac{\sqrt{2}}{2}} = \frac{2 + \sqrt{2}}{\sqrt{2}} = 1 + \sqrt{2}$$

Problem Set 4.4

For Problems 1–8, find the exact values for $\sin 2\theta$, $\cos 2\theta$, and $\tan 2\theta$. Do not use a calculator or a table.

1. $\cos \theta = \frac{4}{5}$ and θ is a first-quadrant angle
2. $\sin \theta = -\frac{4}{5}$ and θ is a third-quadrant angle
3. $\tan \theta = -\frac{12}{5}$ and θ is a second-quadrant angle
4. $\cot \theta = \frac{12}{5}$ and θ is a first-quadrant angle
5. $\sin \theta = -\frac{7}{25}$ and θ is a fourth-quadrant angle
6. $\cos \theta = \frac{15}{17}$ and θ is a fourth-quadrant angle
7. $\tan \theta = \frac{1}{2}$ and θ is a first-quadrant angle
8. $\tan \theta = -\frac{3}{2}$ and θ is a second-quadrant angle

For Problems 9–20, use the half-angle formulas to find exact values. Do not use a calculator or a table.

9. $\sin 15°$
10. $\cos 15°$
11. $\tan 15°$
12. $\sin 67.5°$

13. $\tan 157.5°$
14. $\tan 22.5°$
15. $\cos \dfrac{3\pi}{8}$
16. $\sin \dfrac{5\pi}{12}$

17. $\tan \dfrac{5\pi}{12}$
18. $\tan \dfrac{7\pi}{12}$
19. $\cos \dfrac{7\pi}{12}$
20. $\cos \dfrac{5\pi}{8}$

For Problems 21–28, find the exact values for $\sin \theta/2$, $\cos \theta/2$, and $\tan \theta/2$. Do not use a calculator or a table.

21. $\sin \theta = \frac{3}{5}$ and $0° < \theta < 90°$
22. $\sin \theta = \frac{4}{5}$ and $90° < \theta < 180°$
23. $\cos \theta = -\frac{3}{5}$ and $180° < \theta < 270°$
24. $\tan \theta = -\frac{5}{12}$ and $270° < \theta < 360°$
25. $\tan \theta = -\frac{12}{5}$ and $90° < \theta < 180°$
26. $\cos \theta = \frac{1}{3}$ and $0° < \theta < 90°$
27. $\sec \theta = \frac{3}{2}$ and $270° < \theta < 360°$
28. $\sec \theta = -\frac{4}{3}$ and $180° < \theta < 270°$

For Problems 29–38, solve each equation for θ, where $0° \le \theta < 360°$. Do not use a calculator or a table.

29. $\sin 2\theta = \sin \theta$
30. $\cos 2\theta + \cos \theta = 0$
31. $\cos 2\theta + 3 \sin \theta - 2 = 0$
32. $\tan 2\theta = \tan \theta$
33. $2 - \cos^2 \theta = 4 \sin^2 \dfrac{\theta}{2}$
34. $\sin 2\theta \sin \theta + \cos \theta = 0$
35. $\cos \theta = \cos \dfrac{\theta}{2}$
36. $\sin \dfrac{\theta}{2} + \cos \theta = 1$

37. $\sin 4\theta = \sin 2\theta$ 38. $\cos 4\theta = \cos 2\theta$

For Problems 39–48, solve each equation for x, where $0 \le x < 2\pi$. Do not use a calculator or a table.

39. $\cos x = \sin 2x$ 40. $\sin 2x + \sqrt{2}\cos x = 0$

41. $\cos 2x - 3\sin x - 2 = 0$ 42. $\sin 2x - 2\cos x + \sin x - 1 = 0$

43. $\sin \dfrac{x}{2} + \cos x = 1$ 44. $\cos 2x = 1 - \sin x$

45. $\tan 2x - 2\cos x = 0$ 46. $\tan 2x + \sec 2x = 1$

47. $2 - \sin^2 x = 2\cos^2 \dfrac{x}{2}$ 48. $\cos \dfrac{x}{2} + \cos x = 0$

For Problems 49–62, verify each identity.

49. $\dfrac{\sin 2\theta}{1 - \cos 2\theta} = \cot \theta$ 50. $(\sin \theta + \cos \theta)^2 - \sin 2\theta = 1$

51. $\dfrac{\sin 2\theta \sin \theta}{2\cos \theta} + \cos^2 \theta = 1$ 52. $\csc 2\theta = \dfrac{1}{2}\sec \theta \csc \theta$

53. $2\sin^2 x = \tan x \sin 2x$ 54. $\dfrac{1 - \tan^2 x}{1 + \tan^2 x} = \cos 2x$

55. $2\cos^2 \dfrac{x}{2}\tan x = \tan x + \sin x$ 56. $\dfrac{\tan x}{1 + \tan^2 x} = \dfrac{1}{2}\sin 2x$

57. $\cot \theta \sin 2\theta = 1 + \cos 2\theta$ 58. $\dfrac{\sin 2\theta}{\sin \theta} - \dfrac{\cos 2\theta}{\cos \theta} = \sec \theta$

59. $\sin 4\theta = 4\cos \theta \sin \theta (1 - 2\sin^2 \theta)$ 60. $\cos^4 \theta - \sin^4 \theta = \cos 2\theta$

61. $\sec 2\theta = \dfrac{\sec^2 \theta}{2 - \sec^2 \theta}$ 62. $\cot 2\theta = \dfrac{\cot^2 \theta - 1}{2\cot \theta}$

63. Express $\cos 3\theta$ in terms of $\cos \theta$.   64. Express $\cos 4\theta$ in terms of $\cos \theta$.

65. Express $\cos 6\theta$ in terms of $\cos \theta$.

Chapter Summary

You should have the following basic trigonometric identities at your fingertips.

$$\csc \theta = \frac{1}{\sin \theta}, \qquad \sec \theta = \frac{1}{\cos \theta}, \qquad \cot \theta = \frac{1}{\tan \theta}$$

$$\tan \theta = \frac{\sin \theta}{\cos \theta}, \qquad \cot \theta = \frac{\cos \theta}{\sin \theta}, \qquad \sin^2 \theta + \cos^2 \theta = 1,$$

$$1 + \tan^2 \theta = \sec^2 \theta, \qquad 1 + \cot^2 \theta = \csc^2 \theta$$

These identities can be used to (1) determine other functional values from a given value, (2) simplify trigonometric expressions, (3) verify additional identities (formulas), and (4) help solve trigonometric equations.

Many of the techniques used to solve algebraic equations (such as factoring and applying the property "if $ab = 0$, then $a = 0$ or $b = 0$") also apply to the solving of trigonometric equations.

The following important identities were verified in this chapter.

$$\sin(-\theta) = -\sin\theta$$
$$\cos(-\theta) = \cos\theta$$
$$\tan(-\theta) = -\tan\theta$$
$$\cos(\alpha - \beta) = \cos\alpha\cos\beta + \sin\alpha\sin\beta$$
$$\cos(\alpha + \beta) = \cos\alpha\cos\beta - \sin\alpha\sin\beta$$
$$\cos(90° - \alpha) = \sin\alpha$$
$$\sin(90° - \alpha) = \cos\alpha$$
$$\tan(90° - \alpha) = \cot\alpha$$
$$\sin(\alpha + \beta) = \sin\alpha\cos\beta + \cos\alpha\sin\beta$$
$$\sin(\alpha - \beta) = \sin\alpha\cos\beta - \cos\alpha\sin\beta$$
$$\tan(\alpha + \beta) = \frac{\tan\alpha + \tan\beta}{1 - \tan\alpha\tan\beta}$$
$$\tan(\alpha - \beta) = \frac{\tan\alpha - \tan\beta}{1 + \tan\alpha\tan\beta}$$
$$\sin 2\alpha = 2\sin\alpha\cos\alpha$$
$$\cos 2\alpha = \cos^2\alpha - \sin^2\alpha = 1 - 2\sin^2\alpha = 2\cos^2\alpha - 1$$
$$\tan 2\alpha = \frac{2\tan\alpha}{1 - \tan^2\alpha}$$
$$\sin\frac{\alpha}{2} = \pm\sqrt{\frac{1 - \cos\alpha}{2}}$$
$$\cos\frac{\alpha}{2} = \pm\sqrt{\frac{1 + \cos\alpha}{2}}$$
$$\tan\frac{\alpha}{2} = \pm\sqrt{\frac{1 - \cos\alpha}{1 + \cos\alpha}} = \frac{1 - \cos\alpha}{\sin\alpha} = \frac{\sin\alpha}{1 + \cos\alpha}$$

Chapter 4 Review Problem Set

1. If $\tan\theta = -\frac{5}{12}$ and $\cos\theta > 0$, find $\sin\theta$, $\sec\theta$, and $\cot\theta$.
2. If $\cos\alpha = -\frac{3}{5}$ and $\tan\beta = \frac{5}{12}$, where α is a third-quadrant angle and β is a first-quadrant angle, find $\sin(\alpha + \beta)$, $\cos(\alpha - \beta)$, and $\tan(\alpha + \beta)$.
3. If $\sin\theta = \frac{5}{13}$ and $90° < \theta < 180°$, find $\sin 2\theta$, $\cos 2\theta$, and $\tan 2\theta$.
4. If $\cos x = \frac{4}{5}$ and $\frac{3\pi}{2} < x < 2\pi$, find $\sin\frac{x}{2}$, $\cos\frac{x}{2}$, and $\tan\frac{x}{2}$.

For Problems 5–10, find exact values. Do not use a calculator or a table.

5. $\sin 165°$ 6. $\cos 75°$ 7. $\sin\dfrac{7\pi}{12}$ 8. $\tan\dfrac{\pi}{8}$

9. $\cos\left[\sin^{-1}\dfrac{3}{5} + \tan^{-1}\left(-\dfrac{4}{3}\right)\right]$ ➤   10. $\tan\left(\arcsin\dfrac{4}{5} - \arccos\dfrac{12}{13}\right)$

For Problems 11–24, solve each of the equations. If the variable is θ, find all solutions such that $0° \le \theta < 360°$. If the variable is x, find all solutions such that $0 \le x < 2\pi$. Do not use a calculator or a table.

11. $\sin^2\theta - \sin\theta = 0$

12. $2\cos^2\theta + 5\sin\theta - 4 = 0$

13. $4\sin^2\theta - 4\sin\theta + 1 = 0$

14. $\tan\theta = 2\cos\theta\tan\theta$

15. $\cos 2\theta + 3\cos\theta + 2 = 0$

16. $2\cos^2\dfrac{\theta}{2} - 3\cos\theta = 0$

17. $\cos\theta - \sqrt{3}\sin\theta = 1$

18. $\tan 2\theta + 2\sin\theta = 0$

19. $2\cos x + \tan x = \sec x$

20. $\cos 2x \sin x - \cos 2x = 0$

21. $2\sec x\sin x + 2 = 4\sin x + \sec x$

22. $\sin 2x = \cos 2x$

23. $\cos\dfrac{x}{2} = \dfrac{\sqrt{3}}{2}$

24. $\sin 3x + \sin x = 0$

25. Find *all* solutions of $\sec x - 1 = \tan x$ using radian measure.

26. Find *all* solutions of $2\sin\theta\tan\theta + \tan\theta - 2\sin\theta - 1 = 0$ using degree measure.

27. Solve $8\sin^2\theta + 13\sin\theta - 6 = 0$ for θ, where $0° \le \theta < 360°$. Express the solutions to the nearest tenth of a degree.

28. Solve $2\sin^2\theta - 3\sin\theta - 1 = 0$ for θ, where $0° \le \theta < 360°$. Express the solutions to the nearest tenth of a degree.

29. Solve $2\tan^2 x + 5\tan x - 12 = 0$ for x, where $0 \le x < 2\pi$. Express the solutions to the nearest hundredth of a radian.

30. Solve $2\cos^2(x/2) = \cos^2 x$ for x, where $0 \le x < 2\pi$. Express the solutions to the nearest hundredth of a radian.

For Problems 31–44, verify each of the identities.

31. $(\cot^2 x + 1)(1 - \cos^2 x) = 1$

32. $\tan x = \dfrac{\tan x - 1}{1 - \cot x}$

33. $\dfrac{1}{1 - \sin x} + \dfrac{1}{1 + \sin x} = 2\sec^2 x$

34. $\sin(\theta + 270°) = -\cos\theta$

35. $\tan(x + \pi) = \tan x$

36. $\cos\left(x + \dfrac{\pi}{4}\right) = \dfrac{\sqrt{2}}{2}(\cos x - \sin x)$

37. $\dfrac{\sin(\alpha - \beta)}{\cos(\alpha + \beta)} = \dfrac{\tan\alpha - \tan\beta}{1 - \tan\alpha\tan\beta}$

38. $4\sin^2\dfrac{\theta}{2}\cos^2\dfrac{\theta}{2} = \sin^2\theta$

39. $\dfrac{\cos x}{1 + \sin x} = \sec x - \tan x$

40. $\sin x + \cos x = \dfrac{1 + \cot x}{\csc x}$

41. $2\cot 2\theta = \cot\theta - \tan\theta$

42. $\cos\theta\sin 2\theta = 2\sin\theta - 2\sin^3\theta$

43. $1 - \dfrac{1}{2}\sin 2x = \dfrac{\sin^3 x + \cos^3 x}{\sin x + \cos x}$

44. $\tan\dfrac{\theta}{2} = \csc\theta - \cot\theta$

5 Topics in Trigonometry

© Cleveland W. Bryant, Jr.

Thus far in this text we have had no need for numbers other than the real numbers. However, as you might recall from a previous algebra course, there are applications for which the real numbers are not sufficient. In fact, strictly from a mathematical viewpoint, there are some very simple looking equations that do not have solutions within the set of real numbers. For example, the equation $x^2 = -1$ has no real number solution since any nonzero real number squared is positive. Therefore, we need to extend the real number system, which we will do in a moment by introducing the complex number system. Then in the next two sections we will use some trigonometric ideas to aid in the study of the complex number system. In the last three sections of this chapter we will introduce another plotting system, the polar coordinate system, and take a brief look at vectors from an algebraic viewpoint.

5.1

Complex Numbers

Let's begin by introducing the number i, such that

$$i^2 = -1.$$

The number i is not a real number and it is often called the **imaginary unit**; but the number i^2 is the real number -1. The imaginary unit i is used to define a complex number as follows.

DEFINITION 5.1

> A **complex number** is any number that can be expressed in the form
>
> $a + bi,$
>
> where a and b are real numbers.

The form $a + bi$ is called the **standard form** of a complex number. The real number a is called the **real part** of the complex number and b is called the **imaginary part**. (Note that b is a real number even though it is called the imaginary part.) Each of the following represents a complex number.

$6 + 2i$	It is already expressed in the form $a + bi$. Traditionally, complex numbers where $a \neq 0$ and $b \neq 0$ have been called imaginary numbers.
$5 - 3i$	It can be written as $5 + (-3i)$ even though the form $5 - 3i$ is often used.
$-8 + i\sqrt{2}$	It can be written as $-8 + \sqrt{2}i$. It is easy to mistake $\sqrt{2}i$ for $\sqrt{2i}$. Thus, we commonly write $i\sqrt{2}$ instead of $\sqrt{2}i$ to avoid any difficulties with the radical sign.
$-9i$	It can be written as $0 + (-9i)$. Complex numbers, such as $-9i$, where $a = 0$ and $b \neq 0$ traditionally have been called **pure imaginary numbers**.
5	It can be written as $5 + 0i$.

The set of real numbers is a subset of the set of complex numbers. The following diagram indicates the organizational format of the complex number system.

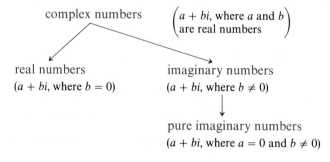

Two complex numbers are said to be *equal* if and only if $a = c$ and $b = d$. In other words, two complex numbers are equal if and only if their real parts are equal and their imaginary parts are equal.

Adding and Subtracting Complex Numbers

The following definition provides the basis for adding complex numbers.

$$(a + bi) + (c + di) = (a + c) + (b + d)i$$

Using this definition, we can find the sum of two complex numbers as follows.

$$(4 + 3i) + (5 + 9i) = (4 + 5) + (3 + 9)i = 9 + 12i$$

$$(-6 + 4i) + (8 - 7i) = (-6 + 8) + (4 - 7)i = 2 - 3i$$

$$\left(\frac{1}{2} + \frac{3}{4}i\right) + \left(\frac{2}{3} + \frac{1}{5}i\right) = \left(\frac{1}{2} + \frac{2}{3}\right) + \left(\frac{3}{4} + \frac{1}{5}\right)i = \left(\frac{3}{6} + \frac{4}{6}\right) + \left(\frac{15}{20} + \frac{4}{20}\right)i$$

$$= \frac{7}{6} + \frac{19}{20}i$$

$$(3 + i\sqrt{2}) + (-4 + i\sqrt{2}) = (3 + (-4)) + (\sqrt{2} + \sqrt{2})i = -1 + 2i\sqrt{2}$$

Note the form for writing $2\sqrt{2}i$.

The set of complex numbers is closed with respect to addition; that is, the sum of two complex numbers is a complex number. Furthermore, the commutative and associative properties of addition hold for all complex numbers. The addition identity element is $0 + 0i$, or simply the real number 0. The additive inverse of $a + bi$ is $-a - bi$ because

$$(a + bi) + (-a - bi) = (a + (-a)) + (b + (-b))i = 0.$$

Therefore, to *subtract* $c + di$ from $a + bi$, we add the additive inverse of $c + di$.

$$(a + bi) - (c + di) = (a + bi) + (-c - di)$$
$$= (a - c) + (b - d)i$$

The following examples illustrate the subtraction of complex numbers.

$$(9 + 8i) - (5 + 3i) = (9 - 5) + (8 - 3)i = 4 + 5i$$

$$(3 - 2i) - (4 - 10i) = (3 - 4) + (-2 - (-10)i = -1 + 8i$$

$$\left(-\frac{1}{2} + \frac{1}{3}i\right) - \left(\frac{3}{4} + \frac{1}{2}i\right) = \left(-\frac{1}{2} - \frac{3}{4}\right) + \left(\frac{1}{3} - \frac{1}{2}\right)i = -\frac{5}{4} - \frac{1}{6}i$$

Multiplying and Dividing Complex Numbers

Since $i^2 = -1$, i is a square root of negative one; so we let $i = \sqrt{-1}$. It should also be evident that $-i$ is a square root of negative one because

$$(-i)^2 = (-i)(-i) = i^2 = -1.$$

Therefore, in the set of complex numbers, -1 has two square roots, namely, i and $-i$. These are symbolically expressed as

$$i = \sqrt{-1} \quad \text{and} \quad -i = -\sqrt{-1}.$$

Let's extend the definition so that in the set of complex numbers every negative real number has two square roots. We define $\sqrt{-b}$, where b is a positive real number, to be the number whose square is $-b$. Therefore,

$$(\sqrt{-b})^2 = -b, \quad \text{for } b > 0.$$

Furthermore, since $(i\sqrt{b})(i\sqrt{b}) = i^2(b) = -1(b) = -b$ we see that

$$\sqrt{-b} = i\sqrt{b}.$$

In other words, a square root of any negative real number can be represented as the product of a real number and the imaginary unit i. Consider the following examples.

$$\sqrt{-4} = i\sqrt{4} = 2i$$
$$\sqrt{-17} = i\sqrt{17}$$
$$\sqrt{-24} = i\sqrt{24} = i\sqrt{4}\sqrt{6} = 2i\sqrt{6} \qquad \text{Note that we simplified the radical } \sqrt{24} \text{ to } 2\sqrt{6}.$$

We should also observe that $-\sqrt{-b}$, where $b > 0$, is a square root of $-b$ because

$$(-\sqrt{-b})^2 = (-i\sqrt{b})^2 = i^2(b) = (-1)b = -b.$$

Thus, in the set of complex numbers, $-b$ (where $b > 0$) has two square roots, $i\sqrt{b}$ and $-i\sqrt{b}$. These are expressed as

$$\sqrt{-b} = i\sqrt{b} \quad \text{and} \quad -\sqrt{-b} = -i\sqrt{b}.$$

We must be careful with the use of the symbol $\sqrt{-b}$, where $b > 0$. Some properties that are true in the set of real numbers involving the square root symbol are not true if the square root symbol does not represent a real number. For example, $\sqrt{a}\sqrt{b} = \sqrt{ab}$ *does not hold* if a and b are both negative numbers.

Correct: $\sqrt{-4}\sqrt{-9} = (2i)(3i) = 6i^2 = 6(-1) = -6$

Incorrect: $\sqrt{-4}\sqrt{-9} = \sqrt{(-4)(-9)} = \sqrt{36} = 6$

To avoid difficulty with this idea, you should rewrite all expressions of the form $\sqrt{-b}$, where $b > 0$, in the form $i\sqrt{b}$ *before* doing any computations. The following examples further illustrate this point.

$$\sqrt{-5}\sqrt{-7} = (i\sqrt{5})(i\sqrt{7}) = i^2\sqrt{35} = (-1)(\sqrt{35}) = -\sqrt{35}$$
$$\sqrt{-2}\sqrt{-8} = (i\sqrt{2})(i\sqrt{8}) = i^2\sqrt{16} = (-1)(4) = -4$$
$$\sqrt{-2}\sqrt{8} = (i\sqrt{2})(\sqrt{8}) = i\sqrt{16} = 4i$$
$$\sqrt{-6}\sqrt{-8} = (i\sqrt{6})(i\sqrt{8}) = i^2\sqrt{48} = i^2\sqrt{16}\sqrt{3} = 4i^2\sqrt{3} = -4\sqrt{3}$$

$$\frac{\sqrt{-2}}{\sqrt{3}} = \frac{i\sqrt{2}}{\sqrt{3}} = \frac{i\sqrt{2}}{\sqrt{3}} \cdot \frac{\sqrt{3}}{\sqrt{3}} = \frac{i\sqrt{6}}{3}$$

$$\frac{\sqrt{-48}}{\sqrt{12}} = \frac{i\sqrt{48}}{\sqrt{12}} = i\sqrt{\frac{48}{12}} = i\sqrt{4} = 2i$$

Since complex numbers have a **binomial form**, we can find the product of two complex numbers in the same way as we find the product of two binomials. Then by replacing i^2 with -1, we are able to simplify and express the final product in the standard form of a complex number. Consider the following examples.

$$
\begin{aligned}
(2 + 3i)(4 + 5i) &= 2(4 + 5i) + 3i(4 + 5i) \\
&= 8 + 10i + 12i + 15i^2 \\
&= 8 + 22i + 15(-1) \\
&= 8 + 22i - 15 \\
&= -7 + 22i
\end{aligned}
$$

$$
\begin{aligned}
(1 - 7i)^2 = (1 - 7i)(1 - 7i) &= 1(1 - 7i) - 7i(1 - 7i) \\
&= 1 - 7i - 7i + 49i^2 \\
&= 1 - 14i + 49(-1) \\
&= 1 - 14i - 49 \\
&= -48 - 14i
\end{aligned}
$$

$$
\begin{aligned}
(2 + 3i)(2 - 3i) &= 2(2 - 3i) + 3i(2 - 3i) \\
&= 4 - 6i + 6i - 9i^2 \\
&= 4 - 9(-1) \\
&= 4 + 9 \\
&= 13
\end{aligned}
$$

The last example above illustrates an important idea. The complex numbers $2 + 3i$ and $2 - 3i$ are called conjugates of each other. In general, the two complex numbers $a + bi$ and $a - bi$ are called **conjugates** of each other and the product of a complex number and its conjugate is a real number. This can be shown as follows.

$$
\begin{aligned}
(a + bi)(a - bi) &= a(a - bi) + bi(a - bi) \\
&= a^2 - abi + abi - b^2i^2 \\
&= a^2 - b^2(-1) \\
&= a^2 + b^2
\end{aligned}
$$

Conjugates are used to simplify an expression such as $3i/(5 + 2i)$, which **indicates the quotient** of two complex numbers. To eliminate i in the denominator and change the indicated quotient to the standard form of a complex number, we can multiply both the numerator and denominator by the conjugate of the denominator as follows.

$$\frac{3i}{5+2i} = \frac{3i}{5+2i} \cdot \frac{5-2i}{5-2i} = \frac{3i(5-2i)}{(5+2i)(5-2i)}$$

$$= \frac{15i - 6i^2}{25 - 4i^2}$$

$$= \frac{15i - 6(-1)}{25 - 4(-1)}$$

$$= \frac{6 + 15i}{29} = \frac{6}{29} + \frac{15}{29}i$$

The following examples further illustrate the process of dividing complex numbers.

$$\frac{2-3i}{4-7i} = \frac{2-3i}{4-7i} \cdot \frac{4+7i}{4+7i} = \frac{(2-3i)(4+7i)}{(4-7i)(4+7i)}$$

$$= \frac{8 + 14i - 12i - 21i^2}{16 - 49i^2}$$

$$= \frac{8 + 2i - 21(-1)}{16 - 49(-1)}$$

$$= \frac{29 + 2i}{65} = \frac{29}{65} + \frac{2}{65}i$$

$$\frac{4-5i}{2i} = \frac{4-5i}{2i} \cdot \frac{-2i}{-2i} = \frac{(4-5i)(-2i)}{(2i)(-2i)}$$

$$= \frac{-8i + 10i^2}{-4i^2}$$

$$= \frac{-8i + 10(-1)}{-4(-1)}$$

$$= \frac{-10 - 8i}{4} = -\frac{5}{2} - 2i$$

For a problem such as the last one above in which the denominator is a pure imaginary number, we can change to standard form by choosing a multiplier other than the conjugate of the denominator. Consider the following alternate approach.

$$\frac{4-5i}{2i} = \frac{4-5i}{2i} \cdot \frac{i}{i} = \frac{(4-5i)(i)}{(2i)(i)}$$

$$= \frac{4i - 5i^2}{2i^2}$$

$$= \frac{4i - 5(-1)}{2(-1)}$$

$$= \frac{5 + 4i}{-2} = -\frac{5}{2} - 2i$$

Let's conclude this section by illustrating the use of the quadratic formula to solve some quadratic equations that contain nonreal complex solutions.

EXAMPLE 1

Solve $x^2 - 2x + 5 = 0$.

Solution Using the quadratic formula we can proceed as follows.

$$x = \frac{-b \pm \sqrt{b^2 - 4ac}}{2a}$$

$$= -(-2) \pm \frac{\sqrt{(-2)^2 - 4(1)(5)}}{2(1)}$$

$$= \frac{2 \pm \sqrt{-16}}{2} = \frac{2 \pm 4i}{2} = \frac{2(1 \pm 2i)}{2} = 1 \pm 2i$$

The solution set is $\{1 \pm 2i\}$.

EXAMPLE 2

Solve $2x^2 - x + 3 = 0$.

Solution

$$x = \frac{-b \pm \sqrt{b^2 - 4ac}}{2a}$$

$$= -(-1) \pm \frac{\sqrt{(-1)^2 - 4(2)(3)}}{2(2)}$$

$$= \frac{1 \pm \sqrt{-23}}{4} = \frac{1 \pm i\sqrt{23}}{4}$$

The solution set is $\{(1 \pm i\sqrt{23})/4\}$.

EXAMPLE 3

Solve $4x^2 - 4x + 7 = 0$.

Solution

$$x = \frac{-b \pm \sqrt{b^2 - 4ac}}{2a}$$

$$= \frac{-(-4) \pm \sqrt{(-4)^2 - 4(4)(7)}}{2(4)}$$

$$= \frac{4 \pm \sqrt{-96}}{8}$$

$$= \frac{4 \pm i\sqrt{96}}{8} = \frac{4 \pm 4i\sqrt{6}}{8}$$

$$= \frac{4(1 \pm i\sqrt{6})}{8} = \frac{1 \pm i\sqrt{6}}{2}$$

The solution set is $\{(1 \pm i\sqrt{6})/2\}$.

Problem Set 5.1

Add or subtract as indicated.

1. $(5 + 2i) + (8 + 6i)$

2. $(-9 + 3i) + (4 + 5i)$

3. $(8 + 6i) - (5 + 2i)$

4. $(-6 + 4i) - (4 + 6i)$

5. $(-7 - 3i) + (-4 + 4i)$

6. $(6 - 7i) - (7 - 6i)$

7. $(-2 - 3i) - (-1 - i)$

8. $\left(\dfrac{1}{3} + \dfrac{2}{5}i\right) + \left(\dfrac{1}{2} + \dfrac{1}{4}i\right)$

9. $\left(-\dfrac{3}{4} - \dfrac{1}{4}i\right) + \left(\dfrac{3}{5} + \dfrac{2}{3}i\right)$

10. $\left(\dfrac{5}{8} + \dfrac{1}{2}i\right) - \left(\dfrac{7}{8} + \dfrac{1}{5}i\right)$

11. $\left(\dfrac{3}{10} - \dfrac{3}{4}i\right) - \left(-\dfrac{2}{5} + \dfrac{1}{6}i\right)$

12. $(4 + i\sqrt{3}) + (-6 - 2i\sqrt{3})$

13. $(5 + 3i) + (7 - 2i) + (-8 - i)$

14. $(5 - 7i) - (6 - 2i) - (-1 - 2i)$

Write each of the following in terms of i and simplify. For example,

$$\sqrt{-20} = i\sqrt{20} = i\sqrt{4}\sqrt{5} = 2i\sqrt{5}.$$

15. $\sqrt{-9}$

16. $\sqrt{-49}$

17. $\sqrt{-19}$

18. $\sqrt{-31}$

19. $\sqrt{-\dfrac{4}{9}}$

20. $\sqrt{-\dfrac{25}{36}}$

21. $\sqrt{-8}$

22. $\sqrt{-18}$

23. $\sqrt{-27}$

24. $\sqrt{-32}$

25. $\sqrt{-54}$

26. $\sqrt{-40}$

27. $3\sqrt{-36}$

28. $5\sqrt{-64}$

29. $4\sqrt{-18}$

30. $6\sqrt{-8}$

Write each of the following in terms of i, perform the indicated operations, and simplify. For example,

$$\sqrt{-9}\sqrt{-16} = (i\sqrt{9})(i\sqrt{16}) = (3i)(4i) = 12i^2 = 12(-1) = -12.$$

31. $\sqrt{-4}\sqrt{-16}$

32. $\sqrt{-25}\sqrt{-9}$

33. $\sqrt{-2}\sqrt{-3}$

34. $\sqrt{-3}\sqrt{-7}$

35. $\sqrt{-5}\sqrt{-4}$

36. $\sqrt{-7}\sqrt{-9}$

37. $\sqrt{-6}\sqrt{-10}$

38. $\sqrt{-2}\sqrt{-12}$

39. $\sqrt{-8}\sqrt{-7}$

40. $\sqrt{-12}\sqrt{-5}$

41. $\dfrac{\sqrt{-36}}{\sqrt{-4}}$

42. $\dfrac{\sqrt{-64}}{\sqrt{-16}}$

43. $\dfrac{\sqrt{-54}}{\sqrt{-9}}$

44. $\dfrac{\sqrt{-18}}{\sqrt{-3}}$

Find each of the following products and express answers in standard form.

45. $(3i)(7i)$

46. $(-5i)(8i)$

47. $(4i)(3 - 2i)$

48. $(5i)(2 + 6i)$

49. $(3 + 2i)(4 + 6i)$

50. $(7 + 3i)(8 + 4i)$

51. $(4 + 5i)(2 - 9i)$

52. $(1 + i)(2 - i)$

53. $(-2 - 3i)(4 + 6i)$

54. $(-3 - 7i)(2 + 10i)$

55. $(6 - 4i)(-1 - 2i)$

56. $(7 - 3i)(-2 - 8i)$

57. $(3 + 4i)^2$

58. $(4 - 2i)^2$

59. $(-1 - 2i)^2$

60. $(-2 + 5i)^2$

61. $(8 - 7i)(8 + 7i)$

62. $(5 + 3i)(5 - 3i)$

63. $(-2 + 3i)(-2 - 3i)$

64. $(-6 - 7i)(-6 + 7i)$

Find each of the following quotients expressing answers in standard form.

65. $\dfrac{4i}{3-2i}$ 66. $\dfrac{3i}{6+2i}$ 67. $\dfrac{2+3i}{3i}$ 68. $\dfrac{3-5i}{4i}$

69. $\dfrac{3}{2i}$ 70. $\dfrac{7}{4i}$ 71. $\dfrac{3+2i}{4+5i}$ 72. $\dfrac{2+5i}{3+7i}$

73. $\dfrac{4+7i}{2-3i}$ 74. $\dfrac{3+9i}{4-i}$ 75. $\dfrac{3-7i}{-2+4i}$ 76. $\dfrac{4-10i}{-3+7i}$

77. $\dfrac{-1-i}{-2-3i}$ 78. $\dfrac{-4+9i}{-3-6i}$

Solve each of the following equations.

79. $x^2 + 4x + 7 = 0$ 80. $x^2 - 2x + 19 = 0$

81. $x^2 + 2x + 5 = 0$ 82. $y^2 - 6y + 10 = 0$

83. $n^2 + 24 = 0$ 84. $2n^2 + 3 = 0$

85. $2t^2 - 3t + 7 = 0$ 86. $3t^2 - 2t + 5 = 0$

87. $3x^2 + 2x + 2 = 0$ 88. $11x^2 + 4x + 1 = 0$

89. Using $a + bi$ and $c + di$ to represent two complex numbers, verify the following properties.

(a) The conjugate of the sum of two complex numbers is equal to the sum of the conjugates of the two numbers.

(b) The conjugate of the product of two complex numbers is equal to the product of the conjugates of the numbers.

5.2

Trigonometric Form of Complex Numbers

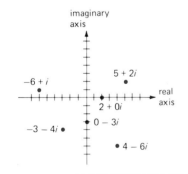

Figure 5.1

As we have seen, real numbers can be represented geometrically by points on a coordinate line. Complex numbers can also be represented geometrically, but by points in a plane. Every complex number $a + bi$ determines an ordered pair of real numbers (a, b), where a is called the **real part** and b the **imaginary part** of the complex number. Therefore, using a rectangular coordinate system with the horizontal axis as the **real axis** and the vertical axis as the **imaginary axis**, we can associate points of the plane with complex numbers as illustrated in Figure 5.1. Note that the complex number $5 + 2i$ is represented by the point $(5, 2)$, the number $-6 + i$ is represented by the point $(-6, 1)$, and so on. A coordinate plane with complex numbers assigned to the points in this manner is called a **complex plane**. In this context, the form $a + bi$ of a complex number is referred to as the **rectangular form** of the number.

The absolute value of a real number can be geometrically interpreted as the distance between 0 and the number on the real number line. It is natural, therefore, to interpret the absolute value of a complex number to be the distance between the origin and the point representing the complex number in the complex plane. Specifically, the absolute value of a complex number is defined as follows.

DEFINITION 5.2

The **absolute value** of a complex number $a + bi$ is denoted by $|a + bi|$ and is defined by

$$|a + bi| = \sqrt{a^2 + b^2}.$$

EXAMPLE 1

Compute (a) $|-5 + 12i|$, (b) $|2 - 4i|$, and (c) $|3i|$.

Solution

(a) $|-5 + 12i| = \sqrt{(-5)^2 + 12^2} = \sqrt{25 + 144} = \sqrt{169} = 13$

(b) $|2 - 4i| = \sqrt{2^2 + (-4)^2} = \sqrt{4 + 16} = \sqrt{20} = 2\sqrt{5}$

(c) $|3i| = |0 + 3i| = \sqrt{0^2 + 3^2} = \sqrt{0 + 9} = 3$

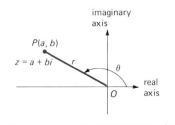

Figure 5.2

The geometric representation of a complex number as a point in a coordinate plane provides the basis for representing a complex number by using trigonometric functions. Consider the complex number $z = a + bi$ as illustrated in Figure 5.2. Let θ be an angle in standard position whose terminal side is OP and let $r = |z|$; that is, $r = \sqrt{a^2 + b^2}$. By the definition of the sine and cosine functions we have

$$\sin \theta = \frac{b}{r} \quad \text{or} \quad b = r \sin \theta$$

and

$$\cos \theta = \frac{a}{r} \quad \text{or} \quad a = r \cos \theta.$$

Therefore, the complex number $a + bi$ can be expressed as

$$a + bi = (r \cos \theta) + (r \sin \theta)i,$$

which can be written in **trigonometric form** as

$$z = a + bi = r(\cos \theta + i \sin \theta).$$

The form $r(\cos \theta + i \sin \theta)$ is also called the **polar form** of a complex number. Remember that r is the absolute value of the complex number and therefore it is always nonnegative. Angle θ can be expressed in either radian or degree measure. Furthermore, θ is not uniquely determined since $\theta \pm 2k\pi$ will also do for k, any integer. We will usually take the smallest positive angle for θ when writing complex numbers in trigonometric form.

Remark: Traditionally, the number r in the trigonometric form of a complex number is also called the **modulus** of the number, and angle θ is called the **argument**. Also, the expression $r(\cos \theta + i \sin \theta)$ can be abbreviated as $r \operatorname{cis} \theta$. However, we will not use this terminology or symbolism in this brief introduction of the complex numbers in trigonometric form.

Since a complex number can be written in either rectangular form ($a + bi$) or trigonometric form ($r(\cos\theta + i\sin\theta)$), we need to be able to switch back and forth between the two forms. Let's consider some examples.

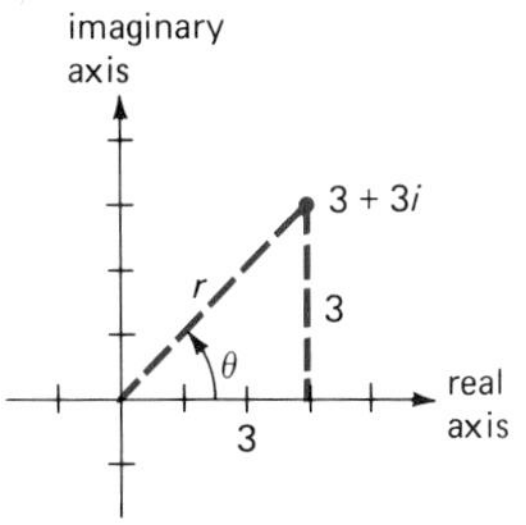

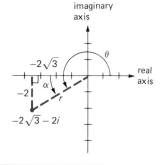

Figure 5.3

EXAMPLE 2

Express the following complex numbers in trigonometric form, for which $0 \le \theta < 2\pi$.

(a) $3 + 3i$ **(b)** $-2\sqrt{3} - 2i$

Solution

(a) The complex number $3 + 3i$ is represented geometrically in Figure 5.3. Since the right triangle indicated in the figure is isosceles, $\theta = \pi/4$. Then using either the Pythagorean Theorem or the distance formula, we can find the value of r.

$$r = \sqrt{3^2 + 3^2} = \sqrt{18} = 3\sqrt{2}.$$

Therefore,

$$3 + 3i = 3\sqrt{2}\left(\cos\frac{\pi}{4} + i\sin\frac{\pi}{4}\right).$$

(b) The complex number $-2\sqrt{3} - 2i$ is graphed in Figure 5.4.

$$r = \sqrt{(-2\sqrt{3})^2 + (-2)^2} = \sqrt{12 + 4} = \sqrt{16} = 4$$

Now we should recognize that $\alpha = \pi/6$ and thus $\theta = \pi + (\pi/6) = 7\pi/6$. Therefore,

$$-2\sqrt{3} - 2i = 4\left(\cos\frac{7\pi}{6} + i\sin\frac{7\pi}{6}\right).$$

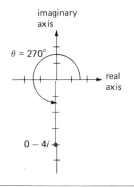

Figure 5.4

EXAMPLE 3

Express the following complex numbers in trigonometric form, for which $0° \le \theta < 360°$.

(a) $0 - 4i$ **(b)** $5 + 2i$

Solution

(a) The complex number $0 - 4i$ is graphed in Figure 5.5. It is evident from the figure that $r = 4$ and $\theta = 270°$. Therefore,

$$0 - 4i = 4(\cos 270° + i\sin 270°).$$

(b) The complex number $5 + 2i$ is represented in Figure 5.6 on page 172.

$$r = \sqrt{5^2 + 2^2} = \sqrt{29}$$

From the figure we see that $\tan\theta = \frac{2}{5} = .4$. Using a calculator or Table A in

Figure 5.5

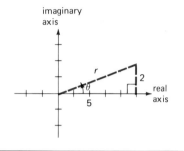

imaginary
axis

real
axis

Figure 5.6

the back of the book, we can find that $\theta = 21.8°$, to the nearest tenth of a degree. Therefore,

$$5 + 2i = \sqrt{29}(\cos 21.8° + i\sin 21.8°).$$

EXAMPLE 4

Change the complex number $8(\cos 60° + i\sin 60°)$ to $a + bi$ form.

Solution

$$8(\cos 60° + i\sin 60°) = 8\left(\frac{1}{2} + i\frac{\sqrt{3}}{2}\right)$$
$$= 4 + 4\sqrt{3}\,i$$

Multiplying Complex Numbers in Trigonometric Form

In Section 5.1 we discussed the multiplication of complex numbers in the form $a + bi$. For example, the product of $2 + 2i$ and $-3 + 3i$ was handled as follows.

$$(2 + 2i)(-3 + 3i) = 2(-3 + 3i) + 2i(-3 + 3i)$$
$$= -6 + 6i - 6i + 6i^2$$
$$= -6 + 6(-1)$$
$$= -12 + 0i$$

Let's consider the product of two complex numbers using their trigonometric forms. Let $z_1 = r_1(\cos\theta_1 + i\sin\theta_1)$ and $z_2 = r_2(\cos\theta_2 + i\sin\theta_2)$ be the two numbers. Therefore,

$$z_1 z_2 = [r_1(\cos\theta_1 + i\sin\theta_1)][r_2(\cos\theta_2 + i\sin\theta_2)]$$
$$= r_1 r_2[\cos\theta_1 \cos\theta_2 + i\cos\theta_1 \sin\theta_2 + i\sin\theta_1 \cos\theta_2$$
$$+ i^2 \sin\theta_1 \sin\theta_2]$$
$$= r_1 r_2[\cos\theta_1 \cos\theta_2 - \sin\theta_1 \sin\theta_2 + i(\cos\theta_1 \sin\theta_2 + \sin\theta_1 \cos\theta_2)].$$

Now applying the sum formulas for $\cos(\theta_1 + \theta_2)$ and $\sin(\theta_1 + \theta_2)$, we obtain the following description for multiplying complex numbers in trigonometric form.

$$z_1 z_2 = r_1 r_2[\cos(\theta_1 + \theta_2) + i\sin(\theta_1 + \theta_2)]$$

EXAMPLE 5

Use their trigonometric forms to find the product of $(2 + 2i)$ and $(-3 + 3i)$.

Solution Let $z_1 = 2 + 2i$ and $z_2 = -3 + 3i$. Their trigonometric forms are as follows.

$$z_1 = 2 + 2i = 2\sqrt{2}\left(\cos\frac{\pi}{4} + i\sin\frac{\pi}{4}\right)$$

$$z_2 = -3 + 3i = 3\sqrt{2}\left(\cos\frac{3\pi}{4} + i\sin\frac{3\pi}{4}\right)$$

Therefore,

$$z_1 z_2 = (2\sqrt{2})(3\sqrt{2})\left[\cos\left(\frac{\pi}{4} + \frac{3\pi}{4}\right) + i\sin\left(\frac{\pi}{4} + \frac{3\pi}{4}\right)\right]$$

$$= 12[\cos\pi + i\sin\pi]$$

$$= 12[-1 + i(0)]$$

$$= -12 + 0i. \qquad \text{This agrees with our earlier result!}$$

As illustrated in Example 5, multiplying complex numbers in trigonometric form is quite easy. Basically, we multiply absolute values and add angles. Most of us would probably agree that multiplying complex numbers in the $a + bi$ form is not very difficult. Therefore, it may seem as if the trigonometric form has not provided us with much extra fire power. That may be true at this time; however, in the next section we will find the trigonometric form very convenient for finding powers and roots of complex numbers.

Problem Set 5.2

For Problems 1–12, plot each complex number and find its absolute value.

1. $3 + 4i$	2. $-4 + 3i$	3. $-5 - 12i$
4. $12 - 5i$	5. $0 - 5i$	6. $-4 + 0i$
7. $1 - 2i$	8. $-1 - i$	9. $-2 + 3i$
10. $3 - 2i$	11. $\frac{3}{5} - \frac{4}{5}i$	12. $-\frac{5}{13} - \frac{12}{13}i$

For Problems 13–22, express each complex number in trigonometric form, where $0 \le \theta < 2\pi$.

13. $-2 + 2i$	14. $-4 - 4i$	15. $0 - 3i$	16. $-4 + 0i$
17. $2\sqrt{3} - 2i$	18. $-3\sqrt{3} + 3i$	19. $-1 - i$	20. $1 - i$
21. $-1 + \sqrt{3}i$	22. $2 - 2\sqrt{3}i$		

For Problems 23–32, express each complex number in trigonometric form, where $0 \le \theta < 360°$.

23. $5 - 5i$	24. $6 + 6i$	25. $-2 + 0i$	26. $0 + 7i$
27. $\sqrt{3} + i$	28. $-\sqrt{3} - i$	29. $-4 - 4\sqrt{3}i$	30. $5 - 5\sqrt{3}i$
31. $6\sqrt{3} - 6i$	32. $-7\sqrt{3} + 7i$		

For Problems 33–40, express each complex number in trigonometric form, where $0° \le \theta < 360°$. Express θ to the nearest tenth of a degree.

33. $2 + 3i$	34. $4 - 3i$	35. $-2 - i$	36. $-5 + 3i$

37. $4 - i$ 38. $2 + i$ 39. $-2 + 4i$ 40. $-6 - 3i$

For Problems 41–50, change the given complex number from trigonometric form to $a + bi$ form.

41. $4(\cos 30° + i \sin 30°)$

42. $5(\cos 120° + i \sin 120°)$

43. $3\left(\cos \dfrac{5\pi}{4} + i \sin \dfrac{5\pi}{4}\right)$

44. $1\left(\cos \dfrac{11\pi}{6} + i \sin \dfrac{11\pi}{6}\right)$

45. $2\left(\cos \dfrac{4\pi}{3} + i \sin \dfrac{4\pi}{3}\right)$

46. $\dfrac{1}{2}\left(\cos \dfrac{5\pi}{6} + i \sin \dfrac{5\pi}{6}\right)$

47. $\dfrac{2}{3}\left(\cos \dfrac{5\pi}{3} + i \sin \dfrac{5\pi}{3}\right)$

48. $7(\cos \pi + i \sin \pi)$

49. $6(\cos 0° + i \sin 0°)$

50. $8(\cos 60° + i \sin 60°)$

For Problems 51–60, find the product $z_1 z_2$ by using the trigonometric forms of the numbers. Express the final result in $a + bi$ form. Check by using the methods of Section 5.1.

51. $z_1 = \sqrt{3} + i, \quad z_2 = -2\sqrt{3} - 2i$

52. $z_1 = -3\sqrt{3} - 3i, \quad z_2 = 2\sqrt{3} + 2i$

53. $z_1 = 5\sqrt{3} + 5i, \quad z_2 = 6\sqrt{3} + 6i$

54. $z_1 = 1 + \sqrt{3}i, \quad z_2 = -\dfrac{1}{2} - \dfrac{\sqrt{3}}{2}i$

55. $z_1 = \dfrac{1}{2} + \dfrac{\sqrt{3}}{2}i, \quad z_2 = -\dfrac{3}{2} - \dfrac{3\sqrt{3}}{2}i$

56. $z_1 = -1 + i, \quad z_2 = 1 + i$

57. $z_1 = -2 - 2\sqrt{3}i, \quad z_2 = 0 + 5i$

58. $z_1 = 0 + 4i, \quad z_2 = 0 - 7i$

59. $z_1 = 8 + 0i, \quad z_2 = 0 - 3i$

60. $z_1 = -\dfrac{5\sqrt{2}}{2} + \dfrac{5\sqrt{2}}{2}i, \quad z_2 = -3 + 0i$

For Problems 61–66, find the product $z_1 z_2$ by using the trigonometric forms of the numbers. Express the final results in trigonometric form.

61. $z_1 = 5(\cos 20° + i \sin 20°), \quad z_2 = 4(\cos 55° + i \sin 55°)$

62. $z_1 = 3(\cos 110° + i \sin 110°), \quad z_2 = (\cos 28° + i \sin 28°)$

63. $z_1 = \sqrt{2}(\cos 120° + i \sin 120°), \quad z_2 = 3\sqrt{2}(\cos 310° + i \sin 310°)$

64. $z_1 = 2\sqrt{3}(\cos 260° + i \sin 260°), \quad z_2 = 4\sqrt{3}(\cos 320° + i \sin 320°)$

65. $z_1 = 5\left(\cos \dfrac{3\pi}{5} + i \sin \dfrac{3\pi}{5}\right), \quad z_2 = 7\left(\cos \dfrac{\pi}{2} + i \sin \dfrac{\pi}{2}\right)$

66. $z_1 = 6\left(\cos \dfrac{3\pi}{4} + i \sin \dfrac{3\pi}{4}\right), \quad z_2 = 4\left(\cos \dfrac{2\pi}{3} + i \sin \dfrac{2\pi}{3}\right)$

Miscellaneous Problems

67. If $z_1 = r_1(\cos \theta_1 + i \sin \theta_1)$ and $z_2 = r_2(\cos \theta_2 + i \sin \theta_2)$, then verify that $\dfrac{z_1}{z_2} =$

$\dfrac{r_1}{r_2}[\cos(\theta_1 - \theta_2) + i \sin(\theta_1 - \theta_2)]$.

For Problems 68–73, find the quotient z_1/z_2 using the trigonometric forms of the numbers (see Problem 67). Express the final quotient in $a + bi$ form.

68. $z_1 = 1 + i, \quad z_2 = 0 + i$

69. $z_1 = 2 - 2i, \quad z_2 = 0 + 3i$

70. $z_1 = -1 + i, \quad z_2 = 1 + i$

71. $z_1 = 1 - i, \quad z_2 = -1 - i$

72. $z_1 = -1 + \sqrt{3}\,i, \quad z_2 = -1 - \sqrt{3}\,i$

73. $z_1 = 3 + 3\sqrt{3}\,i, \quad z_2 = -\dfrac{3\sqrt{3}}{2} - \dfrac{3}{2}i$

5.3

Powers and Roots of Complex Numbers

By repeated application of the principle for multiplying complex numbers in trigonometric form, the powers of a complex number $z = r(\cos\theta + i\sin\theta)$ can be easily obtained.

$$z^2 = z \cdot z = [r(\cos\theta + i\sin\theta)][r(\cos\theta + i\sin\theta)]$$
$$= r^2(\cos 2\theta + i\sin 2\theta)$$

$$z^3 = z^2 \cdot z = [r^2(\cos 2\theta + i\sin 2\theta)][r(\cos\theta + i\sin\theta)]$$
$$= r^3(\cos 3\theta + i\sin 3\theta)$$

$$z^4 = z^3 \cdot z = [r^3(\cos 3\theta + i\sin 3\theta)][r(\cos\theta + i\sin\theta)]$$
$$= r^4(\cos 4\theta + i\sin 4\theta)$$

In general, the following result, named after the French mathematician Abraham De Moivre (1667–1754) can be stated.

De Moivre's Theorem

For every positive integer n,

$$[r(\cos\theta + i\sin\theta)]^n = r^n(\cos n\theta + i\sin n\theta).$$

EXAMPLE 1

Find $(1 + i)^{16}$.

Solution First, let's change $1 + i$ to trigonometric form.

$$1 + i = \sqrt{2}\left(\cos\frac{\pi}{4} + i\sin\frac{\pi}{4}\right)$$

Next, we can apply De Moivre's Theorem.

$$(1 + i)^{16} = (2^{1/2})^{16}\left[\cos 16\left(\frac{\pi}{4}\right) + i\sin 16\left(\frac{\pi}{4}\right)\right]$$
$$= 2^8[\cos 4\pi + i\sin 4\pi]$$
$$= 256[\cos 4\pi + i\sin 4\pi]$$

Finally, we can change back to $a + bi$ form.

$$(1 + i)^{16} = 256[\cos 4\pi + i\sin 4\pi]$$
$$= 256(1) + 256(i)(0)$$
$$= 256 + 0i$$

EXAMPLE 2

Find $\left(-\dfrac{\sqrt{3}}{2}-\dfrac{1}{2}i\right)^{20}$

Solution

$$-\dfrac{\sqrt{3}}{2}-\dfrac{1}{2}i = 1(\cos 210° + i\sin 210°)$$

Therefore,

$$\left(-\dfrac{\sqrt{3}}{2}-\dfrac{1}{2}i\right)^{20} = 1^{20}[\cos 20(210°) + i\sin 20(210°)]$$

$$= 1[\cos 4200° + i\sin 4200°]$$
$$= 1[\cos 240° + i\sin 240°]$$
$$= 1\left(-\dfrac{1}{2}\right) + (1)(i)\left(-\dfrac{\sqrt{3}}{2}\right)$$
$$= -\dfrac{1}{2} - \dfrac{\sqrt{3}}{2}i.$$

Finding Roots of Complex Numbers

If the complex number w is an nth root of the complex number z, then $w^n = z$. Substituting the trigonometric forms $s(\cos \alpha + i\sin \alpha)$ for w and $r(\cos \theta + i\sin \theta)$ for z, we obtain

$$[s(\cos \alpha + i\sin \alpha)]^n = r(\cos \theta + i\sin \theta).$$

Now applying De Moivre's Theorem to the left side produces

$$s^n(\cos n\alpha + i\sin n\alpha) = r(\cos \theta + i\sin \theta). \tag{1}$$

Equal complex numbers have equal absolute values. Therefore, $s^n = r$ and because s and r are nonnegative

$$s = \sqrt[n]{r}.$$

Furthermore, for equation (1) to hold,

$$\cos n\alpha + i\sin n\alpha = \cos \theta + i\sin \theta.$$

It follows that

$$\cos n\alpha = \cos \theta \quad \text{and} \quad \sin n\alpha = \sin \theta.$$

Since both the sine and cosine functions have a period of 2π, the last two equations are true if and only if $n\alpha$ and θ differ by a multiple of 2π. Therefore,

$$n\alpha = \theta + 2\pi k, \quad \text{where } k \text{ is an integer.}$$

Solving for α produces

$$\alpha = \dfrac{\theta + 2\pi k}{n}.$$

Substituting $\sqrt[n]{r}$ for s and $(\theta + 2\pi k)/n$ for α in the trigonometric form for w produces

$$w = \sqrt[n]{r}\left[\cos\left(\frac{\theta + 2\pi k}{n}\right) + i\sin\left(\frac{\theta + 2\pi k}{n}\right)\right].$$

If we let $k = 0, 1, 2, \ldots, n - 1$ successively, we obtain n distinct values for w, that is, n distinct nth roots of z. Furthermore, no other value of k will produce a new value for w. For example, if $k = n$, then $\alpha = (\theta + 2\pi n)/n = (\theta/n) + 2\pi$, which produces the same value for w as when $k = 0$. Similarly, $k = n + 1$ yields the same value for w as $k = 1$, and so on. Likewise, negative values of k will merely produce repeat values for w.

The previous discussion has verified the following property.

PROPERTY 5.1

If $z = r(\cos\theta + i\sin\theta)$ is any nonzero complex number and n is any positive integer, then z has precisely n distinct nth roots that are given by

$$\sqrt[n]{r}\left[\cos\left(\frac{\theta + 2\pi k}{n}\right) + i\sin\left(\frac{\theta + 2\pi k}{n}\right)\right]$$

where $k = 0, 1, 2, \ldots, n - 1$.

Property 5.1 can also be stated in terms of degree measure. It then becomes

$$\sqrt[n]{r}\left[\cos\left(\frac{\theta + k \cdot 360°}{n}\right) + i\sin\left(\frac{\theta + k \cdot 360°}{n}\right)\right]$$

where $k = 0, 1, 2, \ldots, n - 1$.

EXAMPLE 3

Find the four fourth roots of $-8 - 8\sqrt{3}\,i$.

Solution First, let's express the given number in trigonometric form.

$$-8 - 8\sqrt{3}\,i = 16(\cos 240° + i\sin 240°)$$

With $n = 4$, the fourth roots are given by

$$\sqrt[4]{16}\left[\cos\left(\frac{240° + k \cdot 360°}{4}\right) + i\sin\left(\frac{240° + k \cdot 360°}{4}\right)\right],$$

which simplifies to

$$2[\cos(60° + k \cdot 90°) + i\sin(60° + k \cdot 90°)].$$

Substituting 0, 1, 2, and 3 for k yields the following fourth roots.

$$2(\cos 60° + i\sin 60°) = 2\left(\frac{1}{2}\right) + 2i\left(\frac{\sqrt{3}}{2}\right) = 1 + \sqrt{3}\,i$$

$$2(\cos 150° + i\sin 150°) = 2\left(-\frac{\sqrt{3}}{2}\right) + 2i\left(\frac{1}{2}\right) = -\sqrt{3} + i$$

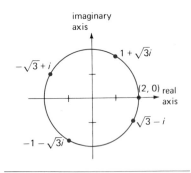

imaginary axis

$1 + \sqrt{3}i$

$-\sqrt{3} + i$

$(2, 0)$ real axis

$\sqrt{3} - i$

$-1 - \sqrt{3}i$

Figure 5.7

$$2(\cos 240° + i\sin 240°) = 2\left(-\frac{1}{2}\right) + 2i\left(-\frac{\sqrt{3}}{2}\right) = -1 - \sqrt{3}\,i$$

$$2(\cos 330° + i\sin 330°) = 2\left(\frac{\sqrt{3}}{2}\right) + 2i\left(-\frac{1}{2}\right) = \sqrt{3} - i$$

Note in Example 3 that each root has an absolute value of 2. Therefore, geometrically they are all located on a circle with its center at the origin and having a radius of length 2 units as illustrated in Figure 5.7. Furthermore, the points associated with the roots are equally spaced around the circle. In general, the nth roots of a complex number $z = r(\cos\theta + i\sin\theta)$ are equally spaced around a circle with a radius of length $\sqrt[n]{r}$ that has its center at the origin.

EXAMPLE 4

Find the three cube roots of $8i$.

Solution In trigonometric form

$$0 + 8i = 8\left(\cos\frac{\pi}{2} + i\sin\frac{\pi}{2}\right).$$

With $n = 3$, the cube roots are given by

$$\sqrt[3]{8}\left[\cos\left(\frac{\frac{\pi}{2} + 2\pi k}{3}\right) + i\sin\left(\frac{\frac{\pi}{2} + 2\pi k}{3}\right)\right],$$

which simplifies to

$$2\left[\cos\left(\frac{\pi}{6} + \frac{2\pi k}{3}\right) + i\sin\left(\frac{\pi}{6} + \frac{2\pi k}{3}\right)\right]$$

Substituting 0, 1, and 2 for k yields the following cube roots.

$$2\left(\cos\frac{\pi}{6} + i\sin\frac{\pi}{6}\right) = 2\left(\frac{\sqrt{3}}{2}\right) + 2i\left(\frac{1}{2}\right) = \sqrt{3} + i$$

$$2\left(\cos\frac{5\pi}{6} + i\sin\frac{5\pi}{6}\right) = 2\left(-\frac{\sqrt{3}}{2}\right) + 2i\left(\frac{1}{2}\right) = -\sqrt{3} + i$$

$$2\left(\cos\frac{3\pi}{2} + i\sin\frac{3\pi}{2}\right) = 2(0) + 2i(-1) = 0 - 2i$$

EXAMPLE 5

Find the five fifth roots of -243.

Solution In trigonometric form $-243 + 0i = 243(\cos 180° + i\sin 180°)$. With $n = 5$, the fifth roots are given by

$$\sqrt[5]{243}\left[\cos\left(\frac{180° + k \cdot 360°}{5}\right) + i\sin\left(\frac{180° + k \cdot 360°}{5}\right)\right],$$

which simplifies to

$$3[\cos(36° + k \cdot 72°) + i\sin(36° + k \cdot 72°)].$$

Substituting 0, 1, 2, 3, and 4 for k produces the following fifth roots.

$$3(\cos 36° + i\sin 36°)$$
$$3(\cos 108° + i\sin 108°)$$
$$3(\cos 180° + i\sin 180°) = -3 + 0i$$
$$3(\cos 252° + i\sin 252°)$$
$$3(\cos 324° + i\sin 324°)$$

There is one exact root, namely, $-3 + 0i$. Approximations for the other four roots could be obtained by using a calculator or a table. We will leave them in trigonometric form.

Problem Set 5.3

For Problems 1–16, use De Moivre's Theorem to find the indicated powers. Express results in $a + bi$ form.

1. $(1 + i)^{20}$
2. $(1 - i)^{12}$
3. $(-1 + i)^{10}$
4. $(-2 - 2i)^4$
5. $(3 + 3i)^5$
6. $(\sqrt{3} + i)^7$
7. $(-1 + \sqrt{3}i)^4$
8. $(-2\sqrt{3} + 2i)^5$
9. $\left(\dfrac{\sqrt{3}}{2} - \dfrac{1}{2}i\right)^{14}$
10. $\left(\dfrac{1}{2} - \dfrac{\sqrt{3}}{2}i\right)^{11}$
11. $\left(-\dfrac{\sqrt{2}}{2} + \dfrac{\sqrt{2}}{2}i\right)^{15}$
12. $\left(-\dfrac{\sqrt{2}}{2} - \dfrac{\sqrt{2}}{2}i\right)^{13}$
13. $[2(\cos 15° + i\sin 15°)]^4$
14. $[2(\cos 50° + i\sin 50°)]^6$
15. $\left(\cos\dfrac{\pi}{8} + i\sin\dfrac{\pi}{8}\right)^{10}$
16. $\left(\cos\dfrac{\pi}{12} + i\sin\dfrac{\pi}{12}\right)^8$

For Problems 17–32, find the indicated roots. Express the roots in $a + bi$ form if they are exact. Otherwise, leave them in trigonometric form.

17. The three cube roots of 8
18. The three cube roots of -27
19. The two square roots of $-16i$
20. The two square roots of $9i$
21. The four fourth roots of $-8 + 8\sqrt{3}i$
22. The four fourth roots of $-8 - 8\sqrt{3}i$
23. The five fifth roots of $1 + i$
24. The five fifth roots of $1 - i$
25. The six sixth roots of 1
26. The eight eighth roots of 1
27. The three cube roots of $-1 + \sqrt{3}i$
28. The three cube roots of $1 - \sqrt{3}i$
29. The five fifth roots of $-\sqrt{2} + \sqrt{2}i$

30. The five fifth roots of $\sqrt{2} - \sqrt{2}\,i$

31. The two square roots of $\dfrac{9}{2} + \dfrac{9\sqrt{3}}{2}\,i$

32. The two square roots of $-2 - 2\sqrt{3}\,i$

5.4

The Polar Coordinate System

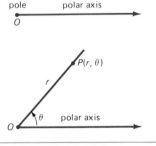

Figure 5.8

Some problems in analytic geometry, especially those involving motion about a point, are difficult to solve using the rectangular coordinate system. In fact, unwieldy equations such as $x^2 + y^2 - 2x = 2\sqrt{x^2 + y^2}$ are generated from relatively simple motion problems. However, this same equation can be transformed into a much more workable form using the variables r and θ of a plotting system called the **polar coordinate system**.

To set up the polar coordinate system in a plane, we begin with a fixed point O (called the **origin** or **pole**) and a directed half-line (called the **polar axis**) with its endpoint at O (Figure 5.8). The polar axis is usually drawn horizontal and directed to the right. To each point P in the plane we assign the **polar coordinates** (r, θ) as indicated in Figure 5.8. The angle θ has the polar axis as its initial side and the half-line OP as its terminal side. As usual, θ is considered positive if the angle is generated by a counterclockwise rotation of the polar axis and negative if the rotation is clockwise. Either radians or degrees can be used to express the measure of θ. The number r indicates the distance from the pole to the point P. If r is positive, then the distance is measured along the terminal side of θ. If r is negative, then the distance is measured along the half-line from O in the opposite direction of the terminal side of θ. The points associated with the ordered pairs $(5, \pi/3)$, $(-5, \pi/3)$, $(4, -30°)$, and $(-4, -135°)$ are plotted in Figure 5.9.

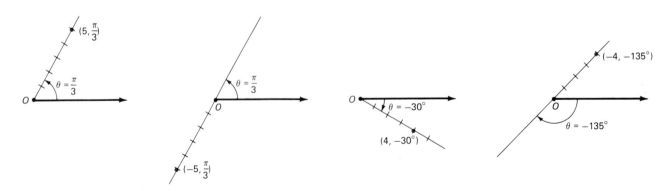

Figure 5.9

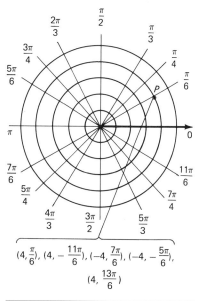

$$\left(4, \frac{\pi}{6}\right), \left(4, -\frac{11\pi}{6}\right), \left(-4, \frac{7\pi}{6}\right), \left(-4, -\frac{5\pi}{6}\right),$$
$$\left(4, \frac{13\pi}{6}\right)$$

Figure 5.10

In Figure 5.10 we have drawn a model of polar coordinate paper. The concentric circles having point O as a common center and the rays emanating from O at intervals corresponding to some special angles facilitate the plotting of points. It should be evident that the polar coordinates of a point are not unique. Note in Figure 5.10 that we have assigned five different ordered pairs to the point P. Actually, every point has infinitely many ordered pairs that can be assigned to it. The pole has polar coordinates $(0, \theta)$ for any angle θ.

Relationships between Rectangular and Polar Coordinates

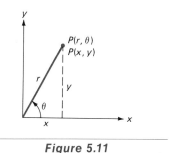

Figure 5.11

In Figure 5.11 we have superimposed an xy-plane on an $r\theta$-plane with the positive x-axis coinciding with the polar axis. The following polar-rectangular relationships can be easily deduced.

$$x = r \cos \theta, \qquad y = r \sin \theta$$
$$\tan \theta = \frac{y}{x}, \qquad r^2 = x^2 + y^2$$

EXAMPLE 1

Find the rectangular coordinates of the point P whose polar coordinates are $(4, 210°)$.

Solution Let's substitute 4 for r and $210°$ for θ in the equations $x = r\cos\theta$ and $y = r\sin\theta$.

$$x = 4\cos 210° = 4\left(-\frac{\sqrt{3}}{2}\right) = -2\sqrt{3}$$

$$y = 4\sin 210° = 4\left(-\frac{1}{2}\right) = -2$$

Thus, the rectangular coordinates of point P are $(-2\sqrt{3}, -2)$.

EXAMPLE 2

Suppose that point P has rectangular coordinates $(3, -3)$. Find polar coordinates (r, θ), such that $r > 0$ and $0 \le \theta < 2\pi$, for point P.

Solution Since r is to be positive, we can use $r = \sqrt{x^2 + y^2}$.

$$r = \sqrt{3^2 + (-3)^2} = \sqrt{9 + 9} = \sqrt{18} = 3\sqrt{2}$$

Also, we can use the equation $\tan\theta = y/x$.

$$\tan\theta = \frac{-3}{3} = -1$$

From this and the fact that $(3, -3)$ lies in the fourth quadrant, it follows that

$$\theta = \frac{7\pi}{4}.$$

Thus, $(3\sqrt{2}, 7\pi/4)$ are polar coordinates for point P.

The polar-rectangular relationships also provide the basis for changing polar equations to equations in rectangular form and vice versa. The next two examples illustrate this process.

EXAMPLE 3

Change $x^2 + y^2 - 2x = 2\sqrt{x^2 + y^2}$ to polar form.

Solution Substituting r^2 for $x^2 + y^2$, $r\cos\theta$ for x, and r for $\sqrt{x^2 + y^2}$, the equation $x^2 + y^2 - 2x = 2\sqrt{x^2 + y^2}$ becomes

$$r^2 - 2r\cos\theta = 2r.$$

This equation simplifies to

$$r^2 - 2r\cos\theta - 2r = 0$$

$$r(r - 2\cos\theta - 2) = 0$$

$$r = 0 \quad \text{or} \quad r - 2\cos\theta - 2 = 0.$$

The graph of $r = 0$ is the pole and since the pole is also included in the graph of $r - 2\cos\theta - 2 = 0$ (let $\theta = \pi$), we can discard $r = 0$ and keep only

$$r - 2\cos\theta - 2 = 0.$$

This could also be written as

$$r = 2 + 2\cos\theta.$$

Notice in Example 3 that the original complicated equation in rectangular form produced a fairly simple polar equation. Furthermore, in the next section we will see that the polar equation $r = 2 + 2\cos$ is easy to graph. So in a case like this, changing from rectangular form to polar form can be very beneficial.

EXAMPLE 4

Change $r = \cos\theta + \sin\theta$ to rectangular form.

Solution Substituting x/r for $\cos\theta$ and y/r for $\sin\theta$, the given equation $r = \cos\theta + \sin\theta$ becomes

$$r = \frac{x}{r} + \frac{y}{r}.$$

Now we can multiply both sides by r. This, in effect, adds $r = 0$ (the pole) to the graph. But the pole is already a part of the graph of $r = \cos\theta + \sin\theta$ (let $\theta = 3\pi/4$), so an equivalent equation

$$r^2 = x + y$$

is produced. Finally, by substituting $x^2 + y^2$ for r^2, we obtain

$$x^2 + y^2 = x + y,$$

which can be written as

$$x^2 + y^2 - x - y = 0.$$

In Example 4, the switch from polar form to rectangular form produced an equation $(x^2 + y^2 - x - y = 0)$ that should look familiar to us. Its graph is a circle and by completing the square, we could find its center and the length of a radius. In other words, in this case the switch from polar to rectangular form was beneficial.

Together Examples 3 and 4 illustrate that for some problems the rectangular system is most appropriate; however, for other problems it may be easier to use the polar coordinate system. Having both systems provides us with more flexibility to solve problems.

Graphing Polar Equations

In Chapter 1 when introducing the rectangular coordinate system, we remarked that there are basically two kinds of problems to solve in analytic geometry, namely, (1) given an algebraic equation, find its geometric graph, and (2) given a set of conditions pertaining to a geometric figure, find its algebraic equation. The polar coordinate system provides another basis for solving those same two kinds of problems. However, in this brief introduction to the polar coordinate system,

r	θ
2	0
$\frac{4\sqrt{3}}{3} \approx 2.3$	$\pi/6$
$2\sqrt{2} \approx 2.8$	$\pi/4$
4	$\pi/3$
-4	$2\pi/3$
$-2\sqrt{2} \approx -2.8$	$3\pi/4$
$\frac{-4\sqrt{3}}{3} \approx -2.3$	$5\pi/6$
-2	π

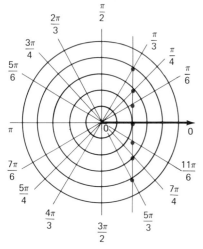

Figure 5.12

we will limit our study to problems of type (1), that is, sketching the graph of a given polar equation.

A **polar equation** is an equation involving the variables r and θ. An ordered pair (a, b) is said to be a *solution* of a polar equation if a substituted for r and b substituted for θ produces a true numerical statement. For example, $(1, \pi/6)$ is a solution of $r = 2 \sin \theta$ because 1 substituted for r and $\pi/6$ substituted for θ produces the true numerical statement $1 = 2(\frac{1}{2})$. The **graph** of a polar equation is the set of all points (in the $r\theta$-plane) that correspond to the set of all solutions of the equation.

EXAMPLE 5

Graph the polar equation $r \cos \theta = 2$.

Solution Let's change the form of the given equation by solving for r.

$$r \cos \theta = 2$$

$$r = \frac{2}{\cos \theta}, \qquad \cos \theta \neq 0$$

For each value assigned to θ, starting with 0 and using special positive angles, r takes on a corresponding value. (Since $\cos \theta$ cannot equal zero, θ cannot equal $\pi/2$.) The accompanying table contains eight solutions of the equation. Plotting the points associated with the ordered pairs (r, θ) and connecting them produces the line in Figure 5.12.

Notice that the last entry, $(-2, \pi)$, in the table for Example 5 determines the same point as the first entry $(2, 0)$. This fact alerted us to the realization that in this case there was no need to allow θ to vary from π to 2π because the same points would be determined again. For example, if $\theta = 7\pi/6$ we get the ordered pair $(-4\sqrt{3}/3, 7\pi/6)$ which determines the same point as $(4\sqrt{3}/3, \pi/6)$. In other words, by paying special attention to the nature of the trigonometric functions, you can often circumvent the need for a large table of values.

EXAMPLE 6

Graph the polar equation $r = 4 \sin \theta$.

Solution Let's set up a table of values, plot the points associated with the ordered pairs, and draw the graph. (Figure 5.13)

Examples 5 and 6 illustrate the fact that the graphs of some polar equations are familiar geometric figures. In fact, the polar equation $r \cos \theta = 2$ of Example 5 can be easily changed to the rectangular form $x = 2$, where the graph in Figure 5.12 is obvious. Likewise, the polar equation $r = 4 \sin \theta$ of Example 6 can be changed to the rectangular form $x^2 + y^2 = 4y$, which is equivalent to $x^2 + (y - 2)^2 = 4$. So in the xy-plane it is a circle with its center at $(0, 2)$ that

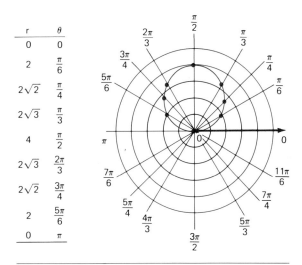

r	θ
0	0
2	$\frac{\pi}{6}$
$2\sqrt{2}$	$\frac{\pi}{4}$
$2\sqrt{3}$	$\frac{\pi}{3}$
4	$\frac{\pi}{2}$
$2\sqrt{3}$	$\frac{2\pi}{3}$
$2\sqrt{2}$	$\frac{3\pi}{4}$
2	$\frac{5\pi}{6}$
0	π

Figure 5.13

has a radius of 2 units. This agrees with our graph in Figure 5.13. However, at this time our primary objective is to give you some practice graphing polar equations by plotting a sufficient number of points to determine the figure. This will help you in the next section when we encounter some "not so familiar" geometric figures.

Problem Set 5.4

For Problems 1–12, plot the indicated ordered pairs as points in a polar coordinate system.

1. $A\left(3, \frac{\pi}{4}\right)$

2. $B\left(4, \frac{\pi}{3}\right)$

3. $C\left(-2, \frac{2\pi}{3}\right)$

4. $D\left(-3, \frac{5\pi}{6}\right)$

5. $E\left(5, -\frac{3\pi}{4}\right)$

6. $F\left(5, -\frac{5\pi}{4}\right)$

7. $G\left(-4, -\frac{\pi}{6}\right)$

8. $H\left(-4, -\frac{23\pi}{6}\right)$

9. $I(5, 270°)$

10. $J(4, -180°)$

11. $K(-2, -510°)$

12. $L(-2, 150°)$

For Problems 13–30, find the rectangular coordinates of the points whose polar coordinates are given.

13. $(3, 30°)$

14. $(6, 150°)$

15. $(-4, 225°)$

16. $(-2, 315°)$

17. $(2, 420°)$

18. $(5, 570°)$

19. $\left(-3, \frac{\pi}{3}\right)$

20. $\left(-7, \frac{5\pi}{6}\right)$

21. $\left(4, \frac{4\pi}{3}\right)$

22. $\left(6, \frac{5\pi}{3}\right)$

23. $\left(1, -\frac{2\pi}{3}\right)$

24. $\left(3, -\frac{11\pi}{6}\right)$

25. $\left(-7, \frac{9\pi}{4}\right)$

26. $\left(-4, \frac{11\pi}{4}\right)$

27. $\left(-2, -\frac{17\pi}{6}\right)$

28. $\left(-1, -\dfrac{11\pi}{3}\right)$ 29. $\left(-3, -\dfrac{3\pi}{2}\right)$ 30. $\left(8, -\dfrac{5\pi}{2}\right)$

For Problems 31–40, the rectangular coordinates of a point P are given. Find a pair of polar coordinates (r, θ) for P such that $r > 0$ and $0 \le \theta < 2\pi$.

31. $(-\sqrt{2}, \sqrt{2})$ 32. $(2\sqrt{2}, -2\sqrt{2})$ 33. $\left(-\dfrac{5\sqrt{3}}{2}, -\dfrac{5}{2}\right)$

34. $(-3\sqrt{3}, 3)$ 35. $(3, -3\sqrt{3})$ 36. $\left(-\dfrac{1}{2}, -\dfrac{\sqrt{3}}{2}\right)$

37. $(-4, 0)$ 38. $(0, -3)$ 39. $\left(\dfrac{3\sqrt{3}}{2}, \dfrac{3}{2}\right)$ 40. $(\sqrt{3}, 1)$

For Problems 41–46, the rectangular coordinates of a point P are given. Find a pair of polar coordinates (r, θ) for P such that $r < 0$ and $0 \le \theta < 2\pi$.

41. $(\sqrt{2}, \sqrt{2})$ 42. $\left(-\dfrac{3\sqrt{2}}{2}, \dfrac{3\sqrt{2}}{2}\right)$ 43. $(2, -2\sqrt{3})$

44. $\left(\dfrac{1}{2}, \dfrac{\sqrt{3}}{2}\right)$ 45. $\left(-\dfrac{5\sqrt{3}}{2}, \dfrac{5}{2}\right)$ 46. $(-\sqrt{3}, -1)$

For Problems 47–52, the rectangular coordinates of a point P are given. Find a pair of polar coordinates (r, θ) for P such that $r > 0$ and $0° \le \theta < 360°$. Express θ to the nearest tenth of a degree.

47. $(3, 2)$ 48. $(2, 5)$ 49. $(-4, 3)$
50. $(6, -2)$ 51. $(-4, -1)$ 52. $(-3, -4)$

For Problems 53–64, change each equation to polar form.

53. $y = 2$ 54. $x = 7$ 55. $3x - 2y = 4$
56. $5x + 4y = 10$ 57. $y = x$ 58. $y = -2x$
59. $x^2 + y^2 - 8x = 0$ 60. $x^2 + y^2 + 6y = 0$
61. $x^2 + y^2 + x = \sqrt{x^2 + y^2}$ 62. $x^2 + y^2 - 2y = 2\sqrt{x^2 + y^2}$
63. $x^2 = 4y$ 64. $y^2 = x$

For Problems 65–76, change each polar equation to rectangular form.

65. $r \sin\theta = -4$ 66. $r \cos\theta = 6$
67. $r - 3\cos\theta = 0$ 68. $r = 2\sin\theta$
69. $r = 2\cos\theta + 3\sin\theta$ 70. $r = 3\cos\theta - 4\sin\theta$
71. $r(\sin\theta + 4\cos\theta) = 5$ 72. $r(2\sin\theta - 3\cos\theta) = -4$

73. $r = \dfrac{4}{2 + \cos\theta}$ 74. $r = \dfrac{5}{2 - 3\sin\theta}$

75. $r = 2 + 3\sin\theta$ 76. $-3 - 2\cos\theta$

For Problems 77–92, sketch the graph of each of the polar equations. These graphs should be the familiar figures straight line, circle, parabola, ellipse, or hyperbola.

77. $r = 4$ 78. $r = -3$ 79. $\theta = \dfrac{\pi}{6}$

80. $\theta = -\dfrac{\pi}{4}$ 81. $r = -4\sin\theta$ 82. $r = -3\cos\theta$

83. $r \sin\theta = 3$ 84. $r \cos\theta = -2$

85. $r = 3\cos\theta + 4\sin\theta$

86. $r = 2\cos\theta - 3\sin\theta$

87. $r = \dfrac{4}{1 + \sin\theta}$

88. $r = \dfrac{3}{1 - \sin\theta}$

89. $r = \dfrac{5}{3 + 2\cos\theta}$

90. $r = \dfrac{5}{3 - 2\cos\theta}$

91. $r = \dfrac{5}{2 + 3\sin\theta}$

92. $r = \dfrac{5}{2 - 3\cos\theta}$

Miscellaneous Problems

93. The formula $d = \sqrt{(r_1)^2 + (r_2)^2 - 2r_1r_2\cos(\theta_2 - \theta_1)}$ can be used to find the distance between two points (r_1, θ_1) and (r_2, θ_2) in the polar coordinate system. Use the following figure and the law of cosines to develop the formula.

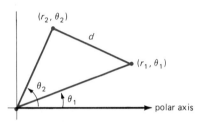

94. Use the distance formula from Problem 93 to find the distance between each of the following pairs of points.

(a) $\left(4, \dfrac{\pi}{2}\right)$ and $\left(3, \dfrac{\pi}{6}\right)$

(b) $\left(6, \dfrac{3\pi}{4}\right)$ and $\left(8, \dfrac{\pi}{4}\right)$

(c) $\left(10, \dfrac{7\pi}{6}\right)$ and $\left(2, \dfrac{2\pi}{3}\right)$

(d) $\left(3, \dfrac{5\pi}{6}\right)$ and $\left(6, \dfrac{\pi}{6}\right)$

5.5

More on Graphing Polar Equations

In the previous section we graphed some polar equations by plotting a sufficient number of points to determine the curve. Now let's discuss how the concept of **symmetry** can decrease the number of points that we need to plot and increase our efficiency in graphing polar equations.

In Figure 5.14(a) we have indicated that the polar axis reflection of point (r, θ) can be named $(r, -\theta)$ or $(-r, \pi - \theta)$. Likewise, in parts (b) and (c) of Figure 5.14 we have indicated the $(\pi/2)$-axis reflection and the pole reflection of (r, θ). From this information the following tests for symmetry can be stated.

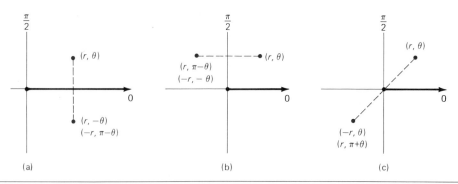

Figure 5.14

r	θ
4	0
$2 + \sqrt{3} \approx 3.7$	$\pi/6$
$2 + \sqrt{2} \approx 3.4$	$\pi/4$
3	$\pi/3$
2	$\pi/2$
1	$2\pi/3$
$2 - \sqrt{2} \approx 0.6$	$3\pi/4$
$2 - \sqrt{3} \approx 0.3$	$5\pi/6$
0	π

Polar axis: A polar equation exhibits polar axis symmetry if replacing θ by $-\theta$ or replacing r by $-r$ and θ by $\pi - \theta$ produces an equivalent equation.

$\dfrac{\pi}{2}$-axis: A polar equation exhibits $(\pi/2)$-axis symmetry if replacing r by $-r$ and θ by $-\theta$ or replacing θ by $\pi - \theta$ produces an equivalent equation.

Pole: A polar equation exhibits pole symmetry if replacing r by $-r$ or replacing θ by $\pi + \theta$ produces an equivalent equation.

A few comments about the tests for symmetry should be made. We refer to polar axis symmetry, but technically we mean symmetry with respect to the line determined by the polar axis. Likewise, $\pi/2$-axis symmetry means symmetry with respect to the line determined by the $\pi/2$-axis. Also, note that we have stated more than one test for each kind of symmetry. This is due to the fact that different polar equations produce the same set of points. For example, $r = 2$ and $r = -2$ would both produce a circle of radius 2 with the center at the pole. Finally, we suggest that as you begin to use the tests for symmetry you retain a mental picture of Figure 5.14. It may help you recall the specific tests.

EXAMPLE 1

Graph $r = 2 + 2\cos\theta$.

Solution First, since $\cos(-\theta) = \cos\theta$, we know that replacing θ by $-\theta$ will produce an equivalent equation. Thus, this curve is symmetric with respect to the polar axis. So in our table of values we can let θ vary from 0 to π. Then $\cos\theta$ decreases from 1 to -1 and $2 + 2\cos\theta$ decreases from 4 to 0. Plotting the points represented in the table and connecting them with a smooth curve produces the upper half of Figure 5.15. Then reflecting this across the polar axis completes the figure.

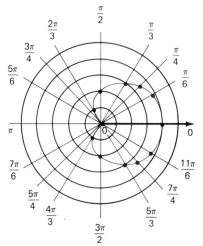

Figure 5.15

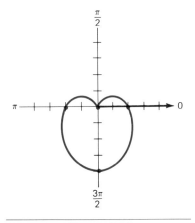

Figure 5.16

r	θ
6	0
$2 + 2\sqrt{3} \approx 5.4$	$\pi/6$
$2 + 2\sqrt{2} \approx 4.8$	$\pi/4$
4	$\pi/3$
2	$\pi/2$
0	$2\pi/3$
$2 - 2\sqrt{2} \approx -0.8$	$3\pi/4$
$2 - 2\sqrt{3} \approx -1.4$	$5\pi/6$
-2	π

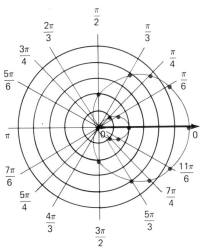

Figure 5.17

The heart-shaped graph in Figure 5.15 is called a **cardioid**. In general, the graph of a polar equation of the form

$$r = a(1 \pm \cos\theta) \quad \text{or} \quad r = a(1 \pm \sin\theta)$$

where a is a nonzero real number, is called a cardioid. Therefore, by recognizing the general form of the equation of a cardioid, we can sketch a rough graph of one by simply plotting points for $\theta = 0$, $\pi/2$, π, and $3\pi/2$. Let's consider an example.

EXAMPLE 2

Sketch the graph of $r = 2 - 2\sin\theta$.

Solution Letting $\theta = 0, \pi/2, \pi$, and $3\pi/2$ produces the ordered pairs $(2, 0), (0, \pi/2)$, $(2, \pi)$, and $(4, 3\pi/2)$. These points along with the knowledge that the graph is a cardioid allow us to sketch a rough graph, as in Figure 5.16.

The graph of a polar equation of the form

$$r = a \pm b\cos\theta \quad \text{or} \quad r = a \pm b\sin\theta,$$

where $a \neq b$, is called a **limacon**. The graph of a limacon is similar in shape to a cardioid, but depending upon the relative sizes of a and b, it may contain an additional "loop," as illustrated by the next example.

EXAMPLE 3

Graph $r = 2 + 4\cos\theta$.

Solution Again, since $\cos(-\theta) = \cos\theta$, this equation exhibits polar axis symmetry. So in the table we have allowed θ to vary from 0 to π. Notice in the table that $r = 0$ when $\theta = 2\pi/3$. Then r becomes negative for $2\pi/3 < \theta \leq \pi$. So the points from the table determine the upper half of the large loop and the lower half of the small loop in Figure 5.17. Then because of symmetry, the complete figure is determined.

EXAMPLE 4

Graph $r = 5\sin 2\theta$.

Solution To test for symmetry, it might be easier to replace $\sin 2\theta$ by $2\sin\theta\cos\theta$. Then the given equation becomes $r = 10\sin\theta\cos\theta$. Using the identities $\sin(\pi - \theta) = \sin\theta$, $\cos(\pi - \theta) = -\cos\theta$, $\sin(\pi + \theta) = -\sin\theta$, and $\cos(\pi + \theta) = -\cos\theta$, we can verify that this curve is symmetric with respect to the polar axis, $(\pi/2)$-axis, and the pole. (We shall leave the details of applying the tests for symmetry for you to complete.) Thus, we can concentrate on values of θ from 0 to $\pi/2$. As θ increases from 0 to $\pi/4$, the value of r increases from 0 to 5. Then as θ continues to increase from $\pi/4$ to $\pi/2$, the value of r decreases from 5 to 0. By keeping these facts in mind and plotting the points $(5\sqrt{3}/2, \pi/6)$, $(5, \pi/4)$, and $(5\sqrt{3}/2, \pi/3)$, we can sketch the upper right-hand part of Figure 5.18. Then the concept of symmetry allows us to complete the figure. It is called a **four-leafed rose**.

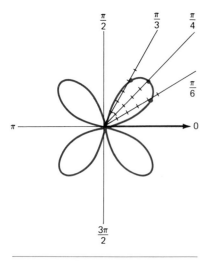

Figure 5.18

Another interesting type of curve is produced by equations of the form $r = a\theta$. These curves "wind around the pole" infinitely many times in such a way that r increases (or decreases) steadily as θ increases (or decreases). They are called **Archimedean spirals**. Let's consider one specific example.

EXAMPLE 5

Sketch the curve $r = \theta$ for $\theta \geq 0$.

Solution A reasonably accurate sketch can be obtained by plotting some points on the axes and using the fact that r increases steadily as θ increases. In Figure 5.19 we have plotted the points $(0,0)$, $(\pi/2, \pi/2)$, (π, π), $(3\pi/2, 3\pi/2)$, $(2\pi, 2\pi)$, $(5\pi/2, 5\pi/2)$, $(3\pi, 3\pi)$, $(7\pi/2, 7\pi/2)$, and $(4\pi, 4\pi)$, and sketched the curve.

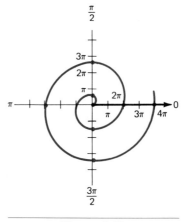

Figure 5.19

Problem Set 5.5

Determine the symmetry (polar axis, $(\pi/2)$-axis, pole, or none) that each of the following equations exhibits. Do not sketch the graphs.

1. $r\cos\theta = -6$
2. $r\sin\theta = 8$
3. $r = \dfrac{3}{1 - \sin\theta}$

4. $r = \dfrac{2}{1 + \cos\theta}$
5. $r = \dfrac{4}{2 + 3\cos\theta}$
6. $r = \dfrac{3}{3 - 2\sin\theta}$

7. $r = 4\sin\theta$
8. $r = 6\cos\theta$
9. $r = 3\cos\theta + 2\sin\theta$

10. $r = 5\sin\theta + 3\cos\theta$
11. $r = \sec\theta + 2$
12. $r = \csc\theta - 3$

13. $r = 10\tan\theta\sin\theta$
14. $r = 4\cot\theta\cos\theta$
15. $r^2 = \sin 2\theta$

16. $r^2 = \cos 2\theta$

Graph each of the following polar equations.

17. $r = 3 + 3\sin\theta$
18. $r = 2 - 2\cos\theta$
19. $r = 1 - \cos\theta$

20. $r = 3 - 3 \sin \theta$ 21. $r = 2 + 4 \sin \theta$ 22. $r = 3 - 4 \sin \theta$

23. $r = 4 - 2 \cos \theta$ 24. $r = 4 + 2 \cos \theta$ 25. $r = 2 - 4 \cos \theta$

26. $r = 1 - 3 \cos \theta$ 27. $r = 3 + \sin \theta$ 28. $r = 3 - \sin \theta$

29. $r = 4 \cos 2\theta$ 30. $r = 3 \sin 2\theta$ 31. $r = 3 \sin 3\theta$

32. $r = 3 \cos 2\theta$ 33. $r^2 = 9 \cos 2\theta$ 34. $r^2 = -16 \cos 2\theta$

35. $r^2 = -16 \sin 2\theta$ 36. $r^2 = 9 \sin 2\theta$ 37. $r = 3 \sin \theta \tan \theta$

38. $r = 2 \cos \theta \cot \theta$ 39. $r = \theta,\ \theta \le 0$ 40. $r = 2\theta,\ \theta \ge 0$

5.6

Vectors as Ordered Pairs of Real Numbers

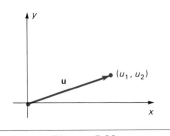

Figure 5.20

Our study of vectors in Section 2.6 relied primarily on geometric ideas. Now let's switch to more of an algebraic approach by introducing the rectangular coordinate system to our study of vectors. To each vector **u** having its initial point at the origin of a rectangular coordinate system, we can assign an ordered pair of real numbers (u_1, u_2) to its terminal point, as indicated in Figure 5.20. Likewise, to each ordered pair of real numbers (u_1, u_2), we can associate a vector that has its initial point at the origin and its terminal point at (u_1, u_2). Thus, we obtain a one-to-one correspondence between vectors and ordered pairs of real numbers. This allows us to regard a vector in a plane as an ordered pair of real numbers.

We will use the symbol $\langle u_1, u_2 \rangle$ for an ordered pair of real numbers that represents a vector **u**. Thus, we can write $\mathbf{u} = \langle u_1, u_2 \rangle$ and the numbers u_1 and u_2 are called the **components** of the vector. Two vectors $\mathbf{u} = \langle u_1, u_2 \rangle$ and $\mathbf{v} = \langle v_1, v_2 \rangle$ are equal if and only if $u_1 = v_1$ and $u_2 = v_2$. It should also be evident from Figure 5.20 and the distance formula that the magnitude $\|\mathbf{u}\|$ of the vector $\mathbf{u} = \langle u_1, u_2 \rangle$ is given by

$$\|\mathbf{u}\| = \sqrt{(u_1)^2 + (u_2)^2}.$$

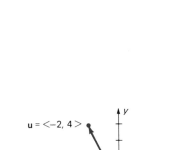

Figure 5.21

EXAMPLE 1

Sketch the vector $\mathbf{u} = \langle -2, 4 \rangle$ and find its magnitude.

Solution The initial point of **u** is at the origin and the terminal point is at $(-2, 4)$ as shown in Figure 5.21. The magnitude of **u** is

$$\|\mathbf{u}\| = \sqrt{(-2)^2 + 4^2} = \sqrt{20} = 2\sqrt{5}.$$

One advantage of considering vectors as ordered pairs becomes evident immediately. The basic operations of vector addition and multiplication by scalars are easy to define and perform in terms of components. If $\mathbf{u} = \langle u_1, u_2 \rangle$ and $\mathbf{v} = \langle v_1, v_2 \rangle$ and k is any scalar (real number), then

$$\mathbf{u} + \mathbf{v} = \langle u_1 + v_1, u_2 + v_2 \rangle \qquad \text{Vector addition}$$

$$k\mathbf{u} = \langle ku_1, ku_2 \rangle. \qquad \text{Scalar multiplication}$$

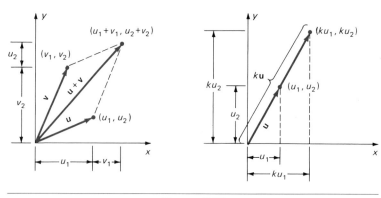

Figure 5.22

Using Figure 5.22 we could show that these definitions for vector addition and scalar multiplication are consistent with those given in Section 2.6. We will leave the details of this argument for you to complete in the next problem set. Since $\mathbf{u} - \mathbf{v} = \mathbf{u} + (-1)\mathbf{v}$, vector subtraction can be defined as follows.

$$\mathbf{u} - \mathbf{v} = \langle u_1 - v_1, u_2 - v_2 \rangle$$

EXAMPLE 2

If $\mathbf{u} = \langle 3, 5 \rangle$ and $\mathbf{v} = \langle -1, -4 \rangle$, determine

(a) $\mathbf{u} + \mathbf{v}$,   (b) $\mathbf{u} - \mathbf{v}$,   (c) $-4\mathbf{u}$,   and   (d) $2\mathbf{v} - 3\mathbf{u}$.

Solutions

(a) $\mathbf{u} + \mathbf{v} = \langle 3, 5 \rangle + \langle -1, -4 \rangle = \langle 2, 1 \rangle$

(b) $\mathbf{u} - \mathbf{v} = \langle 3, 5 \rangle - \langle -1, -4 \rangle = \langle 4, 9 \rangle$

(c) $-4\mathbf{u} = (-4)\langle 3, 5 \rangle = \langle -12, -20 \rangle$

(d) $2\mathbf{v} - 3\mathbf{u} = (2)\langle -1, -4 \rangle - (3)\langle 3, 5 \rangle$
$\qquad\qquad = \langle -2, -8 \rangle - \langle 9, 15 \rangle$
$\qquad\qquad = \langle -11, -23 \rangle$

Sometimes a vector may be positioned so that its initial point is not at the origin. If the coordinates of the initial and terminal points of a vector are known, then we can determine a vector that has its initial point at the origin and has the same magnitude and same direction as the given vector. For example, in Figure 5.23 consider the vector that has an initial point at $(4, 1)$ and a terminal point at $(2, 4)$. By subtracting the coordinates of the initial point from the coordinates of the terminal point we obtain $(2 - 4, 4 - 1) = (-2, 3)$. The vector $\langle -2, 3 \rangle$ in Figure 5.23 has the same magnitude and direction as the given vector.

By defining vectors as ordered pairs of real numbers, some of the basic properties of vector addition and scalar multiplication can be easily verified using the properties of real numbers. For example, to verify that vector addition is a commutative operation, we can reason as follows.

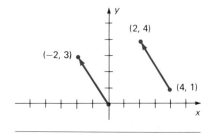

Figure 5.23

$$\mathbf{u} + \mathbf{v} = \langle u_1, u_2 \rangle + \langle v_1, v_2 \rangle$$
$$= \langle u_1 + v_1, u_2 + v_2 \rangle$$
$$= \langle v_1 + u_1, v_2 + u_2 \rangle$$
$$= \langle v_1, v_2 \rangle + \langle u_1, u_2 \rangle$$
$$= \mathbf{v} + \mathbf{u}$$

For any vectors $\mathbf{u}$, $\mathbf{v}$, and $\mathbf{w}$ and any scalars k and l, the following properties hold:

1. $\mathbf{u} + \mathbf{v} = \mathbf{v} + \mathbf{u}$
2. $(\mathbf{u} + \mathbf{v}) + \mathbf{w} = \mathbf{u} + (\mathbf{v} + \mathbf{w})$
3. $\mathbf{u} + \mathbf{0} = \mathbf{0} + \mathbf{u} = \mathbf{u}$
4. $\mathbf{u} + (-\mathbf{u}) = \mathbf{0}$
5. $k(l\mathbf{u}) = (kl)\mathbf{u}$
6. $k(\mathbf{u} + \mathbf{v}) = k\mathbf{u} + k\mathbf{v}$
7. $(k + l)\mathbf{u} = k\mathbf{u} + l\mathbf{u}$
8. $1(\mathbf{u}) = \mathbf{u}$

Dot Product

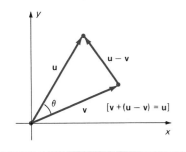

Figure 5.24

To motivate a definition for another operation involving vectors, consider two vectors $\mathbf{u} = \langle u_1, u_2 \rangle$, $\mathbf{v} = \langle v_1, v_2 \rangle$, and the difference vector $\mathbf{u} - \mathbf{v} = \langle u_1 - v_1, u_2 - v_2 \rangle$ as shown in Figure 5.24. Let θ be the smallest positive angle between $\mathbf{u}$ and $\mathbf{v}$. By applying the law of cosines we obtain

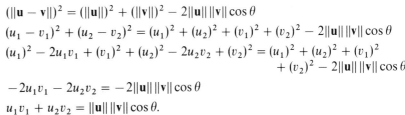

$$(\|\mathbf{u} - \mathbf{v}\|)^2 = (\|\mathbf{u}\|)^2 + (\|\mathbf{v}\|)^2 - 2\|\mathbf{u}\|\,\|\mathbf{v}\| \cos \theta$$
$$(u_1 - v_1)^2 + (u_2 - v_2)^2 = (u_1)^2 + (u_2)^2 + (v_1)^2 + (v_2)^2 - 2\|\mathbf{u}\|\,\|\mathbf{v}\| \cos \theta$$
$$(u_1)^2 - 2u_1 v_1 + (v_1)^2 + (u_2)^2 - 2u_2 v_2 + (v_2)^2 = (u_1)^2 + (u_2)^2 + (v_1)^2 + (v_2)^2 - 2\|\mathbf{u}\|\,\|\mathbf{v}\| \cos \theta$$

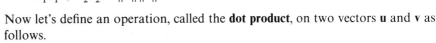

$$-2u_1 v_1 - 2u_2 v_2 = -2\|\mathbf{u}\|\,\|\mathbf{v}\| \cos \theta$$
$$u_1 v_1 + u_2 v_2 = \|\mathbf{u}\|\,\|\mathbf{v}\| \cos \theta.$$

Now let's define an operation, called the **dot product**, on two vectors $\mathbf{u}$ and $\mathbf{v}$ as follows.

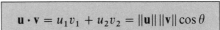

$$\mathbf{u} \cdot \mathbf{v} = u_1 v_1 + u_2 v_2 = \|\mathbf{u}\|\,\|\mathbf{v}\| \cos \theta$$

EXAMPLE 3

Find the dot product of $\langle -4, -2 \rangle$ and $\langle 3, 7 \rangle$.

Solution Using $\mathbf{u} \cdot \mathbf{v} = u_1 v_1 + u_2 v_2$, we obtain

$$\langle -4, -2 \rangle \cdot \langle 3, 7 \rangle = -4(3) + (-2)(7)$$
$$= -12 - 14 = -26.$$

Note that the dot product of two vectors is a real number (scalar), not another vector. Furthermore, it should be evident that the expression $u_1 v_1 + u_2 v_2$

provides a convenient way of computing the dot product when the components of the two vectors are known. However, the equation $\mathbf{u} \cdot \mathbf{v} = \|\mathbf{u}\|\,\|\mathbf{v}\| \cos \theta$ gives a geometric interpretation to the operation of dot product. Solving for $\cos \theta$, we obtain

$$\cos \theta = \frac{\mathbf{u} \cdot \mathbf{v}}{\|\mathbf{u}\|\,\|\mathbf{v}\|}.$$

In other words, the cosine of the angle of intersection of two vectors is the quotient of the dot product and the product of the magnitudes of the two vectors. Let's consider some examples using this idea.

EXAMPLE 4

Find the angle of intersection of the two vectors $\langle -3, 2 \rangle$ and $\langle 2, 3 \rangle$.

Solution

$$\cos \theta = \frac{(-3)(2) + 2(3)}{\sqrt{(-3)^2 + 2^2} \sqrt{2^2 + 3^2}} = \frac{0}{\sqrt{13} \sqrt{13}} = 0$$

Therefore, $\theta = 90°$, and the vectors are perpendicular.

Example 4 illustrates that the dot product is very convenient for determining perpendicular vectors. **In general, two nonzero vectors are perpendicular if and only if their dot product is zero.**

EXAMPLE 5

Find the angle of intersection of the two vectors $\langle -3, 1 \rangle$ and $\langle 5, 2 \rangle$.

Solution

$$\cos \theta = \frac{-3(5) + 1(2)}{\sqrt{(-3)^2 + 1^2} \sqrt{5^2 + 2^2}} = \frac{-13}{\sqrt{10} \sqrt{29}} = -0.7634$$

Therefore,

$$\theta = 139.8° \text{ to the nearest tenth of a degree.}$$

Problem Set 5.6

In Problems 1–8, sketch the vector and find its magnitude.

1. $\langle 3, 4 \rangle$ 2. $\langle -4, 3 \rangle$ 3. $\langle -1, -3 \rangle$ 4. $\langle 4, -1 \rangle$

5. $\langle 6, -2 \rangle$ 6. $\langle -2, -4 \rangle$ 7. $\langle -6, 4 \rangle$ 8. $\langle 2, 6 \rangle$

In Problems 9–16, find $\mathbf{u} + \mathbf{v}$, $\mathbf{u} - \mathbf{v}$, $3\mathbf{u} + 4\mathbf{v}$, and $2\mathbf{u} - 5\mathbf{v}$.

9. $\mathbf{u} = \langle 1, 2 \rangle, \mathbf{v} = \langle 3, 5 \rangle$ 10. $\mathbf{u} = \langle 5, 6 \rangle, \mathbf{v} = \langle 4, -2 \rangle$

11. $\mathbf{u} = \langle -4, -3 \rangle, \mathbf{v} = \langle -1, 6 \rangle$ 12. $\mathbf{u} = \langle 0, -4 \rangle, \mathbf{v} = \langle -3, -7 \rangle$

13. $\mathbf{u} = \langle 7, -1 \rangle, \mathbf{v} = \langle -4, 0 \rangle$ 14. $\mathbf{u} = \langle -2, 8 \rangle, \mathbf{v} = \langle -1, -6 \rangle$

15. $\mathbf{u} = \langle -3, -6 \rangle, \mathbf{v} = \langle -2, -4 \rangle$ 16. $\mathbf{u} = \langle 4, -4 \rangle, \mathbf{v} = \langle -4, 4 \rangle$

In Problems 17–24, (a) sketch the vector that has its initial point at P and its terminal point at Q, (b) determine a vector $\mathbf{u}$ that has its initial point at the origin and has the same magnitude and direction as $\overrightarrow{PQ}$, and (c) sketch the vector $\mathbf{u}$.

17. $P(2, 3)$ and $Q(4, 9)$ 18. $P(4, 9)$ and $Q(2, 3)$

19. $P(-1, 3)$ and $Q(-5, 6)$ 20. $P(-2, 4)$ and $Q(1, -3)$

21. $P(-1, -5)$ and $Q(-1, 4)$ 22. $P(3, -2)$ and $Q(-4, -1)$

23. $P(2, -1)$ and $Q(5, -7)$ 24. $P(-3, -4)$ and $Q(1, 2)$

In Problems 25–30, find the dot product of the two given vectors.

25. $\langle 4, 3 \rangle$ and $\langle -2, 6 \rangle$ 26. $\langle -1, 2 \rangle$ and $\langle 3, -4 \rangle$

27. $\langle -4, -2 \rangle$ and $\langle -1, -6 \rangle$ 28. $\langle 3, -4 \rangle$ and $\langle 5, 4 \rangle$

29. $\langle -2, 7 \rangle$ and $\langle -6, 1 \rangle$ 30. $\langle 0, 4 \rangle$ and $\langle 3, -6 \rangle$

In Problems 31–36, determine if the two vectors are perpendicular.

31. $\langle -1, -3 \rangle$ and $\langle 3, -1 \rangle$ 32. $\langle 4, 3 \rangle$ and $\langle 3, -4 \rangle$

33. $\langle 4, 5 \rangle$ and $\langle -4, 5 \rangle$ 34. $\langle -2, 7 \rangle$ and $\langle -2, -7 \rangle$

35. $\langle -2, -6 \rangle$ and $\langle 6, -2 \rangle$ 36. $\langle 5, -4 \rangle$ and $\langle 2, 5 \rangle$

In Problems 37–46, find the angle of intersection of the two vectors. Express answers to the nearest tenth of a degree.

37. $\langle 2, 5 \rangle$ and $\langle 6, 2 \rangle$ 38. $\langle 4, 4 \rangle$ and $\langle 5, 2 \rangle$

39. $\langle -5, 2 \rangle$ and $\langle -6, -1 \rangle$ 40. $\langle -2, -7 \rangle$ and $\langle 2, -1 \rangle$

41. $\langle -5, 1 \rangle$ and $\langle 5, 1 \rangle$ 42. $\langle -2, 6 \rangle$ and $\langle -2, -6 \rangle$

43. $\langle -1, \sqrt{3} \rangle$ and $\langle 3, 3 \rangle$ 44. $\langle \sqrt{3}, 1 \rangle$ and $\langle \sqrt{3}, -1 \rangle$

45. $\langle -4, -1 \rangle$ and $\langle 3, -6 \rangle$ 46. $\langle -3, 5 \rangle$ and $\langle 7, -2 \rangle$

47. Show that $k(\mathbf{u} + \mathbf{v}) = k\mathbf{u} + k\mathbf{v}$ for $k = 3$, $\mathbf{u} = \langle 4, 6 \rangle$, and $\mathbf{v} = \langle -3, 9 \rangle$.

48. Show that $(k + l)\mathbf{u} = k\mathbf{u} + l\mathbf{u}$ for $k = -2$, $l = 5$, and $\mathbf{u} = \langle -3, -6 \rangle$.

49. Show that $\mathbf{u} \cdot (\mathbf{v} + \mathbf{w}) = \mathbf{u} \cdot \mathbf{v} + \mathbf{u} \cdot \mathbf{w}$ for $\mathbf{u} = \langle 3, -2 \rangle$, $\mathbf{v} = \langle -4, 5 \rangle$, and $\mathbf{w} = \langle -1, -4 \rangle$.

50. Show that $k(\mathbf{u} \cdot \mathbf{v}) = (k\mathbf{u}) \cdot \mathbf{v} = \mathbf{u} \cdot (k\mathbf{v})$ for $k = 5$, $\mathbf{u} = \langle 2, 7 \rangle$, and $\mathbf{v} = \langle -3, -5 \rangle$.

Miscellaneous Problems

51. Prove Properties (2) through (8) on page 193.

52. Prove that $\mathbf{u} \cdot \mathbf{v} = \mathbf{v} \cdot \mathbf{u}$.

53. Prove that $\mathbf{u} \cdot (\mathbf{v} + \mathbf{w}) = \mathbf{u} \cdot \mathbf{v} + \mathbf{u} \cdot \mathbf{w}$.

54. Using Figure 5.22, verify that the definitions for vector addition and scalar multiplication given in this section are consistent with those given in Section 2.6.

Chapter Summary

Let's summarize this chapter in terms of the three topics, complex numbers, polar coordinate system, and algebra of vectors.

Complex Numbers

A number of the form $a + bi$, where a and b are real numbers and i is the imaginary unit such that $i^2 = -1$, is a **complex number**.

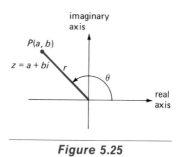

Figure 5.25

Two complex numbers $a + bi$ and $c + di$ are said to be equal if and only if $a = c$ and $b = d$.

Addition and subtraction of complex numbers are described as follows:

$$(a + bi) + (c + di) = (a + c) + (b + d)i;$$
$$(a + bi) - (c + di) = (a - c) + (b - d)i.$$

A square root of any negative real number can be represented as the product of a real number and the imaginary unit i. That is,

$$\sqrt{-b} = i\sqrt{b}, \text{ where } b \text{ is a positive real number.}$$

The product of two complex numbers is defined to conform with the product of two binomials.

The **conjugate** of $a + bi$ is $a - bi$. The product of a complex number and its conjugate is a real number. Therefore, conjugates are used to simplify expressions, such as $\dfrac{4 + 3i}{5 - 2i}$, that indicate the quotient of two complex numbers.

The following relationships are apparent from Figure 5.25.

$$b = r \sin \theta$$
$$a = r \cos \theta$$
$$r = \sqrt{a^2 + b^2}$$
$$a + bi = r(\cos \theta + i \sin \theta) \qquad \text{trigonometric form of a complex number}$$

The product of two complex numbers z_1 and z_2, expressed in trigonometric form, is given by

$$z_1 z_2 = r_1 r_2 [\cos(\theta_1 + \theta_2) + i \sin(\theta_1 + \theta_2)].$$

De Moivre's theorem is the basis for raising a complex number to a power.

$$[r(\cos \theta + i \sin \theta)]^n = r^n(\cos n\theta + i \sin n\theta)$$

The nth roots of a complex number $z = r(\cos \theta + i \sin \theta)$ are given by

$$\sqrt[n]{r}\left[\cos\left(\frac{\theta + 2\pi k}{n}\right) + i \sin\left(\frac{\theta + 2\pi k}{n}\right)\right]$$

where $k = 0, 1, 2, \ldots, n - 1$.

Polar Coordinates

The polar coordinate system provides another way of naming points in a coordinatized plane. The polar coordinates (r, θ) of a point P measure its distance r from a fixed point O and the angle θ that ray OP makes with a horizontal ray OA directed to the right. The point O is called the **pole** and the ray OA is called the **polar axis**.

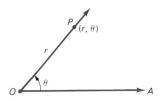

The following equations express relationships between the polar coordinates (r, θ) and the rectangular coordinates (x, y).

$$x = r \cos \theta \qquad y = r \sin \theta$$

$$\tan \theta = \frac{y}{x} \qquad r^2 = x^2 + y^2$$

The graph of a polar equation of the form

$$r = a(1 + \cos \theta), \qquad r = a(1 - \cos \theta),$$

$$r = a(1 + \sin \theta), \quad \text{or} \quad r = a(1 - \sin \theta)$$

is a **cardioid**.

The graph of a polar equation of the form

$$r = a \pm b \cos \theta \quad \text{or} \quad r = a \pm b \sin \theta,$$

where $a \neq b$, is a **limacon**.

Vectors

Treating 2-space vectors as ordered pairs of real numbers provides an algebraic setting for the study of vectors. Some basic operations can be defined in terms of real numbers.

Vector addition: $\mathbf{u} + \mathbf{v} = \langle u_1 + v_1, u_2 + v_2 \rangle$

Scalar multiplication: $k\mathbf{u} = \langle ku_1, ku_2 \rangle$

The basic properties of vector addition and scalar multiplication can be reviewed on page 193.

The magnitude or length of a vector is given by

$$\|\mathbf{u}\| = \sqrt{(u_1)^2 + (u_2)^2}.$$

The operation of dot product is defined as

$$\mathbf{u} \cdot \mathbf{v} = u_1 v_2 + u_2 v_2 = \|\mathbf{u}\|\,\|\mathbf{v}\|\cos\theta.$$

From the equation $\mathbf{u} \cdot \mathbf{v} = \|\mathbf{u}\|\,\|\mathbf{v}\|\cos\theta$, we get

$$\cos\theta = \frac{\mathbf{u} \cdot \mathbf{v}}{\|\mathbf{u}\|\,\|\mathbf{v}\|}$$

which provides a form for finding the angle of intersection, θ, of two vectors. In a special case, two vectors are perpendicular if and only if their dot product is zero.

Chapter 5 Review Problem Set

For Problems 1–12, perform the indicated operations and express the resulting complex number in standard form.

1. $(-7 + 3i) + (-4 - 9i)$
2. $(2 - 10i) - (3 - 8i)$
3. $(-1 + 4i) - (-2 + 6i)$
4. $(3i)(-7i)$
5. $(2 - 5i)(3 + 4i)$
6. $(-3 - i)(6 - 7i)$
7. $(4 + 2i)(-4 - i)$
8. $(5 - 2i)(5 + 2i)$
9. $\dfrac{5}{3i}$
10. $\dfrac{2 + 3i}{3 - 4i}$
11. $\dfrac{-1 - 2i}{-2 + i}$
12. $\dfrac{-6i}{5 + 2i}$

For Problems 13–18, write each radical expression in terms of i and simplify.

13. $\sqrt{-100}$
14. $\sqrt{-40}$
15. $4\sqrt{-80}$
16. $(-\sqrt{9})(\sqrt{-16})$
17. $(\sqrt{-6})(\sqrt{-8})$
18. $\dfrac{\sqrt{-24}}{\sqrt{-3}}$

19. Plot the complex number $2 - 4i$ and find its absolute value.
20. Express the complex number $\sqrt{3} - i$ in trigonometric form, where $0 \le \theta < 2\pi$.
21. Express the complex number $-3\sqrt{2} - 3\sqrt{2}i$ in trigonometric form, where $0° \le \theta < 360°$.
22. Express the complex number $5(\cos(3\pi/2) + i\sin(3\pi/2))$ in $a + bi$ form.
23. Express the complex number $8(\cos 300° + i\sin 300°)$ in $a + bi$ form.

For Problems 24–26, use De Moivre's theorem to find the indicated powers. Express results in $a + bi$ form.

24. $(-1 - i)^8$
25. $(1 - \sqrt{3}i)^{10}$
26. $\left(\dfrac{\sqrt{2}}{2} + \dfrac{\sqrt{2}}{2}i\right)^{17}$

For Problems 27–29, find the indicated roots and express them in $a + bi$ form.

27. The three cube roots of $-27i$

28. The four fourth roots of $-2 + 2\sqrt{3}\,i$

29. The two square roots of $8 - 8\sqrt{3}\,i$

For Problems 30–35, sketch the graph of each of the polar equations.

30. $r = \dfrac{2}{1 + \cos\theta}$ 31. $r = 2\cos\theta$ 32. $r = 1 + \cos\theta$

33. $r = 1 - \sin\theta$ 34. $r = 2 - 4\sin\theta$ 35. $r = 3 - 2\cos\theta$

For Problems 36–39, if the equation is given in rectangular form, change it to polar form. If the equation is given in polar form, change it to rectangular form. Also identify each curve using either of the two forms.

36. $r = 1 - \cos\theta$ 37. $y = -\frac{1}{3}x^2$

38. $r = 3\sin\theta$ 39. $x^2 + y^2 - 3y = 2\sqrt{x^2 + y^2}$

40. Sketch the vector and find its magnitude.

    (a) $\langle -3, 3 \rangle$ (b) $\langle -2, -4 \rangle$

41. Find $\mathbf{v} - \mathbf{u}$, $2\mathbf{u} + 5\mathbf{v}$, and $3\mathbf{u} - 2\mathbf{v}$.

    (a) $\mathbf{u} = \langle -5, 4 \rangle$, $\mathbf{v} = \langle -3, -2 \rangle$ (b) $\mathbf{u} = \langle 2, -6 \rangle$, $\mathbf{v} = \langle 4, 6 \rangle$

42. Find the angle of intersection of the two vectors. Express answers to the nearest tenth of a degree.

    (a) $\langle 4, 1 \rangle$ and $\langle 5, -3 \rangle$ (b) $\langle -3, -4 \rangle$ and $\langle -1, 6 \rangle$

6 Exponential and Logarithmic Functions

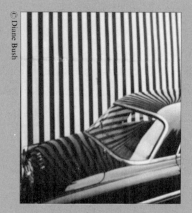

© Diane Bush

In this chapter we will (1) review the concept of an exponent, (2) work with some exponential functions, (3) review the concept of a logarithm, (4) work with some logarithmic functions, and (5) use the concepts of exponent and logarithm to expand our problem solving skills. Your calculator will be a valuable tool throughout this chapter.

6.1

Exponents

Let's begin by reviewing the usual sequential development of the use of real numbers as exponents. Positive integers are used as exponents to indicate repeated multiplication. For example, $4 \cdot 4 \cdot 4$ can be written as 4^3, where the "raised 3" indicates that 4 is to be used as a factor three times. The following general definition can be given.

DEFINITION 6.1

> If n is a positive integer and b is any real number, then
>
> $$b^n = \underbrace{b\,b\,b\ldots b.}_{n \text{ factors of } b}$$

The b is referred to as the **base** and n as the **exponent**. The expression b^n can be read as "b to the nth power." The terms **squared** and **cubed** are commonly associated with exponents of 2 and 3, respectively. For example, b^2 is read "b squared" and b^3 as "b cubed." An exponent of 1 is usually not written, so b^1 is written as b.

The following examples illustrate Definition 6.1.

$$2^3 = 2 \cdot 2 \cdot 2 = 8; \qquad \left(\frac{1}{2}\right)^5 = \frac{1}{2} \cdot \frac{1}{2} \cdot \frac{1}{2} \cdot \frac{1}{2} \cdot \frac{1}{2} = \frac{1}{32};$$

$$3^4 = 3 \cdot 3 \cdot 3 \cdot 3 = 81; \qquad (.7)^2 = (.7)(.7) = .49;$$

$$(-5)^2 = (-5)(-5) = 25; \qquad -5^2 = -(5 \cdot 5) = -25$$

We especially want to call your attention to the last two examples. Note that $(-5)^2$ means -5 is the base and is to be used as a factor twice. However, -5^2 means that 5 is the base and that after it is squared, we take the opposite of that result.

There are various properties that pertain to the use of exponents, but for now we shall state only one of these that is then used to motivate further use of exponents. The property states that: **If a and b are real numbers, and m and n are positive integers, then $b^n \cdot b^m = b^{n+m}$**. Thus, we can write $x^3 \cdot x^4 = x^{3+4} = x^7$, $y^2 \cdot y^8 = y^{2+8} = y^{10}$, $2^5 \cdot 2^6 = 2^{5+6} = 2^{11}$, and so on.

Now we can extend the concept of an exponent to include the use of zero and negative integers as exponents. First, let's consider the use of zero as an exponent. We want to use zero in such a way that the previous property will continue to hold. For example, if $b^n \cdot b^m = b^{n+m}$ is to hold, then $x^4 \cdot x^0$ should equal x^{4+0}, which equals x^4. In other words, x^0 **acts like 1** because $x^4 \cdot x^0 = x^4$. This line of reasoning suggests the following definition.

DEFINITION 6.2

> If b is a nonzero real number, then
>
> $$b^0 = 1.$$

Therefore, according to Definition 6.2 the following statements are all true.

$$5^0 = 1, \qquad (-413)^0 = 1, \qquad \left(\frac{3}{11}\right)^0 = 1,$$

$$(x^3 y^4)^0 = 1 \qquad x \neq 0 \text{ and } y \neq 0$$

A similar line of reasoning can be used to motivate a definition for the use of negative integers as exponents. Consider the example $x^4 \cdot x^{-4}$. If $b^n \cdot b^m = b^{n+m}$ is to hold, then $x^4 \cdot x^{-4}$ should equal $x^{4+(-4)}$, which equals x^0, which equals 1. Therefore, x^{-4} must be the reciprocal of x^4, since their product is 1. That is to say, $x^{-4} = 1/x^4$. This suggests the following definition.

DEFINITION 6.3

If n is a positive integer and b is a nonzero real number, then

$$b^{-n} = \frac{1}{b^n}.$$

According to Definition 6.3 the following statements are true.

$$x^{-5} = \frac{1}{x^5}, \qquad 2^{-4} = \frac{1}{2^4} = \frac{1}{16},$$

$$\left(\frac{3}{4}\right)^{-2} = \frac{1}{\left(\frac{3}{4}\right)^2} = \frac{1}{\frac{9}{16}} = \frac{16}{9}, \qquad \frac{2}{x^{-3}} = \frac{2}{\frac{1}{x^3}} = 2x^3$$

Now let's state a general property that includes the various uses of exponents. We will include "name tags" for easy reference.

PROPERTY 6.1

If m and n are integers and a and b are real numbers, except $b \neq 0$ whenever it appears in a denominator, then

1. $b^n \cdot b^m = b^{n+m}$ Product of two powers
2. $(b^n)^m = b^{mn}$ Power of a power
3. $(ab)^n = a^n b^n$ Power of a product
4. $\left(\dfrac{a}{b}\right)^n = \dfrac{a^n}{b^n}$ Power of a quotient
5. $\dfrac{b^n}{b^m} = b^{n-m}$ Quotient of two powers

Having the use of all integers as exponents allows us to work with a large variety of numerical and algebraic expressions. Let's consider some examples illustrating the various parts of Property 6.1.

EXAMPLE 1

Evaluate each of the following numerical expressions.

(a) $(2^{-1} \cdot 3^2)^{-1}$

(b) $\left(\dfrac{2^{-3}}{3^{-2}}\right)^{-2}$

(c) $(4^{-1} - 3^{-2})^{-1}$

Solution

(a) $(2^{-1} \cdot 3^2)^{-1} = (2^{-1})^{-1}(3^2)^{-1}$ Power of a product
$$= (2^1)(3^{-2})$$ Power of a power
$$= (2)\left(\frac{1}{3^2}\right)$$
$$= 2\left(\frac{1}{9}\right) = \frac{2}{9}$$

(b) $\left(\dfrac{2^{-3}}{3^{-2}}\right)^{-2} = \dfrac{(2^{-3})^{-2}}{(3^{-2})^{-2}}$ Power of a quotient
$$= \frac{2^6}{3^4}$$ Power of a power
$$= \frac{64}{81}$$

(c) $(4^{-1} - 3^{-2})^{-1} = \left(\dfrac{1}{4^1} - \dfrac{1}{3^2}\right)^{-1}$
$$= \left(\frac{1}{4} - \frac{1}{9}\right)^{-1}$$
$$= \left(\frac{9}{36} - \frac{4}{36}\right)^{-1}$$
$$= \left(\frac{5}{36}\right)^{-1} = \frac{1}{\left(\frac{5}{36}\right)^1} = \frac{36}{5}$$

EXAMPLE 2

Find the indicated products and quotients expressing final results using positive integral exponents only.

(a) $(3x^2y^{-4})(4x^{-3}y)$

(b) $\dfrac{12a^3b^2}{-3a^{-1}b^5}$

(c) $\left(\dfrac{15x^{-1}y^2}{5xy^{-4}}\right)^{-1}$

Solutions

(a) $(3x^2y^{-4})(4x^{-3}y) = 12x^{2+(-3)}y^{-4+1}$ Product of powers
$$= 12x^{-1}y^{-3}$$
$$= \frac{12}{xy^3}$$

(b) $\dfrac{12a^3b^2}{-3a^{-1}b^5} = -4a^{3-(-1)}b^{2-5}$ Quotient of powers

$= -4a^4b^{-3}$

$= -\dfrac{4a^4}{b^3}$

(c) $\left(\dfrac{15x^{-1}y^2}{5xy^{-4}}\right)^{-1} = [3x^{-1-1}y^{2-(-4)}]^{-1}$ Notice that we are first simplifying inside the parentheses.

$= [3x^{-2}y^6]^{-1}$

$= 3^{-1}x^2y^{-6}$

$= \dfrac{x^2}{3y^6}$

Now the following definition provides the basis for the use of all rational numbers as exponents.

DEFINITION 6.4

If m/n is a rational number, where n is a positive integer greater than one and m is any integer, and if b is a real number such that $\sqrt[n]{b}$ exists, then

$$b^{m/n} = \sqrt[n]{b^m} = (\sqrt[n]{b})^m.$$

In Definition 6.4, whether we use the form $\sqrt[n]{b^m}$ or $(\sqrt[n]{b})^m$ for computational purposes depends somewhat on the magnitude of the problem. Let's use both forms on two problems to illustrate this point.

$$8^{2/3} = \sqrt[3]{8^2} = \sqrt[3]{64} = 4 \quad \text{or} \quad 8^{2/3} = (\sqrt[3]{8})^2 = 2^2 = 4$$
$$27^{2/3} = \sqrt[3]{27^2} = \sqrt[3]{729} = 9 \quad \text{or} \quad 27^{2/3} = (\sqrt[3]{27})^2 = 3^2 = 9$$

To compute $8^{2/3}$, both forms seem to work equally well. However, to compute $27^{2/3}$, the form $(\sqrt[3]{27})^2$ is much easier to handle. The following examples further illustrate Definition 6.4.

$$25^{3/2} = (\sqrt{25})^3 = 5^3 = 125$$
$$(32)^{-2/5} = \dfrac{1}{(32)^{2/5}} = \dfrac{1}{(\sqrt[5]{32})^2} = \dfrac{1}{2^2} = \dfrac{1}{4}$$
$$(-64)^{2/3} = (\sqrt[3]{-64})^2 = (-4)^2 = 16$$
$$-8^{4/3} = -(\sqrt[3]{8})^4 = -(2)^4 = -16$$

It can be shown that Property 6.1 holds for all rational exponents. Let's consider some examples to illustrate the use of Property 6.1.

$$x^{1/2} \cdot x^{2/3} = x^{1/2+2/3}$$
$$= x^{3/6+4/6} \qquad\qquad b^n \cdot b^m = b^{n+m}$$
$$= x^{7/6}$$

$$(a^{2/3})^{3/2} = a^{(3/2)(2/3)} \qquad\qquad (b^n)^m = b^{mn}$$
$$= a^1 = a$$

$$(16y^{2/3})^{1/2} = (16)^{1/2}(y^{2/3})^{1/2} \qquad (ab)^n = a^n b^n$$
$$= 4y^{1/3}$$

$$\frac{y^{3/4}}{y^{1/2}} = y^{3/4 - 1/2} \qquad\qquad \frac{b^n}{b^m} = b^{n-m}$$
$$= y^{3/4 - 2/4}$$
$$= y^{1/4}$$

$$\left(\frac{x^{1/2}}{y^{1/3}}\right)^6 = \frac{(x^{1/2})^6}{(y^{1/3})^6} \qquad\qquad \left(\frac{a}{b}\right)^n = \frac{a^n}{b^n}$$
$$= \frac{x^3}{y^2}$$

To formally extend the concept of an exponent to include the use of irrational numbers requires some ideas from calculus and is therefore beyond the scope of this text. However, we can give you a brief glimpse at the general idea involved. Consider the number $2^{\sqrt{3}}$. By using the nonterminating and nonrepeating decimal representation $1.73205\ldots$ for $\sqrt{3}$, we can form the sequence of numbers $2^1, 2^{1.7}, 2^{1.73}, 2^{1.732}, 2^{1.7320}, 2^{1.73205}\ldots$. It should seem reasonable that each successive power gets closer to $2^{\sqrt{3}}$. This is precisely what happens if b^n, where n is irrational, is properly defined by using the concept of a limit. Furthermore, this will insure that an expression such as 2^x will yield exactly one value for each value of x. Thus, we can use any real number as an exponent and Property 6.1 remains true.

Finally, another property that can be used to solve certain types of equations involving exponents can be stated as follows.

PROPERTY 6.2

If $b > 0$, $b \neq 1$, and m and n are real numbers, then

$$b^n = b^m \quad \text{if and only if} \quad n = m.$$

The following three examples illustrate the use of Property 6.2.

EXAMPLE 1

Solve $2^x = 32$.

Solution

$$2^x = 32$$
$$2^x = 2^5 \qquad 32 = 2^5$$
$$x = 5 \qquad \text{Apply Property 6.2.}$$

The solution set is $\{5\}$.

EXAMPLE 2

Solve $2^{3x} = \frac{1}{64}$.

Solution

$$2^{3x} = \frac{1}{64}$$

$$2^{3x} = \frac{1}{2^6}$$

$$2^{3x} = 2^{-6}$$

$$3x = -6 \qquad \text{Apply Property 6.2.}$$

$$x = -2$$

The solution set is $\{-2\}$.

EXAMPLE 3

Solve $(\frac{1}{5})^{x-4} = \frac{1}{125}$.

Solution

$$\left(\frac{1}{5}\right)^{x-4} = \frac{1}{125}$$

$$\left(\frac{1}{5}\right)^{x-4} = \left(\frac{1}{5}\right)^3$$

$$x - 4 = 3 \qquad \text{Apply Property 6.2.}$$

$$x = 7$$

The solution set is $\{7\}$.

Problem Set 6.1

For Problems 1–58, evaluate each of the numerical expressions.

1. 2^{-3}

2. 3^{-2}

3. -10^{-3}

4. 10^{-4}

5. $\frac{1}{3^{-3}}$

6. $\frac{1}{2^{-5}}$

7. $\left(\frac{1}{2}\right)^{-2}$

8. $-\left(\frac{1}{3}\right)^{-2}$

9. $\left(-\frac{2}{3}\right)^{-3}$

10. $\left(\frac{5}{6}\right)^{-2}$

11. $\left(-\frac{1}{5}\right)^{0}$

12. $\frac{1}{\left(\frac{3}{5}\right)^{-2}}$

13. $\frac{1}{\left(\frac{4}{5}\right)^{-2}}$

14. $\left(\frac{4}{5}\right)^{0}$

15. $2^5 \cdot 2^{-3}$

16. $3^{-2} \cdot 3^5$

17. $10^{-6} \cdot 10^4$

18. $10^6 \cdot 10^{-9}$

19. $10^{-2} \cdot 10^{-3}$

20. $10^{-1} \cdot 10^{-5}$

21. $(3^{-2})^{-2}$

22. $((-2)^{-1})^{-3}$

23. $(4^2)^{-1}$

24. $(3^{-1})^3$

25. $(3^{-1} \cdot 2^2)^{-1}$

26. $(2^3 \cdot 3^{-2})^{-2}$

27. $(4^2 \cdot 5^{-1})^2$

28. $(2^{-2} \cdot 4^{-1})^3$

29. $\left(\dfrac{2^{-2}}{5^{-1}}\right)^{-2}$

30. $\left(\dfrac{3^{-1}}{2^{-3}}\right)^{-2}$

31. $\left(\dfrac{3^{-2}}{8^{-1}}\right)^2$

32. $\left(\dfrac{4^2}{5^{-1}}\right)^{-1}$

33. $\dfrac{2^3}{2^{-3}}$

34. $\dfrac{2^{-3}}{2^3}$

35. $\dfrac{10^{-1}}{10^4}$

36. $\dfrac{10^{-3}}{10^{-7}}$

37. $3^{-2} + 2^{-3}$

38. $2^{-3} + 5^{-1}$

39. $\left(\dfrac{2}{3}\right)^{-1} - \left(\dfrac{3}{4}\right)^{-1}$

40. $3^{-2} - 2^3$

41. $(2^{-4} + 3^{-1})^{-1}$

42. $(3^{-2} - 5^{-1})^{-1}$

43. $49^{1/2}$

44. $64^{1/3}$

45. $32^{3/5}$

46. $(-8)^{1/3}$

47. $-8^{2/3}$

48. $64^{-1/2}$

49. $\left(\dfrac{1}{4}\right)^{-1/2}$

50. $\left(-\dfrac{27}{8}\right)^{-1/3}$

51. $16^{3/2}$

52. $(.008)^{1/3}$

53. $(.01)^{3/2}$

54. $\left(\dfrac{1}{27}\right)^{-2/3}$

55. $64^{-5/6}$

56. $-16^{5/4}$

57. $\left(\dfrac{1}{8}\right)^{-1/3}$

58. $\left(-\dfrac{1}{8}\right)^{-2/3}$

For Problems 59–84, find the indicated products and quotients; express results using positive exponents only.

59. $(2x^{-1}y^2)(3x^{-2}y^{-3})$

60. $(4x^{-2}y^3)(-5x^3y^{-4})$

61. $(-6a^5y^{-4})(-a^{-7}y)$

62. $(-8a^{-4}b^{-5})(-6a^{-1}b^8)$

63. $\dfrac{24x^{-1}y^{-2}}{6x^{-4}y^3}$

64. $\dfrac{56xy^{-3}}{8x^2y^2}$

65. $\dfrac{-35a^3b^{-2}}{7a^5b^{-1}}$

66. $\dfrac{27a^{-4}b^{-5}}{-3a^{-2}b^{-4}}$

67. $\left(\dfrac{14x^{-2}y^{-4}}{7x^{-3}y^{-6}}\right)^{-2}$

68. $\left(\dfrac{24x^5y^{-3}}{-8x^6y^{-1}}\right)^{-3}$

69. $(3x^{1/4})(5x^{1/3})$

70. $(2x^{2/5})(6x^{1/4})$

71. $(y^{2/3})(y^{-1/4})$

72. $(2x^{1/3})(x^{-1/2})$

73. $(4x^{1/4}y^{1/2})^3$

74. $(5x^{1/2}y)^2$

75. $\dfrac{24x^{3/5}}{6x^{1/3}}$

76. $\dfrac{18x^{1/2}}{9x^{1/3}}$

77. $\dfrac{56a^{1/6}}{8a^{1/4}}$

78. $\dfrac{48b^{1/3}}{12b^{3/4}}$

79. $\left(\dfrac{2x^{1/3}}{3y^{1/4}}\right)^4$

80. $\left(\dfrac{6x^{2/5}}{7y^{2/3}}\right)^2$

81. $\left(\dfrac{x^2}{y^3}\right)^{-1/2}$

82. $\left(\dfrac{a^3}{b^{-2}}\right)^{-1/3}$

83. $\left(\dfrac{4a^2x}{2a^{1/2}x^{1/3}}\right)^3$

84. $\left(\dfrac{3ax^{-1}}{a^{1/2}x^{-2}}\right)^2$

For Problems 85–100, solve each of the equations.

85. $3^x = 27$

86. $2^x = 64$

87. $\left(\dfrac{1}{2}\right)^x = \dfrac{1}{8}$

88. $\left(\dfrac{1}{2}\right)^n = 4$

89. $3^{-x} = \dfrac{1}{81}$

90. $3^{x+1} = 9$

91. $5^{2n-1} = 125$

92. $2^{3-n} = 8$

93. $\left(\dfrac{2}{3}\right)^t = \dfrac{9}{4}$

94. $\left(\dfrac{3}{4}\right)^n = \dfrac{64}{27}$

95. $4^{3x-1} = 256$

96. $16^x = 64$

97. $4^n = 8$

98. $27^{4x} = 9^{x+1}$

99. $32^x = 16^{1-x}$

100. $\left(\dfrac{1}{8}\right)^{-2t} = 2^{t+3}$

6.2

Exponential Functions and Applications

If b is any positive number, then the expression b^x designates exactly one real number for every real value of x. Therefore, the equation $f(x) = b^x$ defines a function whose domain is the set of real numbers. Furthermore, if we place the additional restriction $b \neq 1$, then any equation of the form $f(x) = b^x$ describes a one-to-one function and is called an **exponential function**. This leads to the following definition.

DEFINITION 6.5

If $b > 0$ and $b \neq 1$, then the function f defined by

$$f(x) = b^x,$$

where x is any real number, is called the **exponential function with base b**.

Remark: The function $f(x) = 1^x$ is a constant function and therefore it is not a one-to-one function. Remember from Chapter 3 that one-to-one functions have inverses; this becomes a key issue in a later section.

Now let's consider graphing some exponential functions.

EXAMPLE 1

Graph the function $y = 2^x$.

Solution Let's set up a table of values keeping in mind that x can be any real number. Plotting these points and connecting them with a smooth curve produces Figure 6.1.

x	2^x
-2	$\frac{1}{4}$
-1	$\frac{1}{2}$
0	1
1	2
2	4
3	8

Figure 6.1

In the table for Example 1 we chose integral values for x to keep the computation simple. However, with the use of a calculator we could easily acquire some values for 2^x by using nonintegral exponents. Consider the following additional values rounded to the nearest hundredth.

$$2^{.5} = 1.41, \qquad 2^{1.7} = 3.25,$$
$$2^{-.5} = .71, \qquad 2^{-2.6} = .16$$

Notice that the points generated by these values do fit the graph in Figure 6.1.

EXAMPLE 2

Graph $y = (1/2)^x$.

Solution Again, let's set up a table of values, plot the points, and connect them with a smooth curve. The graph is shown in Figure 6.2.

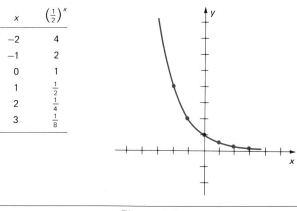

x	$\left(\frac{1}{2}\right)^x$
-2	4
-1	2
0	1
1	$\frac{1}{2}$
2	$\frac{1}{4}$
3	$\frac{1}{8}$

Figure 6.2

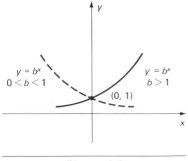

Figure 6.3

The graphs in Figures 6.1 and 6.2 illustrate a "general behavior" pattern of exponential functions; that is to say, if $b > 1$, then the graph of $y = b^x$ "goes up to the right" and the function is called an **increasing function**. If $0 < b < 1$, then the graph of $y = b^x$ "goes down to the right" and the function is called a **decreasing function**. These facts are illustrated in Figure 6.3. Notice that $b^0 = 1$ for any $b > 0$; thus, all graphs of the form $y = b^x$ contain the point $(0, 1)$.

As you graph exponential functions, don't forget the use of previous graphing experiences. Consider the following variations of the basic exponential curve $y = 2^x$.

1. The graph of $y = 2^x + 3$ is the graph of $y = 2^x$ *moved up 3 units.*

2. The graph of $y = 2^{x-4}$ is the graph of $y = 2^x$ *moved to the right 4 units.*

3. The graph of $y = -2^x$ is the graph of $y = 2^x$ *reflected across the x-axis.*

Applications of Exponential Functions

Many real-world situations that exhibit growth or decay can be represented by equations that describe exponential functions. For example, suppose that an economist predicts an annual inflation rate of 5% per year for the next 10 years. This means that an item that presently costs \$8 will cost $8(105\%) = 8(1.05) = \$8.40$ in a year from now. The same item will cost $[8(105\%)](105\%) = 8(1.05)^2 = \$8.82$ in 2 years. In general, the equation

$$P = P_0(1.05)^t$$

yields the predicted price P of an item in t years if the present cost is P_0 and the annual inflation rate is 5%. By using this equation, we can look at some future prices based on the prediction of a 5% inflation rate.

1. A \$3.27 container of hot cocoa mix will cost $\$3.27(1.05)^3 = \$3.79$ in 3 years;

2. A \$4.07 jar of coffee will cost $\$4.07(1.05)^5 = \$5.19$ in 5 years;

3. A \$9500 car will cost $\$9500(1.05)^7 = \$13,367$ (nearest dollar) in 7 years.

Suppose that it is estimated that the value of a car depreciates 15% per year for the first five years. Therefore, a car costing \$9500 will be worth $\$9500(100\% - 15\%) = \$9500(85\%) = \$9500(.85) = \$8075$ in one year. In two years the value of the car will have depreciated to $9500(.85)^2 = \$6864$ (nearest dollar). The equation

$$V = V_0(.85)^t$$

yields the value V of a car in t years if the initial cost is V_0 and it depreciates 15% per year. Therefore, we can estimate some car values to the nearest dollar as follows.

1. A \$6900 car will be worth $\$6900(.85)^3 = \$4237$ in 3 years.

2. A \$10,900 car will be worth $\$10900(.85)^4 = \$5690$ in 4 years.

3. A \$13,000 car will be worth $\$13000(.85)^5 = \$5768$ in 5 years.

Compound Interest

Compound interest provides another illustration of exponential growth. Suppose that $500 (called the **principal**) is invested at an interest rate of 8% **compounded annually**. The interest earned the first year is $500(.08) = $40 and this amount is added to the original $500 to form a new principal of $540 for the second year. The interest earned during the second year is $540(.08) = $43.20 and this amount is added to $540 to form a new principal of $583.20 for the third year. Each year a new principal is formed by reinvesting the interest earned during that year.

In general, suppose that a sum of money P (called the principal) is invested at an interest rate of r percent compounded annually. The interest earned the first year is Pr and the new principal for the second year is $P + Pr$ or $P(1 + r)$. Note that the new principal for the second year can be found by multiplying the original principal P by $(1 + r)$. In a like fashion, the new principal for the third year can be found by multiplying the previous principal $P(1 + r)$ by $1 + r$, thus obtaining $P(1 + r)^2$. If this process is continued, then after t years the total amount of money accumulated, A, is given by

$$A = P(1 + r)^t.$$

Consider the following examples of investments made at a certain rate of interest compounded annually.

1. $750 invested for 5 years at 9% compounded annually produces

$$A = \$750(1.09)^5 = \$1153.97.$$

2. $1000 invested for 10 years at 11% compounded annually produces

$$A = \$1000(1.11)^{10} = \$2839.42.$$

3. $5000 invested for 20 years at 12% compounded annually produces

$$A = \$5000(1.12)^{20} = \$48{,}231.47.$$

If money invested at a certain rate of interest is to be compounded more than once a year, then the basic formula, $A = P(1 + r)^t$, can be adjusted according to the number of compounding periods in a year. For example, when **compounding semiannually**, the formula becomes $A = P[1 + (r/2)]^{2t}$ and when **compounding quarterly**, the formula becomes $A = P[1 + (r/4)]^{4t}$. In general, if n represents the number of compounding periods in a year, the formula becomes

$$A = P\left(1 + \frac{r}{n}\right)^{nt}.$$

The following examples illustrate the use of the formula.

1. $750 invested for 5 years at 9% compounded semiannually produces

$$A = \$750\left(1 + \frac{.09}{2}\right)^{2(5)} = \$750(1.045)^{10} = \$1164.73.$$

2. $1000 invested for 10 years at 11% compounded quarterly produces

$$A = \$1000\left(1 + \frac{.11}{4}\right)^{4(10)} = \$1000(1.0275)^{40} = \$2959.87.$$

3. $5000 invested for 20 years at 12% compounded monthly produces

$$A = \$5000\left(1 + \frac{.12}{12}\right)^{12(20)} = \$5000(1.01)^{240} = \$54,462.77.$$

You may find it interesting to compare these results with those obtained earlier for compounding annually.

The Number e

n	$\left(1 + \frac{1}{n}\right)^n$
5	2.4883200
10	2.5937425
100	2.7048138
1000	2.7169236
10,000	2.7181459
100,000	2.7182818

An interesting situation occurs if we consider the compound interest formula for $P = \$1, r = 100\%$, and $t = 1$ year. The formula becomes $A = 1[1 + (1/n)]^n$. The accompanying table shows some values, rounded to seven decimals, for different values of n. The table suggests that as n increases, the value of $[1 + (1/n)]^n$ gets closer to some fixed number. This does happen and the fixed number is called e. To five decimal places,

$$e = 2.71828.$$

Exponential expressions using e as a base are found in many real-world applications. Before considering some of these applications, let's take a look at the graph of the basic exponential function using e as a base, namely, the function $y = e^x$.

EXAMPLE 3

Graph $y = e^x$.

Solution For graphing purposes, let's use 2.72 as an approximation for e and express the functional values to the nearest tenth. The table on the left can be easily obtained by using a calculator. The graph of $y = e^x$ is shown in Figure 6.4.

x	$(2.72)^x$
0	1
1	2.7
2	7.4
−1	0.4
−2	0.1

Let's return to the concept of compound interest. If the number of compounding periods in a year is increased indefinitely, we arrive at the concept of **compounding continuously**. Mathematically, this can be accomplished by applying the limit concept to the expression $P[1 + (r/n)]^{nt}$. We will not show the details here, but the following result is obtained. The formula

$$A = Pe^{rt}$$

yields the accumulated value (A) of a sum of money (P) that has been invested for t years at a rate of r percent compounded continuously. The following examples illustrate the use of this formula. (We are using 2.718 as an approximation for e in these calculations.)

1. $750 invested for 5 years at 9% compounded continuously produces

$$A = \$750(2.718)^{(.09)(5)} = \$750(2.718)^{.45} = \$1176.18.$$

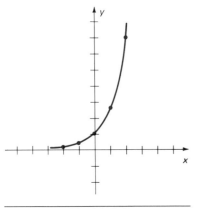

Figure 6.4

2. $1000 invested for 10 years at 11% compounded continuously produces

$$A = \$1000(2.718)^{(.11)(10)} = \$1000(2.718)^{1.1} = \$3003.82.$$

3. $5000 invested for 20 years at 12% compounded continuously produces

$$A = \$5000(2.718)^{(.12)(20)} = \$5000(2.718)^{2.4} = \$55,102.17.$$

Again you may find it interesting to compare these results to those obtained earlier using a different number of compounding periods.

Law of Exponential Growth

The ideas behind compounding continuously carry over to other growth situations. The equation

$$Q(t) = Q_0 e^{kt} \qquad \text{Law of exponential growth}$$

is used as a mathematical model for numerous growth and decay applications. In this equation, $Q(t)$ represents the quantity of a given substance at any time t, Q_0 is the initial amount of the substance (when $t = 0$), and k is a constant that depends on the particular application. If $k < 0$, then $Q(t)$ decreases as t increases and the model is referred to as the **law of decay**.

Let's consider some growth and decay applications.

EXAMPLE 4

Suppose that in a certain culture, the equation $Q(t) = 15{,}000e^{.3t}$ expresses the number of bacteria present as a function of the time, where t is expressed in hours. How many bacteria are present at the end of 3 hours?

Solution Using $Q(t) = 15{,}000e^{.3t}$, we obtain

$$\begin{aligned}
Q(3) &= 15{,}000(2.718)^{.3(3)} \\
&= 15{,}000(2.718)^{.9} \\
&= 36{,}981, \text{ to the nearest whole number.}
\end{aligned}$$

EXAMPLE 5

Suppose the number of bacteria present in a certain culture after t minutes is given by the equation $Q(t) = Q_0 e^{.05t}$, where Q_0 represents the initial number of bacteria. If 5000 bacteria are present after 20 minutes, how many bacteria were present initially?

Solution Since $Q(20) = 5000$, we obtain

$$5000 = Q_0(2.718)^{(.05)(20)}$$

$$5000 = Q_0(2.718)^{1}$$

$$\frac{5000}{2.718} = Q_0$$

$$1840 = Q_0, \text{ to the nearest whole number.}$$

Therefore, there were approximately 1840 bacteria present initially.

EXAMPLE 6

The number of grams of a certain radioactive substance present after t seconds is given by the equation $Q(t) = 200e^{-.3t}$. How many grams remain after 7 seconds?

Solution Using $Q(t) = 200e^{-.3t}$, we obtain

$$Q(7) = 200e^{(-.3)(7)}$$
$$= 200e^{-2.1}$$
$$= 200(2.718)^{-2.1}$$
$$= 24.5, \text{ to the nearest tenth.}$$

Thus, approximately 24.5 grams remain after 7 seconds.

The concept of **half-life** plays an important role in many decay type problems. The **half-life** of a substance is the time required for one-half of the substance to disintegrate or disappear. For example, the half-life of radium is approximately 1600 years. Thus, if 2000 grams of radium are present now, 1600 years from now 1000 grams will be present.

EXAMPLE 7

Suppose that a certain substance has a half-life of 5 years. If there are presently 100 grams of the substance, then the equation $Q(t) = 100(2)^{-t/5}$ yields the amount remaining after t years. How much remains after 10 years? 18 years?

Solution

$$Q(10) = 100(2)^{-10/5}$$
$$= 100(2)^{-2}$$
$$= 100\left(\frac{1}{4}\right)$$
$$= 25$$

Therefore, 25 grams remain after 10 years.

$$Q(18) = 100(2)^{-18/5}$$
$$= 100(2)^{-3.6}$$
$$= 8, \text{ to the nearest whole number}$$

Therefore, approximately 8 grams remain after 18 years.

Problem Set 6.2

For Problems 1–26, graph each of the exponential functions.

1. $y = 3^x$
2. $y = \left(\frac{1}{3}\right)^x$
3. $y = 4^x$
4. $y = \left(\frac{1}{4}\right)^x$
5. $y = \left(\frac{2}{3}\right)^x$
6. $y = \left(\frac{3}{2}\right)^x$
7. $y = 2^x + 1$
8. $y = 2^x - 3$
9. $y = 2^{x-1}$

10. $y = 2^{x+2}$ 11. $y = -3^x$ 12. $y = -2^x$

13. $y = 2^{-x+1}$ 14. $y = 2^{-x-2}$ 15. $y = 2^x + 2^{-x}$

16. $y = 2^{x^2}$ 17. $y = 3^{1-x^2}$ 18. $y = 2^{|x|}$

19. $y = 2^{-|x|}$ 20. $y = 2^x - 2^{-x}$ 21. $y = e^x + 1$

22. $y = e^x - 2$ 23. $y = 2e^x$ 24. $y = -e^x$

25. $y = e^{2x}$ 26. $y = e^{-x}$

27. Assuming that the rate of inflation is 7% per year, the equation $P = P_0(1.07)^t$ yields the predicted price P of an item in t years that presently costs P_0. Find the predicted price of each of the following items for the indicated years ahead.

(a) \$.55 can of soup in 3 years

(b) \$3.43 container of cocoa mix in 5 years

(c) \$1.76 jar of coffee-creamer in 4 years

(d) \$.44 can of beans and bacon in 10 years

(e) \$9000 car in 5 years (nearest dollar)

(f) \$50,000 house in 8 years (nearest dollar)

(g) \$500 TV set in 7 years (nearest dollar)

28. Suppose that it is estimated that the value of a car depreciates 20% per year for the first 5 years. The equation $A = P_0(.8)^t$ yields the value (A) of a car after t years if the original price is P_0. Find the value (to the nearest dollar) of each of the following priced cars after the indicated time.

(a) \$9000 car after 4 years

(b) \$5295 car after 2 years

(c) \$6395 car after 5 years

(d) \$15,595 car after 3 years

For Problems 29–38, use the formula $A = P[1 + (r/n)]^{nt}$ to find the total amount of money accumulated at the end of the indicated time period for each of the following investments.

29. \$250 for 5 years at 9% compounded annually

30. \$350 for 7 years at 11% compounded annually

31. \$300 for 6 years at 8% compounded semiannually

32. \$450 for 10 years at 10% compounded semiannually

33. \$600 for 12 years at 12% compounded quarterly

34. \$750 for 15 years at 9% compounded quarterly

35. \$1000 for 5 years at 12% compounded monthly

36. \$1250 for 8 years at 9% compounded monthly

37. \$600 for 10 years at $8\frac{1}{2}$% compounded annually

38. \$1500 for 15 years at $9\frac{1}{4}$% compounded semiannually

For Problems 39–44, use the formula $A = Pe^{rt}$ to find the total amount of money accumulated at the end of the indicated time period by compounding continuously. Use 2.718 as an approximation for e.

39. \$500 for 5 years at 8%   40. \$750 for 7 years at 9%

41. \$800 for 10 years at 10%   42. \$1000 for 10 years at $9\frac{1}{2}$%

43. \$1500 for 12 years at 9%   44. \$2000 for 20 years at 11%

45. Complete the following chart, which will illustrate what happens to $1000 in 10 years based on different rates of interest and different numbers of compounding periods. Round your answers to the nearest dollar.

$1000 for 10 Years

	8%	10%	12%	14%
COMPOUNDED ANNUALLY				
COMPOUNDED SEMIANNUALLY				
COMPOUNDED QUARTERLY				
COMPOUNDED MONTHLY				
COMPOUNDED CONTINUOUSLY				

46. Complete the following chart, which will illustrate what happens to $1000 invested at 12% for different lengths of time and different numbers of compounding periods. Round all of your answers to the nearest dollar.

$1000 at 12%

	1 YEAR	5 YEARS	10 YEARS	20 YEARS
COMPOUNDED ANNUALLY				
COMPOUNDED SEMIANNUALLY				
COMPOUNDED QUARTERLY				
COMPOUNDED MONTHLY				
COMPOUNDED CONTINUOUSLY				

47. Complete the following chart, which will illustrate what happens to $1000 invested at various rates of interest for different lengths of time, always compounded continuously.

$1000 Compounded Continuously

	8%	10%	12%	14%
5 years				
10 years				
15 years				
20 years				
25 years				

48. Suppose that in a certain culture, the equation $Q(t) = 1000e^{.4t}$ expresses the number of bacteria present as a function of the time t, where t is expressed in hours. How many bacteria are present at the end of 2 hours?   3 hours?   5 hours?

49. The number of bacteria present at a given time under certain conditions is given by the equation $Q(t) = 5000e^{.05t}$, where t is expressed in minutes. How many bacteria are present at the end of 10 minutes?   30 minutes?   1 hour?

50. The number of bacteria present in a certain culture after t hours is given by the equation $Q(t) = Q_0e^{.3t}$, where Q_0 represents the initial number of bacteria. If 6640 bacteria are present after 4 hours, how many bacteria were present initially?

51. The number of grams of a certain radioactive substance present after t seconds is given by the equation $Q(t) = 1500e^{-.4t}$. How many grams remain after 5 seconds? 10 seconds?   20 seconds?

52. The atmospheric pressure, measured in pounds per square inch, is a function of the

altitude above sea level. The equation $P(a) = 14.7e^{-.21a}$, where a is the altitude measured in miles, can be used to approximate atmospheric pressure. Find the atmospheric pressure at each of the following locations.

(a) Mount McKinley in Alaska—altitude of 3.85 miles

(b) Denver, Colorado—"the mile-high" city

(c) Asheville, North Carolina—altitude of 1985 feet

(d) Phoenix, Arizona—altitude of 1090 feet

53. Suppose that the present population of a city is 75,000 and the equation $P(t) = 75000e^{.01t}$ is used to estimate future growth. Estimate the population (a) 10 years from now, (b) 15 years from now, and (c) 25 years from now.

54. Suppose that a certain substance has a half-life of 20 years. If there are presently 2500 milligrams of the substance, then the equation $Q(t) = 2500(2)^{-t/20}$ yields the amount remaining after t years. How much remains after 40 years?   50 years?

55. Assume that the half-life of a certain radioactive substance is 1200 years and that the amount of the substance remaining after t years is given by the equation $Q(t) = 500(2)^{-t/1200}$. If the substance is measured in grams, how many grams are present now? How many grams will remain after 1500 years?

56. The half-life of radium is approximately 1600 years. If the initial amount is Q_0 milligrams, then the quantity $Q(t)$ remaining after t years is given by $Q(t) = Q_0 2^{kt}$. Find the value of k.

57. The half-life of a certain substance is 25 years. If the initial amount is Q_0 grams, then the quantity $Q(t)$ remaining after t years is given by $Q(t) = Q_0(2)^{kt}$. Find the value of k.

6.3

Logarithms

In Sections 6.1 and 6.2, (1) we gave meaning to exponential expressions of the form b^n, where b is any positive real number and n is any real number, (2) we used exponential expressions of the form b^n to define exponential functions, and (3) we used exponential functions to help solve problems. In the next three sections we will follow the same basic pattern with respect to a new concept, that of a **logarithm**. Let's begin with the following definition.

DEFINITION 6.6

If r is any positive real number, then the unique exponent t such that $b^t = r$ is called the **logarithm of r with base b** and is denoted by $\log_b r$.

According to Definition 6.6, the logarithm of 16 base 2 is the exponent t such that $2^t = 16$; thus, we can write $\log_2 16 = 4$. Likewise, we can write $\log_{10} 1000 = 3$ because $10^3 = 1000$. In general, Definition 6.6 can be remembered in terms of the statement

$$\log_b r = t \quad \text{is equivalent to} \quad b^t = r.$$

Therefore, we can easily switch back and forth between exponential and logarithmic forms of equations, as the next examples illustrate.

$$\log_2 8 = 3 \quad \text{is equivalent to} \quad 2^3 = 8$$

$$\log_{10} 100 = 2 \quad \text{is equivalent to} \quad 10^2 = 100$$

$$\log_3 81 = 4 \quad \text{is equivalent to} \quad 3^4 = 81$$

$$\log_{10}.001 = -3 \quad \text{is equivalent to} \quad 10^{-3} = .001$$

$$2^7 = 128 \quad \text{is equivalent to} \quad \log_2 128 = 7$$

$$5^3 = 125 \quad \text{is equivalent to} \quad \log_5 125 = 3$$

$$\left(\frac{1}{2}\right)^4 = \frac{1}{16} \quad \text{is equivalent to} \quad \log_{1/2}\left(\frac{1}{16}\right) = 4$$

$$10^{-2} = .01 \quad \text{is equivalent to} \quad \log_{10}.01 = -2$$

Some logarithms can be determined by changing to exponential form and using the properties of exponents, as the next two examples illustrate.

EXAMPLE 1

Evaluate $\log_{10}.0001$.

Solution Let $\log_{10}.0001 = x$. Then by changing to exponential form we have $10^x = .0001$, which can be solved as follows.

$$10^x = .0001 = 10^{-4} \qquad \left(.0001 = \frac{1}{10000} = \frac{1}{10^4} = 10^{-4}\right)$$

$$x = -4$$

Thus, we have $\log_{10}.0001 = -4$.

EXAMPLE 2

Evaluate $\log_9(\sqrt[5]{27}/3)$.

Solution Let $\log_9(\sqrt[5]{27}/3) = x$. Then by changing to exponential form we have $9^x = \sqrt[5]{27}/3$, which can be solved as follows.

$$9^x = \frac{(27)^{1/5}}{3}$$

$$(3^2)^x = \frac{(3^3)^{1/5}}{3}$$

$$3^{2x} = \frac{3^{3/5}}{3}$$

$$3^{2x} = 3^{-2/5}$$

$$2x = -\tfrac{2}{5}$$

$$x = -\tfrac{1}{5}$$

Therefore, we have $\log_9(\sqrt[5]{27}/3) = -\tfrac{1}{5}$.

Some equations involving logarithms can also be solved by changing to exponential form and using our knowledge of exponents.

EXAMPLE 3

Solve $\log_8 x = \frac{2}{3}$.

Solution Changing $\log_8 x = \frac{2}{3}$ to exponential form we obtain

$$8^{2/3} = x.$$

Therefore,

$$x = (\sqrt[3]{8})^2$$
$$= 2^2$$
$$= 4.$$

The solution set is $\{4\}$.

EXAMPLE 4

Solve $\log_b(\frac{27}{64}) = 3$.

Solution Changing $\log_b(\frac{27}{64}) = 3$ to exponential form we obtain

$$b^3 = \frac{27}{64}.$$

Therefore,

$$b = \sqrt[3]{\frac{27}{64}}$$
$$= \frac{3}{4}.$$

The solution set is $\{\frac{3}{4}\}$.

Properties of Logarithms

There are some properties of logarithms that are a direct consequence of Definition 6.6 and the properties of exponents. For example, by writing the exponential equations $b^1 = b$ and $b^0 = 1$ in logarithmic form, the following property is obtained.

PROPERTY 6.3

For $b > 0$ and $b \neq 1$,

1. $\log_b b = 1$ 2. $\log_b 1 = 0$.

Therefore, according to Property 6.3 we can write

$$\log_{10} 10 = 1, \quad \log_4 4 = 1,$$
$$\log_{10} 1 = 0, \quad \text{and} \quad \log_5 1 = 0.$$

Also from Definition 6.6 we know that $\log_b r$ is the exponent t such that $b^t = r$. Therefore, raising b to the $\log_b r$ power must produce r. This fact is stated in Property 6.4.

PROPERTY 6.4

> For $b > 0$, $b \neq 1$, and $r > 0$
>
> $$b^{\log_b r} = r.$$

Therefore, according to Property 6.4 we can write

$$10^{\log_{10} 72} = 72, \quad 3^{\log_3 85} = 85,$$

and $\quad e^{\log_e 7} = 7.$

Because a logarithm is by definition an exponent, it would seem reasonable to predict that there are some properties of logarithms that correspond to the basic exponential properties. This is an accurate prediction; these properties provide a basis for computational work with logarithms. Let's state the first of these properties and show how it can be verified by using our knowledge of exponents.

PROPERTY 6.5

> For positive numbers b, r, and s where $b \neq 1$,
>
> $$\log_b rs = \log_b r + \log_b s.$$

To verify Property 6.5 we can proceed as follows.

Let $m = \log_b r$ and $n = \log_b s$. Change each of these equations to exponential form.

$$m = \log_b r \quad \text{becomes} \quad r = b^m;$$
$$n = \log_b s \quad \text{becomes} \quad s = b^n.$$

Thus, the product rs becomes

$$rs = b^m \cdot b^n = b^{m+n}.$$

Now, by changing $rs = b^{m+n}$ back to logarithmic form, we obtain

$$\log_b rs = m + n.$$

Replacing m with $\log_b r$ and n with $\log_b s$ yields

$$\log_b rs = \log_b r + \log_b s.$$

The following two examples illustrate one use of Property 6.5.

EXAMPLE 5

If $\log_2 5 = 2.3222$ and $\log_2 3 = 1.5850$, evaluate $\log_2 15$.

Solution Because $15 = 5 \cdot 3$, we can apply Property 6.5 as follows:

$$\log_2 15 = \log_2(5 \cdot 3)$$
$$= \log_2 5 + \log_2 3$$
$$= 2.3222 + 1.5850 = 3.9072.$$

EXAMPLE 6

If $\log_{10} 178 = 2.2504$ and $\log_{10} 89 = 1.9494$, evaluate $\log_{10}(178 \cdot 89)$.

Solution

$$\log_{10}(178 \cdot 89) = \log_{10} 178 + \log_{10} 89$$
$$= 2.2504 + 1.9494$$
$$= 4.1998.$$

Since $b^m / b^n = b^{m-n}$, we would expect a corresponding property pertaining to logarithms. Property 6.6 is that property. It can be verified by using an approach similar to the one used to verify Property 6.5. This verification is left for you to do as an exercise in the next problem set.

PROPERTY 6.6

For positive numbers b, r, and s where $b \neq 1$,

$$\log_b\left(\frac{r}{s}\right) = \log_b r - \log_b s.$$

Property 6.6 can be used to change a division problem into an equivalent subtraction problem as the next two examples illustrate.

EXAMPLE 7

If $\log_5 36 = 2.2265$ and $\log_5 4 = .8614$, evaluate $\log_5 9$.

Solution Since $9 = \frac{36}{4}$, we can use Property 6.6 as follows:

$$\log_5 9 = \log_5\left(\frac{36}{4}\right)$$
$$= \log_5 36 - \log_5 4$$
$$= 2.2265 - .8614 = 1.3651.$$

EXAMPLE 8

Evaluate $\log_{10}(\frac{379}{86})$ given that $\log_{10} 379 = 2.5786$ and $\log_{10} 86 = 1.9345$.

Solution

$$\log_{10}(\tfrac{379}{86}) = \log_{10} 379 - \log_{10} 86$$
$$= 2.5786 - 1.9345$$
$$= .6441.$$

Another property of exponents states that $(b^n)^m = b^{mn}$. The corresponding property of logarithms is stated in Property 6.7. Again we will leave the verification of this property as an exercise for you to do in the next set of problems.

PROPERTY 6.7

If r is a positive real number, b is a positive real number other than 1, and p is any real number, then

$$\log_b r^p = p(\log_b r).$$

The next two examples illustrate a use of Property 6.7.

EXAMPLE 9

Evaluate $\log_2 22^{1/3}$ given that $\log_2 22 = 4.4598$.
Solution

$$\log_2 22^{1/3} = \frac{1}{3}\log_2 22 \qquad \text{Property 6.7}$$

$$= \frac{1}{3}(4.4598)$$

$$= 1.4866.$$

EXAMPLE 10

Evaluate $\log_{10}(8540)^{3/5}$ given that $\log_{10} 8540 = 3.9315$.
Solution

$$\log_{10}(8540)^{3/5} = \frac{3}{5}\log_{10} 8540$$

$$= \frac{3}{5}(3.9315)$$

$$= 2.3589.$$

Working together, the properties of logarithms allow us to change the forms of various logarithmic expressions. For example, an expression such as $\log_b \sqrt{xy/z}$ can be rewritten in terms of sums and differences of simpler logarithmic quantities as follows.

$$\log_b \sqrt{\frac{xy}{z}} = \log_b \left(\frac{xy}{z}\right)^{1/2}$$

$$= \frac{1}{2}\log_b \left(\frac{xy}{z}\right) \qquad \text{Property 6.7}$$

$$= \frac{1}{2}(\log_b xy - \log_b z) \qquad \text{Property 6.6}$$

$$= \frac{1}{2}(\log_b x + \log_b y - \log_b z) \qquad \text{Property 6.5}$$

The properties of logarithms along with the link between logarithmic form and exponential form provide the basis for solving certain types of equations involving logarithms. Our final example of this section illustrates this idea.

EXAMPLE 11

Solve $\log_{10}x + \log_{10}(x + 9) = 1$.

Solution

$$\log_{10}x + \log_{10}(x + 9) = 1.$$

$$\log_{10}[x(x + 9)] = 1 \qquad\qquad \text{Property 6.5}$$

$$10^1 = x(x + 9) \qquad\qquad \text{Change to exponential form.}$$

$$10 = x^2 + 9x$$

$$0 = x^2 + 9x - 10$$

$$0 = (x + 10)(x - 1)$$

$$x + 10 = 0 \qquad \text{or} \quad x - 1 = 0$$

$$x = -10 \quad \text{or} \qquad x = 1.$$

Since the left-hand number of the original equation is meaningful only if $x > 0$ and $x + 9 > 0$, the solution -10 must be discarded. Thus, the solution set is $\{1\}$.

Problem Set 6.3

Write each of the following in logarithmic form. For example, $2^4 = 16$ becomes $\log_2 16 = 4$.

1. $3^2 = 9$
2. $2^5 = 32$
3. $5^3 = 125$

4. $10^1 = 10$
5. $2^{-4} = \dfrac{1}{16}$
6. $\left(\dfrac{2}{3}\right)^{-3} = \dfrac{27}{8}$

7. $10^{-2} = .01$
8. $10^5 = 100,000$

Write each of the following in exponential form. For example, $\log_2 8 = 3$ becomes $2^3 = 8$.

9. $\log_2 64 = 6$
10. $\log_3 27 = 3$
11. $\log_{10}.1 = -1$

12. $\log_5\left(\dfrac{1}{25}\right) = -2$
13. $\log_2\left(\dfrac{1}{16}\right) = -4$
14. $\log_{10}.00001 = -5$

Evaluate each of the following.

15. $\log_6 36$
16. $\log_3 243$
17. $\log_5\left(\dfrac{1}{5}\right)$
18. $\log_4\left(\dfrac{1}{64}\right)$

19. $\log_{10} 10$
20. $\log_{10} 1$
21. $\log_3 \sqrt{3}$
22. $\log_5 \sqrt[3]{25}$

23. $\log_3\left(\dfrac{\sqrt{27}}{3}\right)$
24. $\log_{1/2}\left(\dfrac{\sqrt[4]{8}}{2}\right)$
25. $\log_{1/4}\left(\dfrac{\sqrt[4]{32}}{2}\right)$
26. $\log_2\left(\dfrac{\sqrt[3]{16}}{4}\right)$

27. $10^{\log_{10}7}$ 28. $5^{\log_5 13}$ 29. $\log_2(\log_5 5)$ 30. $\log_6(\log_2 64)$

Solve each of the following equations.

31. $\log_5 x = 2$ 32. $\log_{10} x = 3$ 33. $\log_8 t = \dfrac{5}{3}$ 34. $\log_4 m = \dfrac{3}{2}$

35. $\log_b 3 = \dfrac{1}{2}$ 36. $\log_b 2 = \dfrac{1}{2}$ 37. $\log_{10} x = 0$ 38. $\log_{10} x = 1$

Given that $\log_{10} 2 = .3010$ and $\log_{10} 7 = .8451$, evaluate each of the following by using Properties 6.5–6.7.

39. $\log_{10} 14$ 40. $\log_{10}\left(\dfrac{7}{2}\right)$ 41. $\log_{10} 4$ 42. $\log_{10} 49$

43. $\log_{10} 343$ 44. $\log_{10} 32$ 45. $\log_{10}\sqrt{2}$ 46. $\log_{10}\sqrt[3]{7}$

47. $\log_{10}(7)^{4/3}$ 48. $\log_{10}(2)^{3/5}$ 49. $\log_{10} 28$ 50. $\log_{10} 56$

51. $\log_{10} 98$ 52. $\log_{10} 20$ 53. $\log_{10} 200$ 54. $\log_{10} 70$

55. $\log_{10} 1400$ 56. $\log_{10} 4900$

Express each of the following as the sum or difference of simpler logarithmic quantities. (Assume that all variables represent positive real numbers.) For example,

$$\log_b\left(\frac{x^3}{y^2}\right) = \log_b x^3 - \log_b y^2$$
$$= 3\log_b x - 2\log_b y.$$

57. $\log_b xyz$ 58. $\log_b\left(\dfrac{x^2}{y}\right)$ 59. $\log_b x^2 y^3$ 60. $\log_b x^{2/3} y^{3/4}$

61. $\log_b\sqrt{xy}$ 62. $\log_b\sqrt[3]{x^2 z}$ 63. $\log_b\sqrt{\dfrac{x}{y}}$ 64. $\log_b\left(x\sqrt{\dfrac{x}{y}}\right)$

Express each of the following as a single logarithm. (Assume that all variables represent positive real numbers.) For example,

$$3\log_b x + 5\log_b y = \log_b x^3 y^5.$$

65. $\log_b x + \log_b y - \log_b z$ 66. $2\log_b x - 4\log_b y$

67. $(\log_b x - \log_b y) - \log_b z$ 68. $\log_b x - (\log_b y - \log_b z)$

69. $\log_b x + \dfrac{1}{2}\log_b y$ 70. $2\log_b x + 4\log_b y - 3\log_b z$

71. $2\log_b x + \dfrac{1}{2}\log_b(x - 1) - 4\log_b(2x + 5)$

72. $\dfrac{1}{2}\log_b x - 3\log_b x + 4\log_b y$

Solve each of the following equations.

73. $\log_{10} 5 + \log_{10} x = 1$ 74. $\log_{10} x + \log_{10} 25 = 2$

75. $\log_{10} 20 - \log_{10} x = 1$ 76. $\log_{10} x + \log_{10}(x - 3) = 1$

77. $\log_{10}(x + 2) - \log_{10} x = 1$ 78. $\log_{10} x + \log_{10}(x - 21) = 2$

79. $\log_{10}(x - 4) + \log_{10}(x - 1) = 1$ 80. $\log_{10}(x + 2) + \log_{10}(x - 1) = 1$

81. Verify Property 6.6.   82. Verify Property 6.7.

6.4

Logarithmic Functions

The concept of a logarithm can now be used to define a logarithmic function as follows.

DEFINITION 6.7

If $b > 0$ and $b \neq 1$, then the function defined by

$$f(x) = \log_b x,$$

where x is any positive real number, is called the **logarithmic function with base b**.

x	y
$\frac{1}{8}$	-3
$\frac{1}{4}$	-2
$\frac{1}{2}$	-1
1	0
2	1
4	2
8	3

We can obtain the graph of a specific logarithmic function in various ways. For example, the equation $y = \log_2 x$ can be changed to the exponential equation $2^y = x$ and a table of values can be determined. We will ask you to graph some logarithmic functions using this approach in the next set of exercises.

The graph of a logarithmic function can also be obtained by setting up a table of values directly from the logarithmic equation. Let's illustrate this approach.

EXAMPLE 1

Graph $y = \log_2 x$.

Solution Let's choose some values for x where the corresponding values for $\log_2 x$ are easily determined. (Remember that logarithms are only defined for the positive real numbers.) Plotting these points and connecting them with a smooth curve produces Figure 6.5.

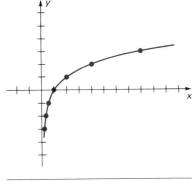

Figure 6.5

Now suppose that we consider two functions f and g as follows:

$f(x) = b^x$ 	 Domain: all real numbers

Range: positive real numbers

$g(x) = \log_b x$ 	 Domain: positive real numbers

Range: all real numbers

Furthermore, suppose that we consider the composition of f and g, and the composition of g and f.

$$(f \circ g)(x) = f(g(x)) = f(\log_b x) = b^{\log_b x} = x$$

$$(g \circ f)(x) = g(f(x)) = g(b^x) = \log_b b^x = x \log_b b = x(1) = x$$

Therefore, because the domain of f is the range of g, the range of f is the domain of g, $f(g(x)) = x$, and $g(f(x)) = x$, the two functions **f and g are inverses of each other**.

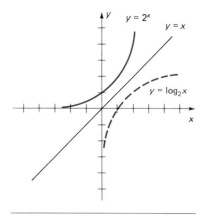

Figure 6.6

Remember also from Chapter 3 that the graphs of a function and its inverse are reflections of each other through the line $y = x$. Thus, the graph of a logarithmic function can also be determined by reflecting the graph of its inverse exponential function through the line $y = x$. This idea is illustrated in Figure 6.6, where the graph of $y = 2^x$ has been reflected across the line $y = x$ to produce the graph of $y = \log_2 x$.

The **general behavior patterns of exponential functions** were illustrated by two graphs back in Figure 6.3. We can now reflect each of these graphs through the line $y = x$ and observe the **general behavior patterns of logarithmic functions** as shown in Figure 6.7.

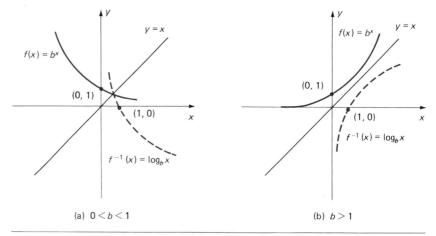

Figure 6.7

Common Logarithms: Base 10

The properties of logarithms discussed in Section 6.3 are true for any valid base. However, since the Hindu-Arabic numeration system that we use is a base 10 system, logarithms to base 10 have historically been used for computational purposes. Base 10 logarithms are called **common logarithms**.

Originally, common logarithms were developed as an aid to numerical calculations. Today they are seldom used for that purpose because the calculator and computer can more effectively handle the messy computational problems. Thus, in this section we will restrict our discussion to evaluating common logarithms and then in an optional section at the end of this chapter we will illustrate a few of their computational characteristics.

As we know from earlier work, the definition of a logarithm provides the basis for evaluating $\log_{10} x$ for values of x that are integral powers of 10. Consider the following examples.

$$\log_{10}1000 = 3 \quad \text{because} \quad 10^3 = 1000$$
$$\log_{10}100 = 2 \quad \text{because} \quad 10^2 = 100$$
$$\log_{10}10 = 1 \quad \text{because} \quad 10^1 = 10$$
$$\log_{10}1 = 0 \quad \text{because} \quad 10^0 = 1$$
$$\log_{10}.1 = -1 \quad \text{because} \quad 10^{-1} = \frac{1}{10} = .1$$
$$\log_{10}.01 = -2 \quad \text{because} \quad 10^{-2} = \frac{1}{10^2} = .01$$
$$\log_{10}.001 = -3 \quad \text{because} \quad 10^{-3} = \frac{1}{10^3} = .001$$

When working exclusively with base 10 logarithms, it is customary to omit writing the numeral 10 to designate the base. Thus, the expression $\log_{10}x$ is written as $\log x$ and a statement such as $\log_{10}1000 = 3$ becomes $\log 1000 = 3$. We will follow this practice from now on in this chapter, but don't forget that the base is understood to be 10.

To find the common logarithm of a positive number that is not an integral power of 10, we can use an appropriately equipped calculator or a table such as the one that appears in Appendix B of this text. Using a calculator equipped with a common logarithm function (ordinarily a key labeled $\boxed{\log}$ is used), we obtained the following results rounded to four decimal places.

$$\log 1.75 = .2430$$
$$\log 23.8 = 1.3766$$
$$\log 134 = 2.1271$$
$$\log .192 = -.7167$$
$$\log .0246 = -1.6091$$

Be sure that you know how to use a calculator and can obtain these results.

Using a table to find a common logarithm is relatively easy but it does require a little more effort than pushing a button as you would with a calculator. Let's consider a small part of the table that appears in the back of the book. Each number in the column headed n represents the first two significant digits of a number between 1 and 10 and each of the column headings 0 through 9 represents the third significant digit. To find the logarithm of a number such as 1.75, we look at the intersection of the row containing 1.7 and the column headed 5. Thus, we obtain

$$\log 1.75 = .2430.$$

Similarly, we can find that

$$\log 2.09 = .3201 \quad \text{and} \quad \log 2.40 = .3802.$$

Keep in mind that these values are rounded to four decimal places. Also, don't lose sight of the meaning of such logarithmic statements. That is to say,

Table of Common Logarithms

n	0	1	2	3	4	5	6	7	8	9
1.0	.0000	.0043	.0086	.0128	.0170	.0212	.0253	.0294	.0334	.0374
1.1	.0414	.0453	.0492	.0531	.0569	.0607	.0645	.0682	.0719	.0755
1.2	.0792	.0828	.0864	.0899	.0934	.0969	.1004	.1038	.1072	.1106
1.3	.1139	.1173	.1206	.1239	.1271	.1303	.1335	.1367	.1399	.1430
1.4	.1461	.1492	.1523	.1553	.1584	.1614	.1644	.1673	.1703	.1732
1.5	.1761	.1790	.1818	.1847	.1875	.1903	.1931	.1959	.1987	.2014
1.6	.2041	.2068	.2095	.2122	.2148	.2175	.2201	.2227	.2253	.2279
1.7	.2304	.2330	.2355	.2380	.2405	.2430	.2455	.2480	.2504	.2529
1.8	.2553	.2577	.2601	.2625	.2648	.2672	.2695	.2718	.2742	.2765
1.9	.2788	.2810	.2833	.2856	.2878	.2900	.2923	.2945	.2967	.2989
2.0	.3010	.3032	.3054	.3075	.3096	.3118	.3139	.3160	.3181	.3201
2.1	.3222	.3243	.3263	.3284	.3304	.3324	.3345	.3365	.3385	.3404
2.2	.3424	.3444	.3464	.3483	.3502	.3522	.3541	.3560	.3579	.3598
2.3	.3617	.3636	.3655	.3674	.3692	.3711	.3729	.3747	.3766	.3784
2.4	.3802	.3820	.3838	.3856	.3874	.3892	.3909	.3927	.3945	.3962

(complete table in back of text)

$\log 1.75 = .2430$ means that 10 raised to the .2430 power is approximately 1.75. (Try it on your calculator!)

Now suppose that we want to use the table to find the logarithm of a positive number greater than 10 or less than 1. This can be accomplished by representing the number in scientific notation and applying the property $\log rs = \log r + \log s$. For example, to find $\log 134$ we can proceed as follows.

$$\log 134 = \log(1.34 \cdot 10^2)$$
$$= \log 1.34 + \log 10^2$$
$$= \log 1.34 + 2\log 10$$
$$= .1271 \quad + \quad 2 \quad = 2.1271$$

from the table

by inspection, since $\log 10 = 1$

The decimal part (.1271) of the logarithm 2.1271 is called the **mantissa**, and the integral part, (2), is called the **characteristic**. Thus, we can find the characteristic of a common logarithm by inspection (since it is the exponent of 10 when the number is written in scientific notation) and the mantissa we can get from a table. Let's consider two more examples.

$$\log 23.8 = \log(2.38 \cdot 10^1)$$
$$= \log 2.38 + \log 10^1$$
$$= .3766 + 1$$

from the table

exponent of 10

$$= 1.3766$$

$$\begin{aligned} \log .192 &= \log(1.92 \cdot 10^{-1}) \\ &= \log 1.92 + \log 10^{-1} \\ &= .2833 + (-1) \end{aligned}$$

from the table   exponent of 10

$$= .2833 + (-1)$$

Notice that in the last example we expressed the logarithm of .192 as .2833 + (−1); we did not add .2833 and −1. This is normal procedure when using a table of common logarithms because the mantissas given in the table are positive numbers. However, you should recognize that adding .2833 and −1 produces −.7167, which agrees with our earlier result obtained with a calculator.

We should realize that the common logarithm table in this text restricts us to reading directly from the table an approximation for the common logarithm of any number between 1.00 and 9.99 with *three* significant digits. A reasonable approximation can also be obtained from the table for the common logarithms of numbers with *four* significant digits by using a process called **linear interpolation**. This process is discussed in a later section of this chapter.

Antilogarithms

It is also necessary to be able to find a number when given the common logarithm of the number; that is, given $\log x$ we need to be able to determine x. In this situation, x is referred to as the **antilogarithm** (abbreviated **antilog**) of $\log x$. Many calculators are equipped to find antilogarithms two ways. Let's consider some examples.

EXAMPLE 2

Determine antilog .2430.

Solution A   The phrase "determine the antilog of .2430" means to find a value for x such that $\log x = .2430$. Thus changing $\log x = .2430$ to exponential form we obtain $10^{.2430} = x$. Now on our calculator we can enter 10, press $\boxed{y^x}$, enter .2430, press $\boxed{=}$, and obtain 1.7498467, which equals 1.75 when rounded to three significant digits.

Solution B   Thinking in terms of functions, the **antilog** function is the inverse of the log function. Therefore, on some calculators we can enter .2430, press $\boxed{\text{INV}}$, press $\boxed{\text{log}}$, and obtain 1.7498467, which rounds to 1.75.

EXAMPLE 3

Determine antilog −1.6091.

Solution   Using the inverse routine we can enter −1.6091, press $\boxed{\text{INV}}$, press $\boxed{\text{log}}$, and obtain .02459801, which equals .0246 when rounded to three significant digits.

The common logarithm table at the back of the book can also be used to determine antilogarithms. Use the procedure illustrated in the next three examples.

EXAMPLE 4

Determine antilog 1.3365.

Solution Finding an antilogarithm simply reverses the process used before for finding a logarithm. Thus, antilog 1.3365 means that 1 is the characteristic and .3365 the mantissa. We look for .3365 in the body of the common logarithm table and we find that it is located at the intersection of the 2.1-row and the 7-column. Therefore, the antilogarithm is

$$2.17 \cdot 10^1 = 21.7.$$

EXAMPLE 5

Determine antilog(.1523 + (−2)).

Solution The mantissa, .1523, is located at the intersection of the 1.4-row and 2-column. The characteristic is −2 and therefore the antilogarithm is

$$1.42 \cdot 10^{-2} = .0142.$$

EXAMPLE 6

Determine antilog −2.6038.

Solution The mantissas given in a table are *positive* numbers. Thus, we need to express −2.6038 in terms of a positive mantissa and this can be done by adding and subtracting 3 as follows.

$$(-2.6038 + 3) - 3 = .3962 + (-3)$$

Now we can look for .3962 and find it at the intersection of the 2.4-row and 9-column. Therefore, the antilogarithm is

$$2.49 \cdot 10^{-3} = .00249.$$

Natural Logarithms

In many practical applications of logarithms, the number e (remember $e \approx 2.71828$) is used as a base. Logarithms with a base of e are called **natural logarithms**, and the symbol **ln** x is commonly used instead of $\log_e x$.

Natural logarithms can be found with an appropriately equipped calculator or with a table of natural logarithms. Using a calculator with a natural logarithm function (ordinarily a key labeled $\boxed{\ln x}$) we obtained the following results rounded to four decimal places.

$$\ln 3.21 = 1.1663$$
$$\ln 47.28 = 3.8561$$
$$\ln 842 = 6.7358$$
$$\ln .21 = -1.5606$$
$$\ln .0046 = -5.3817$$
$$\ln 10 = 2.3026$$

Be sure that you can use a calculator and obtain these results. Also, keep in mind the significance of a statement such as $\ln 3.21 = 1.1663$. We are claiming that e raised to the 1.1663 power is approximately 3.21. Using 2.71828 as an approximation for e we obtain

$$(2.71828)^{1.1663} = 3.2100908.$$

Furthermore, since $f(x) = \ln x$ and $g(x) = e^x$ are inverses of each other, we can enter 1.1663, press $\boxed{\text{INV}}$, press $\boxed{\ln x}$, and obtain 3.2100933 on a calculator. (The last three digits of this result differ from our previous result of 3.2100908 because a better approximation for e is used in this sequence of operations.)

Table 6.1 contains the natural logarithms for numbers between .1 and 10, inclusive, at intervals of .1. Reading directly from the table we obtain $\ln 1.6 = .4700$, $\ln 4.8 = 1.5686$, and $\ln 9.2 = 2.2192$.

Table 6.1/Natural Logarithms

n	$\ln n$	n	$\ln n$	n	$\ln n$	n	$\ln n$
0.1	−2.3026	2.6	0.9555	5.1	1.6292	7.6	2.0281
0.2	−1.6094	2.7	0.9933	5.2	1.6487	7.7	2.0412
0.3	−1.2040	2.8	1.0296	5.3	1.6677	7.8	2.0541
0.4	−0.9163	2.9	1.0647	5.4	1.6864	7.9	2.0669
0.5	−0.6931	3.0	1.0986	5.5	1.7047	8.0	2.0794
0.6	−0.5108	3.1	1.1314	5.6	1.7228	8.1	2.0919
0.7	−0.3567	3.2	1.1632	5.7	1.7405	8.2	2.1041
0.8	−0.2231	3.3	1.1939	5.8	1.7579	8.3	2.1163
0.9	−0.1054	3.4	1.2238	5.9	1.7750	8.4	2.1282
1.0	0.0000	3.5	1.2528	6.0	1.7918	8.5	2.1401
1.1	0.0953	3.6	1.2809	6.1	1.8083	8.6	2.1518
1.2	0.1823	3.7	1.3083	6.2	1.8245	8.7	2.1633
1.3	0.2624	3.8	1.3350	6.3	1.8405	8.8	2.1748
1.4	0.3365	3.9	1.3610	6.4	1.8563	8.9	2.1861
1.5	0.4055	4.0	1.3863	6.5	1.8718	9.0	2.1972
1.6	0.4700	4.1	1.4110	6.6	1.8871	9.1	2.2083
1.7	0.5306	4.2	1.4351	6.7	1.9021	9.2	2.2192
1.8	0.5878	4.3	1.4586	6.8	1.9169	9.3	2.2300
1.9	0.6419	4.4	1.4816	6.9	1.9315	9.4	2.2407
2.0	0.6931	4.5	1.5041	7.0	1.9459	9.5	2.2513
2.1	0.7419	4.6	1.5261	7.1	1.9601	9.6	2.2618
2.2	0.7885	4.7	1.5476	7.2	1.9741	9.7	2.2721
2.3	0.8329	4.8	1.5686	7.3	1.9879	9.8	2.2824
2.4	0.8755	4.9	1.5892	7.4	2.0015	9.9	2.2925
2.5	0.9163	5.0	1.6094	7.5	2.0149	10.0	2.3026

When using Table 6.1 to find the natural logarithm of a positive number less than .1 or greater than 10, we can use the property $\ln rs = \ln r + \ln s$ and proceed as follows.

$$\ln .0084 = \ln(8.4 \cdot 10^{-3})$$
$$= \ln 8.4 + \ln 10^{-3}$$
$$= \ln 8.4 + (-3)(\ln 10)$$
$$= 2.1282 - 3(2.3026)$$

↑ from the table   ↑ from the table

$$= 2.1282 - 6.9078 = -4.7796.$$

$$\ln 190 = \ln(1.9 \cdot 10^{2})$$
$$= \ln 1.9 + \ln 10^{2}$$
$$= \ln 1.9 + 2\ln 10$$
$$= .6419 + 2(2.3026)$$

↑ from the table   ↑ from the table

$$= 5.2471.$$

Problem Set 6.4

For Problems 1–6, graph each of the logarithmic functions.

1. $y = \log_3 x$ 2. $y = \log_4 x$ 3. $y = \log_{10} x$ 4. $y = \log_5 x$
5. $y = \log_{1/2} x$ 6. $y = \log_{1/3} x$

7. Graph $y = \log_3 x$ by changing the equation to exponential form.
8. Graph $y = \log_4 x$ by changing the equation to exponential form.
9. Graph $y = \log_{1/2} x$ by reflecting the graph of $y = (1/2)^x$ across the line $y = x$.
10. Graph $y = \log_{1/3} x$ by reflecting the graph of $y = (1/3)^x$ across the line $y = x$.

For Problems 11–14, graph each of the logarithmic functions. Don't forget that the graph of $f(x) = g(x) + 1$ is the graph of $g(x)$ moved up one unit.

11. $y = 1 + \log_{10} x$ 12. $y = -2 + \log_{10} x$
13. $y = \log_{10}(x - 1)$ 14. $y = \log_{10}(x + 2)$

For Problems 15–24, use a calculator to find each of the **common logarithms**. Express answers to four decimal places.

15. $\log 9.45$ 16. $\log 1.07$ 17. $\log 34.62$ 18. $\log 578.1$
19. $\log 4721.4$ 20. $\log 52,698$ 21. $\log .612$ 22. $\log .08134$
23. $\log .0047$ 24. $\log .000076$

For Problems 25–36, find each **common logarithm** by using the table at the back of the book.

25. $\log 8.72$ 26. $\log 6.04$ 27. $\log 56.9$ 28. $\log 48$
29. $\log 708$ 30. $\log 14,100$ 31. $\log .492$ 32. $\log .023$
33. $\log .00528$ 34. $\log .000415$ 35. $\log 763,000$ 36. $\log 9,180,000$

For Problems 37–46, use a calculator to find each antilogarithm. Express answers to five significant digits of accuracy. For example,

$$\text{antilog } 3.2147 = 1639.4569$$
$$= 1639.5. \quad \text{(5 significant digits)}$$

37. antilog 1.5263   38. antilog 2.7185   39. antilog 3.9335
40. antilog 4.9547   41. antilog 0.5517   42. antilog 1.9006
43. antilog −0.1452   44. antilog −1.3148   45. antilog −2.6542
46. antilog −2.1928

For Problems 47–60, find each antilogarithm by using the table at the back of the book.

47. antilog 0.5502   48. antilog 0.9624   49. antilog 1.4829

50. antilog 1.9170   51. antilog 2.9926   52. antilog 3.4533

53. antilog 5.8062   54. antilog 4.6812   55. antilog$(-1 + .7340)$

56. antilog$(-2 + .7774)$ 57. antilog$(-3 + .8639)$ 58. antilog$(-4 + .0969)$

59. antilog $-.7471$ 60. antilog -1.1232

For Problems 61–72, use a calculator to find each **natural logarithm**. Express answers to four decimal places.

61. ln 2   62. ln 9   63. ln 21.4   64. ln 87.6

65. ln 412   66. ln 384.2   67. ln .32   68. ln .417

69. ln .0715   70. ln .006285   71. ln .0008   72. ln 52,173

For Problems 73–82, find each **natural logarithm** by using the table on page 231.

73. ln 3.7   74. ln 2.8   75. ln 78   76. ln 140

77. ln 620   78. ln 9800   79. ln .42   80. ln .051

81. ln .0085   82. ln .00056

6.5

Exponential Equations, Logarithmic Equations, and Problem Solving

In Section 6.1 we solved exponential equations such as $3^x = 81$ by expressing both sides of the equation as a power of 3 and then applying the property "if $b^n = b^m$, then $n = m$." However, if we try this same approach with an equation such as $3^x = 5$, we face the difficulty of expressing 5 as a power of 3. We can solve this type of problem by using the properties of logarithms and the following property of equality.

PROPERTY 6.8

If $x > 0$, $y > 0$, $b > 0$, and $b \neq 1$, then

$\quad x = y$ if and only if $\log_b x = \log_b y$.

Property 6.8 is stated in terms of any valid base b; however, for most applications either common logarithms or natural logarithms are used. Let's consider some examples.

EXAMPLE 1

Solve $3^x = 5$, to the nearest hundredth.

Solution By using common logarithms we can proceed as follows.

$$3^x = 5$$

$$\log 3^x = \log 5 \qquad \text{Property 6.8}$$

$$x \log 3 = \log 5 \qquad \log r^p = p \log r$$

$$x = \frac{\log 5}{\log 3}$$

$$x = \frac{.6990}{.4771} = 1.47, \text{ to the nearest hundredth}$$

Check: By using a calculator to raise 3 to the 1.47 power, we obtain $3^{1.47} = 5.0276871$. Thus, we say that to the nearest hundredth, the solution set for $3^x = 5$ is $\{1.47\}$.

A word of caution: The expression $\log 5/\log 3$ means that we must *divide*, not subtract, the logarithms. That is, $\log 5/\log 3$ *does not* mean $\log(\frac{5}{3})$.

EXAMPLE 2

Solve $e^{x+1} = 19$, to the nearest hundredth.

Solution Since the base of e is used in the exponential expression, let's use natural logarithms to help solve this equation.

$$e^{x+1} = 5$$

$$\ln e^{x+1} = \ln 5 \qquad \text{Property 6.8}$$

$$(x + 1)\ln e = \ln 5 \qquad \ln r^p = p \ln r$$

$$(x + 1)(1) = \ln 5 \qquad \ln e = 1$$

$$x = \ln 5 - 1$$

$$x = 1.6094 - 1$$

$$x = .6094 = .61, \text{ to the nearest hundredth}$$

The solution set is $\{.61\}$. Check it!

EXAMPLE 3

Solve $2^{3x-2} = 3^{2x+1}$, to the nearest hundredth.

Solution

$$2^{3x-2} = 3^{2x+1}$$

$$\log 2^{3x-2} = \log 3^{2x+1}$$

$$(3x - 2)\log 2 = (2x + 1)\log 3$$

$$3x \log 2 - 2 \log 2 = 2x \log 3 + \log 3$$

$$3x \log 2 - 2x \log 3 = \log 3 + 2 \log 2$$

$$x(3 \log 2 - 2 \log 3) = \log 3 + 2 \log 2$$

$$x = \frac{\log 3 + 2 \log 2}{3 \log 2 - 2 \log 3}$$

At this point we could evaluate x, but instead let's use the properties of logarithms to simplify the expression for x.

$$x = \frac{\log 3 + \log 2^2}{\log 2^3 - \log 3^2}$$

$$= \frac{\log 3 + \log 4}{\log 8 - \log 9}$$

$$= \frac{\log 12}{\log\left(\frac{8}{9}\right)}$$

Now evaluating x we obtain

$$x \approx \frac{1.0792}{-.0512} = -21.08, \text{ to the nearest hundredth.}$$

The solution set is $\{-21.08\}$. Check it!

Logarithmic Equations

In Example 11 of Section 6.3 we solved the logarithmic equation $\log_{10}x + \log_{10}(x + 9) = 1$ by simplifying the left side of the equation to $\log_{10}[x(x + 9)]$, and then changing the equation to exponential form to complete the solution. At this time, using Property 6.8, we can solve such a logarithmic equation another way, and also expand our equation solving capabilities. Let's consider some examples.

EXAMPLE 4

Solve $\log x + \log(x - 15) = 2$.

Solution Since $\log 100 = 2$, the given equation becomes

$$\log x + \log(x - 15) = \log 100.$$

Now simplifying the left side and applying Property 6.8 we can proceed as follows.

$$\log(x)(x - 15) = \log 100$$
$$x(x - 15) = 100$$
$$x^2 - 15x - 100 = 0$$
$$(x - 20)(x + 5) = 0$$
$$x - 20 = 0 \quad \text{or} \quad x + 5 = 0$$
$$x = 20 \quad \text{or} \quad x = -5$$

The domain of a logarithmic function must contain only positive numbers, so x and $x - 15$ must be positive in this problem. Therefore, the solution of -5 is discarded and the solution set is $\{20\}$.

EXAMPLE 5

Solve $\ln(x + 2) = \ln(x - 4) + \ln 3$.

Solution

$$\ln(x + 2) = \ln(x - 4) + \ln 3$$
$$\ln(x + 2) = \ln[3(x - 4)]$$
$$x + 2 = 3(x - 4)$$
$$x + 2 = 3x - 12$$
$$14 = 2x$$
$$7 = x$$

The solution set is $\{7\}$.

EXAMPLE 6

Solve $\log_b(x + 2) + \log_b(2x - 1) = \log_b x$.

Solution

$$\log_b(x + 2) + \log_b(2x - 1) = \log_b x$$
$$\log_b[(x + 2)(2x - 1)] = \log_b x$$
$$(x + 2)(2x - 1) = x$$
$$2x^2 + 3x - 2 = x$$
$$2x^2 + 2x - 2 = 0$$
$$x^2 + x - 1 = 0$$

Using the quadratic formula we obtain

$$x = \frac{-1 \pm \sqrt{1 + 4}}{2}$$
$$= \frac{-1 \pm \sqrt{5}}{2}.$$

Since $x + 2$, $2x - 1$, and x have to be positive, the solution of $(-1 - \sqrt{5})/2$ has to be discarded and the solution set is $\{(-1 + \sqrt{5})/2\}$.

Problem Solving

In Section 6.2 we used the compound interest formula

$$A = P\left(1 + \frac{r}{n}\right)^{nt}$$

to determine the amount of money (A) accumulated at the end of t years if P dollars is invested at r rate of interest compounded n times per year. Now let's use this formula to solve other types of problems that deal with compound interest.

EXAMPLE 7

How long will it take to double $500 if invested at 12% compounded quarterly?

Solution To "double $500" means that the $500 will grow into $1000. Thus,

$$1000 = 500\left(1 + \frac{.12}{4}\right)^{4t}$$
$$= 500(1 + .03)^{4t}$$
$$= 500(1.03)^{4t}.$$

Multiplying both sides of $1000 = 500(1.03)^{4t}$ by $\frac{1}{500}$ yields

$$2 = (1.03)^{4t}.$$

Therefore,

$$\log 2 = \log(1.03)^{4t} \qquad \text{Property 6.8}$$
$$= 4t \log 1.03. \qquad \log r^p = p \log r$$

Solving for t, we obtain

$$\log 2 = 4t \log 1.03$$

$$\frac{\log 2}{\log 1.03} = 4t$$

$$\frac{\log 2}{4 \log 1.03} = t \qquad \text{Multiply both sides by } \frac{1}{4}.$$

$$\frac{.3010}{4(.0128)} = t$$

$$\frac{.3010}{.0512} = t$$

$$t = 5.9, \text{ to the nearest tenth.}$$

Therefore, we are claiming that $500 invested at 12% interest compounded quarterly will double itself in approximately 5.9 years.

Check: $500 invested at 12% compounded quarterly for 5.9 years will produce

$$A = \$500\left(1 + \frac{.12}{4}\right)^{4(5.9)}$$
$$= \$500(1.03)^{23.6}$$
$$= \$1004.45.$$

EXAMPLE 8

What rate of interest (nearest tenth of a percent) is needed so that an investment of $1000 will yield $4000 in 10 years if the money is compounded annually?

Solution Substituting the known facts into the compound interest formula we obtain

$$4000 = 1000(1 + r)^{10}.$$

Multiplying both sides of this equation by $\frac{1}{1000}$ yields

$$4 = (1 + r)^{10}.$$

Therefore,

$$\log 4 = \log(1 + r)^{10} \qquad \text{Property 6.8}$$
$$= 10 \log(1 + r). \qquad \log r^p = p \log r$$

Multiplying both sides by $\frac{1}{10}$ produces

$$\frac{\log 4}{10} = \log(1 + r).$$

Using $\log 4 = .6021$ we obtain

$$\frac{.6021}{10} = \log(1 + r)$$

$$.0602 = \log(1 + r).$$

Finding the antilog of .0602 and solving for r we obtain

$$1.149 = 1 + r$$
$$.149 = r$$
$$14.9\% = r.$$

Therefore, approximately a 14.9% rate of interest is needed.

Check: $1000 invested at 14.9% compounded annually for 10 years will produce

$$A = \$1000(1.149)^{10} = \$4010.52.$$

EXAMPLE 9

How long will it take $100 to triple itself if it is invested at 8% interest compounded continuously?

Solution To **triple itself** means that the $100 will grow into $300. Thus, using the formula for interest that is compounded continuously, we can proceed as follows.

$$A = Pe^{rt}$$
$$300 = 100e^{(.08)t}$$
$$3 = e^{.08t}$$
$$\ln 3 = \ln e^{.08t} \qquad \text{Property 6.8}$$
$$\ln 3 = .08t \ln e \qquad \ln r^p = p \ln r$$
$$\ln 3 = .08t \qquad \ln e = 1$$

$$\frac{\ln 3}{.08} = t$$

$$\frac{1.0986}{.08} = t$$

$$t = 13.7, \text{ to the nearest tenth.}$$

Therefore, in approximately 13.7 years, $100 will triple itself at 8% interest compounded continuously.

> **Check:** $100 invested at 8% compounded continuously for 13.7 years produces
>
> $$A = Pe^{rt}$$
> $$= 100e^{.08(13.7)}$$
> $$= 100(2.718)^{1.096}$$
> $$= 299.18.$$

EXAMPLE 10

Suppose the number of bacteria present in a certain culture after t minutes is given by the equation $Q(t) = Q_0 e^{.04t}$, where Q_0 represents the initial number of bacteria. How long would it take for the bacteria count to grow from 500 to 2000?

Solution Substituting into $Q(t) = Q_0 e^{.04t}$ and solving for t, we obtain

$$2000 = 500e^{.04t}$$

$$4 = e^{.04t}$$

$$\ln 4 = \ln e^{.04t}$$

$$\ln 4 = .04t \ln e$$

$$\ln 4 = .04t$$

$$\frac{\ln 4}{.04} = t$$

$$t = 34.7, \text{ to the nearest tenth.}$$

It should take approximately 34.7 minutes.

The basic approach of applying Property 6.8 and using either common or natural logarithms can also be used to evaluate a logarithm to some base other than 10 or e. The next example illustrates this idea.

EXAMPLE 11

Evaluate $\log_3 41$.

Solution Let $x = \log_3 41$. Changing to exponential form we obtain

$$3^x = 41.$$

Now we can apply Property 6.8 and proceed as follows.

$$\log 3^x = \log 41$$

$$x \log 3 = \log 41$$

$$x = \frac{\log 41}{\log 3}$$

$$x = \frac{1.6128}{.4771}$$

$$x = 3.3804. \qquad \text{Rounded to four decimal places}$$

Therefore, we are claiming that 3 raised to the 3.3804 power will produce approximately 41. Check it!

Using the method of Example 11 to evaluate $\log_a r$ produces the following formula often referred to as the **change-of-base formula for logarithms**.

PROPERTY 6.9

> If a, b, and r are positive numbers with $a \neq 1$ and $b \neq 1$, then
>
> $$\log_a r = \frac{\log_b r}{\log_b a}.$$

By using Property 6.9 we can easily determine a relationship between logarithms of different bases. For example, suppose that in Property 6.9 we let $a = 10$ and $b = e$.

$$\log_a r = \frac{\log_b r}{\log_b a}$$

becomes

$$\log_{10} r = \frac{\log_e r}{\log_e 10}$$

$$\log_e r = (\log_e 10)(\log_{10} r)$$

$$\log_e r = (2.3026)(\log_{10} r).$$

Thus, the natural logarithm of any positive number is approximately equal to the common logarithm of the number times 2.3026.

Problem Set 6.5

Solve each of the following exponential equations and express approximate solutions to the nearest hundredth.

1. $2^x = 9$
2. $3^x = 20$
3. $5^t = 23$
4. $4^t = 12$
5. $2^{x+1} = 7$
6. $3^{x-2} = 11$

7. $7^{2t-1} = 35$ 8. $5^{3t+1} = 9$ 9. $e^x = 4.1$

10. $e^x = 30$ 11. $e^{x-1} = 8.2$ 12. $e^{x-2} = 13.1$

13. $2e^x = 12.4$ 14. $3e^x - 1 = 17$ 15. $3^{x-1} = 2^{x+3}$

16. $5^{2x+1} = 7^{x+3}$ 17. $5^{x-1} = 2^{2x+1}$ 18. $3^{2x+1} = 2^{3x+2}$

Solve each of the following logarithmic equations and express irrational solutions in lowest radical form.

19. $\log x + \log(x + 3) = 1$ 20. $\log x + \log(x + 21) = 2$

21. $\log(2x - 1) - \log(x - 3) = 1$ 22. $\log(3x - 1) = 1 + \log(5x - 2)$

23. $\log(x - 2) = 1 - \log(x + 3)$ 24. $\log(x + 1) = \log 3 - \log(2x - 1)$

25. $\log(x + 1) - \log(x + 2) = \log \dfrac{1}{x}$ 26. $\log(x + 2) - \log(2x + 1) = \log x$

27. $\ln(3t - 4) - \ln(t + 1) = \ln 2$ 28. $\ln(2t + 5) = \ln 3 + \ln(t - 1)$

29. $\log(x^2) = (\log x)^2$ 30. $\log \sqrt{x} = \sqrt{\log x}$

Approximate each of the following logarithms to three decimal places. (Example 11 and/or Property 6.9 should be of some help.)

31. $\log_3 14$ 32. $\log_4 94$ 33. $\log_5 2.1$ 34. $\log_6 .345$

35. $\log_7 176$ 36. $\log_8 296$ 37. $\log_9 14.32$ 38. $\log_7 .024$

Solve each of the following problems.

39. How long will it take $1000 to double itself if it is invested at 9% interest compounded semiannually?

40. How long will it take $750 to be worth $1000 if it is invested at 12% interest compounded quarterly?

41. How long will it take $500 to triple itself if it is invested at 9% interest compounded continuously?

42. How long will it take $2000 to double itself if it is invested at 13% interest compounded continuously?

43. What rate of interest (nearest tenth of a percent) compounded annually is needed so that an investment of $200 will grow to $350 in 5 years?

44. What rate of interest (nearest tenth of a percent) compounded continuously is needed so that an investment of $500 will grow to $900 in 10 years?

45. A piece of machinery valued at $30,000 depreciates at a rate of 10% yearly. How long will it take until it has a value of $15,000?

46. For a certain strain of bacteria, the number present after t hours is given by the equation $Q = Q_0 e^{.34t}$, where Q_0 represents the initial number of bacteria. How long will it take 400 bacteria to increase to 4000 bacteria?

47. The number of grams of a certain radioactive substance present after t hours is given by the equation $Q = Q_0 e^{-.45t}$, where Q_0 represents the initial number of grams. How long would it take 2500 grams to be reduced to 1250 grams?

48. The equation $P(a) = 14.7e^{-.21a}$, where a is the altitude above sea level measured in miles, yields the atmospheric pressure in pounds per square inch. If the atmospheric pressure at Cheyenne, Wyoming is approximately 11.53 pounds per square inch, find its altitude above sea level. Express your answer to the nearest hundred feet.

49. Suppose that the equation $P(t) = P_0 e^{.02t}$, where P_0 represents an initial population and t is the time in years, is used to predict population growth. How long would it take a city of 50,000 to double its population?

50. In a certain culture the equation $Q(t) = Q_0 e^{.4t}$, where Q_0 is an initial number of bacteria and t is time measured in hours, yields the number of bacteria as a function of the time. How long will it take 500 bacteria to increase to 2000?

51. Radon is formed by the radioactive decay of radium. It has a half-life of approximately 4 days and the equation $Q(t) = Q_0 e^{-4k}$ yields the amount of radon that remains after t days if the initial amount is Q_0. Solve the equation $\frac{1}{2}Q_0 = Q_0 e^{-4k}$ for k. Express the answer to the nearest hundredth.

52. Polonium is also formed from the radioactive decay of radium and has a half-life of approximately 140 days. The equation $Q(t) = Q_0 e^{-140k}$ yields the amount of polonium that remains after t days if the initial amount is Q_0. Solve the equation $\frac{1}{2}Q_0 = Q_0 e^{-140k}$ for k and express the answer to the nearest thousandth.

Miscellaneous Problems

53. Use the approach of Example 11 and develop Property 6.9.

54. Let $r = b$ in Property 6.9 and verify that $\log_a b = \dfrac{1}{\log_b a}$.

55. Solve the equation $\dfrac{5^x - 5^{-x}}{2} = 3$. Express your answer to the nearest hundredth.

56. Solve the equation $y = \dfrac{10^x + 10^{-x}}{2}$ for x in terms of y.

57. Solve the equation $y = \dfrac{e^x - e^{-x}}{2}$ for x in terms of y.

6.6

Computation with Common Logarithms (Optional)

As we mentioned earlier, the calculator has replaced the use of common logarithms for most computational purposes. Nevertheless, we feel that a brief look at how common logarithms were used will give you a better insight into the meaning of logarithms and their properties.

Although the computations in this section should be done by using logarithms and the table of common logarithms at the back of the book, we would suggest that you check each problem by using a calculator. This will provide you with some additional practice with your calculator and it will establish the validity of our work with logarithms.

Linear Interpolation

Let's begin by expanding our use of the common logarithm table at the back of the book. Suppose that we try to determine log 2.744 from the table. Because the table contains only logarithms of numbers with, at most, three significant digits, a problem exists. However, by a process called **linear interpolation** we can extend the capabilities of the table to include numbers of four significant digits.

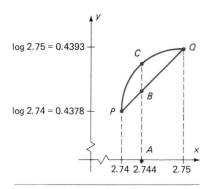

Figure 6.8

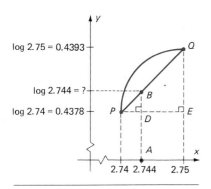

Figure 6.9

First, let's consider a geometric basis of linear interpolation and then we will suggest a systematic procedure for carrying out the necessary calculations. A portion of the graph of $y = \log_{10}x$, with the curvature exaggerated to help illustrate the principle involved, is shown in Figure 6.8. The line segment joining points P and Q is used to approximate the curve from P to Q. The actual value of $\log 2.744$ is the ordinate of the point C, that is, the length of $\overline{AC}$. This cannot be determined from the table. Instead we will use the ordinate of point B (the length of $\overline{AB}$) as an approximation for $\log 2.744$.

Now consider Figure 6.9 where line segments $\overline{DB}$ and $\overline{EQ}$ are drawn perpendicular to $\overline{PE}$. The right triangles formed, $\triangle PDB$ and $\triangle PEQ$, are similar and therefore the lengths of their corresponding sides are proportional. Thus, we can write

$$\frac{PD}{PE} = \frac{DB}{EQ}.$$

Also from Figure 6.9 we see that

$$PD = 2.744 - 2.74 = .004,$$

$$PE = 2.75 - 2.74 = .01,$$

and

$$EQ = .4393 - .4378 = .0015.$$

Therefore, the above proportion becomes

$$\frac{.004}{.01} = \frac{DB}{.0015}.$$

Solving this proportion for DB yields

$$(.01)(DB) = (.004)(.0015)$$

$$DB = .0006.$$

Since $AB = AD + DB$, we have

$$AB = .4378 + .0006$$
$$= .4384.$$

Thus, we obtain $\log 2.744 = .4384$.

Now let's suggest an abbreviated format for carrying out the calculations necessary to find $\log 2.744$.

$$
\begin{array}{cc}
x & \log x \\
4\left\{\begin{array}{l}2.740 \\ 2.744 \\ 2.750\end{array}\right\}10 \quad & k\left\{\begin{array}{l}.4378 \\ ? \\ .4393\end{array}\right\}.0015
\end{array}
$$

Notice that we have used 4 and 10 for the differences for values of x instead of .004

and .01 because the ratio $\frac{.004}{.01}$ equals $\frac{4}{10}$. Setting up a proportion and solving for k yields

$$\frac{4}{10} = \frac{k}{.0015}$$

$$10k = 4(.0015) = .0060$$

$$k = .0006.$$

Thus, $\log 2.744 = .4378 + .0006 = .4384$.

The process of linear interpolation can also be used to approximate an antilogarithm when the mantissa is in between two values in the table. The following example illustrates this procedure.

EXAMPLE 1

Find antilog 1.6157.

Solution From the table we see that the mantissa, .6157, is between .6149 and .6160. We can carry out the interpolation as follows.

$$h\left\{\begin{matrix}4.120 \\ \\ ? \\ \\ 4.130\end{matrix}\right\}.010 \qquad 8\left\{\begin{matrix}.6149 \\ \\ .6157 \\ \\ .6160\end{matrix}\right\}11 \qquad \frac{.0008}{.0011} = \frac{8}{11}$$

$$\frac{h}{.010} = \frac{8}{11}$$

$$11h = 8(.010) = .080$$

$$h = \frac{1}{11}(.080) = .007, \text{ to the nearest thousandth}$$

Thus, antilog $.6157 = 4.120 + .007 = 4.127$. Therefore,

$$\text{antilog } 1.6157 = \text{antilog}(.6157 + 1)$$
$$= 4.127 \cdot 10^1$$
$$= 41.27.$$

Computation with Common Logarithms

Let's first restate the basic properties of logarithms in terms of common logarithms. Remember that we are writing $\log x$ instead of $\log_{10} x$.

If x and y are positive real numbers, then

1. $\log xy = \log x + \log y$

2. $\log \dfrac{x}{y} = \log x - \log y$

3. $\log x^p = p \log x$. p is any real number.

The following two properties of equality pertaining to logarithms will also be used.

    4. If $x = y$ (x and y are positive), then $\log x = \log y$.

    5. If $\log x = \log y$, then $x = y$.

Now let's illustrate how common logarithms can be used for computational purposes.

EXAMPLE 2

Evaluate $\dfrac{(571.4)(8.236)}{71.68}$.

Solution Let $N = \dfrac{(571.4)(8.236)}{71.68}$. Therefore,

$$\begin{aligned}
\log N &= \log \frac{(571.4)(8.236)}{71.68} \\
&= \log 571.4 + \log 8.236 - \log 71.68 \\
&= 2.7569 + 0.9157 - 1.8554 = 1.8172.
\end{aligned}$$

Therefore,

$$N = \text{antilog } 1.8172 = \text{antilog}(.8172 + 1) = 6.564 \cdot 10^1 = 65.64.$$

Check: By using a calculator we obtain

$$N = \frac{(571.4)(8.236)}{71.68} = 65.653605.$$

When using a table of logarithms, it is sometimes necessary to change the form of writing a logarithm so that the mantissa is positive. The next example illustrates this idea.

EXAMPLE 3

Find the quotient $\dfrac{1.73}{5.08}$.

Solution Let $N = \dfrac{1.73}{5.08}$. Therefore,

$$\log N = \log \frac{1.73}{5.08} = \log 1.73 - \log 5.08 = 0.2380 - 0.7059 = -.4679.$$

Now by adding 1 and subtracting 1, which changes the form but not the value, we obtain

$$\log N = -.4679 + 1 - 1 = .5321 - 1 = .5321 + (-1).$$

Therefore,

$$N = \text{antilog}(.5321 + (-1))$$
$$= 3.405 \cdot 10^{-1} = .3405.$$

Check: By using a calculator we obtain

$$N = \frac{1.73}{5.08} = .34055118.$$

Sometimes it is necessary to change the form of a logarithm so that a subsequent calculation will produce an integer for the characteristic part of the logarithm. Let's consider an example to illustrate this idea.

EXAMPLE 4

Evaluate $\sqrt[4]{.0767}$.

Solution Let $N = \sqrt[4]{.0767} = (.0767)^{1/4}$. Therefore,

$$\log N = \log(.0767)^{1/4}$$

$$= \frac{1}{4}\log .0767$$

$$= \frac{1}{4}(.8848 + (-2)) = \frac{1}{4}(-2 + .8848).$$

At this stage we recognize that applying the distributive property will produce a nonintegral characteristic, namely, $-\frac{1}{2}$. Therefore, let's add 4 and subtract 4 inside the parentheses, which will change the form as follows.

$$\log N = \frac{1}{4}(-2 + .8848 + 4 - 4)$$

$$= \frac{1}{4}(4 - 2 + .8848 - 4)$$

$$= \frac{1}{4}(2.8848 - 4)$$

Now applying the distributive property we obtain

$$\log N = \frac{1}{4}(2.8848) - \frac{1}{4}(4)$$

$$= .7212 - 1$$

$$= .7212 + (-1).$$

Therefore,

$$N = \text{antilog}(.7212 + (-1))$$
$$= 5.262 \cdot 10^{-1} = .5262.$$

Check: By using a calculator, we obtain

$$N = \sqrt[4]{.0767} = .52625816.$$

Problem Set 6.6

Use the table at the back of the book and linear interpolation to find each of the following common logarithms.

1. log 4.327
2. log 27.43
3. log 128.9
4. log 3526
5. log .8761
6. log .07692
7. log .005186
8. log .0002558

Use the table at the back of the book and linear interpolation to find each of the following antilogarithms to four significant digits.

9. antilog .4690
10. antilog 1.7971
11. antilog 2.1925
12. antilog 3.7225
13. antilog(.5026 + (−1))
14. antilog(.9397 + (−2))

Use common logarithms and linear interpolation to help evaluate each of the following. Express your answers with four significant digits. Check your answers by using a calculator.

15. $(294)(71.2)$
16. $(192.6)(4.017)$
17. $\dfrac{23.4}{4.07}$

18. $\dfrac{718.5}{8.248}$
19. $(17.3)^5$
20. $(48.02)^3$

21. $\dfrac{(108)(76.2)}{13.4}$
22. $\dfrac{(126.3)(24.32)}{8.019}$
23. $\sqrt[5]{.821}$

24. $\sqrt[4]{645.3}$
25. $(79.3)^{3/5}$
26. $(176.8)^{3/4}$

27. $\sqrt{\dfrac{(7.05)(18.7)}{.521}}$
28. $\sqrt[3]{\dfrac{(41.3)(.271)}{8.05}}$

Chapter Summary

This chapter can be summarized around three main topics, namely, (1) exponents and exponential functions, (2) logarithms and logarithmic functions, and (3) applications of exponential and logarithmic functions.

Exponents and Exponential Functions

If **a** and **b** are positive numbers, and **m** and **n** are real numbers, then

1. $b^n \cdot b^m = b^{n+m}$ Product of two powers
2. $(b^n)^m = b^{mn}$ Power of a power
3. $(ab)^n = a^n b^n$ Power of a product
4. $\left(\dfrac{a}{b}\right)^n = \dfrac{a^n}{b^n}$ Power of a quotient
5. $\dfrac{b^n}{b^m} = b^{n-m}.$ Quotient of two powers

A function defined by an equation of the form

$$y = b^x \qquad b > 0 \text{ and } b \neq 1$$

is called an **exponential function**. The following figure illustrates the general behavior of the graphs of exponential functions of the form $y = b^x$.

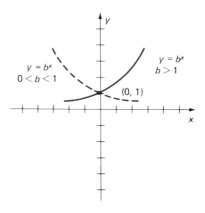

Logarithms and Logarithmic Functions

If r is any positive real number, then the unique exponent t such that $b^t = r$ is called the **logarithm of r with base b** and is denoted by $\log_b r$.

The following properties of logarithms are used frequently.

1. $\log_b b = 1$
2. $\log_b 1 = 0$
3. $b^{\log_b r} = r$
4. $\log_b rs = \log_b r + \log_b s$
5. $\log_b \left(\dfrac{r}{s} \right) = \log_b r - \log_b s$
6. $\log_b (r^p) = p \log_b r$

Logarithms with a base of 10 are called **common logarithms**. The expression $\log_{10} x$ is commonly written as $\log x$.

Many calculators are equipped with a common logarithm function. Often a key labeled $\boxed{\log}$ is used to find common logarithms.

The decimal part (.9425) of the logarithm 1.9425 is called the **mantissa** and the integral part (1) is called the **characteristic.**

Calculators and tables can be used to find x when given $\log x$. The number x is referred to as an **antilogarithm**.

Natural logarithms are logarithms having a base of e, where e is an irrational number whose decimal approximation to eight digits is 2.7182818. Natural logarithms are denoted by $\log_e x$ or $\ln x$.

Many calculators are also equipped with a natural logarithm function. Often a key labeled $\boxed{\ln x}$ is used for this purpose.

A function defined by an equation of the form

$$f(x) = \log_b x \qquad b > 0 \text{ and } b \neq 1$$

is called a **logarithmic function**.

The graph of a logarithmic function, such as $y = \log_2 x$, can be determined by changing the equation to exponential form ($2^y = x$) and plotting points, or by reflecting the graph of $y = 2^x$ across the line $y = x$. This last approach is based on the fact that exponential and logarithmic functions are inverses of each other.

The following figure illustrates the general behavior of the graphs of logarithmic functions of the form $y = \log_b x$.

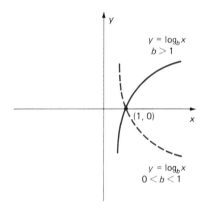

Applications

The following properties of equality are used frequently when solving exponential and logarithmic equations.

1. If $b > 0$, $b \neq 1$ and m and n are real numbers, then

$$b^n = b^m \quad \text{if and only if} \quad n = m.$$

2. If $x > 0$, $y > 0$, $b > 0$, and $b \neq 1$, then

$$x = y \quad \text{if and only if} \quad \log_b x = \log_b y.$$

A general formula for any principal (P) being compounded n times per year for any number (t) of years at a rate (r) is

$$A = P\left(1 + \frac{r}{n}\right)^{nt},$$

where A represents the total amount of money accumulated at the end of the t years.

The value of $(1 + 1/n)^n$, as n gets infinitely large, approaches the number e, where e equals 2.71828 to five decimal places.

The formula

$$A = Pe^{rt}$$

yields the accumulated value, A, of a sum of money, P, that has been invested for t years at a rate of r percent **compounded continuously**. The equation

$$Q(t) = Q_0 e^{kt}$$

is used as a mathematical model for exponential growth and decay problems. The formula

$$\log_a r = \frac{\log_b r}{\log_b a}$$

is often called the **change-of-base** formula.

Chapter 6 Review Problem Set

For Problems 1–16, evaluate each of the expressions.

1. $\left(\frac{3}{4}\right)^{-2}$
2. $-\left(\frac{1}{3}\right)^{-3}$
3. $(2^{-1} \cdot 3^{-1})^{-2}$
4. $\left(\frac{3^{-2}}{4^{-1}}\right)^{-1}$

5. $8^{5/3}$
6. $-25^{3/2}$
7. $(-27)^{4/3}$
8. $\left(\frac{1}{9}\right)^{-1/2}$

9. $\left(-\frac{27}{64}\right)^{-1/3}$
10. $\log_6 216$
11. $\log_7\left(\frac{1}{49}\right)$
12. $\log_2 \sqrt[3]{2}$

13. $\log_2\left(\frac{\sqrt[4]{32}}{2}\right)$
14. $\log_{10}.00001$
15. $\ln e$
16. $7^{\log_7 12}$

For Problems 17–22, find the indicated products and quotients; express results using positive exponents only.

17. $(-3x^{-1}y^{-2})(4xy^5)$
18. $\frac{39x^2 y^{-3}}{13x^{-1}y}$
19. $\left(\frac{18a^{-3}b}{9a^{-1}b^{-2}}\right)^{-1}$

20. $\left(\frac{36ab^{-3}}{-4a^2b^4}\right)^{-2}$
21. $(-3x^{2/3})(7x^{1/4})$
22. $\left(\frac{5ax^{-2}}{a^{1/2}x^{-3}}\right)^2$

For Problems 23–36, solve each of the equations. Express approximate solutions to the nearest hundredth.

23. $\log_{10}2 + \log_{10}x = 1$
24. $\log_3 x = -2$
25. $4^x = 128$
26. $3^t = 42$
27. $\log_2 x = 3$
28. $\left(\frac{1}{27}\right)^{3x} = 3^{2x-1}$
29. $2e^x = 14$
30. $2^{2x+1} = 3^{x+1}$
31. $\ln(x + 4) - \ln(x + 2) = \ln x$
32. $\log x + \log(x - 15) = 2$
33. $\log(\log x) = 2$
34. $\log(7x - 4) - \log(x - 1) = 1$
35. $\ln(2t - 1) = \ln 4 + \ln(t - 3)$
36. $64^{2t+1} = 8^{-t+2}$

For Problems 37–40, if $\log 3 = .4771$ and $\log 7 = .8451$, evaluate each of the following.

37. $\log\left(\frac{7}{3}\right)$
38. $\log 21$
39. $\log 27$
40. $\log(7)^{2/3}$

41. Express each of the following as the sum or difference of simpler logarithmic quantities. Assume that all variables represent positive real numbers.

(a) $\log_b\left(\dfrac{x}{y^2}\right)$ (b) $\log_b\sqrt[4]{xy^2}$ (c) $\log_b\left(\dfrac{\sqrt{x}}{y^3}\right)$

42. Express each of the following as a single logarithm. Assume that all variables represent positive real numbers.

(a) $3\log_b x + 2\log_b y$ (b) $\dfrac{1}{2}\log_b y - 4\log_b x$

(c) $\dfrac{1}{2}(\log_b x + \log_b y) - 2\log_b z$

For Problems 43–46, approximate each of the logarithms to three decimal places.

43. $\log_2 3$ 44. $\log_3 2$ 45. $\log_4 191$ 46. $\log_2 .23$

For Problems 47–54, graph each of the functions.

47. $y = \left(\dfrac{3}{4}\right)^x$ 48. $y = 2^{x+2}$ 49. $y = e^{x-1}$

50. $y = -1 + \log x$ 51. $y = 3^x - 3^{-x}$ 52. $y = e^{-x^2/2}$

53. $y = \log_2(x - 3)$ 54. $y = 3\log_3 x$

For Problems 55–57, use the compound interest formula $A = P(1 + r/n)^{nt}$ to find the total amount of money accumulated at the end of the indicated time period for each of the investments.

55. $750 for 10 years at 11% compounded quarterly

56. $1250 for 15 years at 9% compounded monthly

57. $2500 for 20 years at 9.5% compounded semiannually

58. How long will it take $100 to double itself if it is invested at 14% interest compounded annually?

59. How long will it take $1000 to be worth $3500 if it is invested at 10.5% interest compounded quarterly?

60. What rate of interest (nearest tenth of a percent) compounded continuously is needed so that an investment of $500 will grow to $1000 in 8 years?

61. Suppose that the present population of a city is 50,000 and, furthermore, suppose that the equation $P(t) = P_0 e^{.02t}$, where P_0 represents an initial population, can be used to estimate future populations. Estimate the population of that city in 10 years, 15 years, and 20 years.

62. The number of bacteria present in a certain culture after t hours is given by the equation $Q = Q_0 e^{.29t}$, where Q_0 represents the initial number of bacteria. How long will it take 500 bacteria to increase to 2000 bacteria?

Appendices

The appendices of this text contain a table of trigonometric values and a table of common logarithms. When using tables, we are often confronted with the problem of determining values that are not in the table, but are in between two values in the table. By a process called **linear interpolation** such values can be approximated. Let's briefly explain the *general* process of linear interpolation and then suggest a format for using it with the trigonometric tables. A suggested format for the use of linear interpolation with logarithmic tables is given in Section 6.6.

Suppose that for some function f we know $f(a)$ and $f(b)$, but we want to determine $f(c)$, where c is between a and b as indicated by the following figure. To

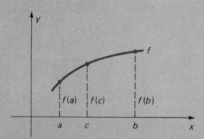

determine a reasonable approximation for $f(c)$ we assume that f is a straight line between $f(a)$ and $f(b)$ as indicated in the following figure.

From similar right triangles we have the proportion

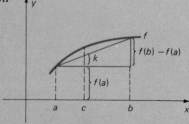

$$\frac{k}{f(b) - f(a)} = \frac{c - a}{b - a}.$$

Solving for k we obtain

$$k = \left(\frac{c - a}{b - a}\right)(f(b) - f(a)).$$

Therefore, an approximation for $f(c)$ is given by

$$f(c) = f(a) + k.$$

Appendix A

Trigonometric Values

Table A contains trigonometric values for angles measured in degrees or radians. Degree measures are given in .1° intervals from .0° to 45.0° in the second column from the left and from 45.0° to 90.0° in the second column from the right. Since .1° ≈ .0017 radians, the radian measures are given in intervals of approximately .0017. The following examples illustrate the use of Table A.

Degree Measure

EXAMPLE 1

Find cos 32.4°.

Locate 32.4° by reading down in the second column from the left. Then read across to the column labeled COS *on top*. Therefore,

cos 32.4° = .8443.

EXAMPLE 2

Find sin 73.8°.

Locate 73.8° by reading up in the second column from the right. Then read across to the column labeled SIN *at the bottom*. Therefore,

sin 73.8° = .9603.

EXAMPLE 3

Find θ if tan θ = 1.076.

Locate 1.076 in the column labeled TAN *at the bottom*. (If 0 < tan θ < 1, then it would be located in the column labeled TAN *at the top*.) Reading across to the *right* in the degree column we obtain

θ = 47.1°.

EXAMPLE 4

Find θ, to the nearest tenth of a degree, if sin θ = .0562.

Reading down in the column labeled SIN *at the top*, we see that .0562 falls between .0558 and .0576. Since .0562 is closer to .0558, we obtain

θ = 3.2°.

EXAMPLE 5

Find θ, to the nearest hundredth of a degree, if cos θ = .7498.

Reading down in the column labeled COS *at the top*, we see that .7498 falls between .7501 and .7490. (The cosine function is a decreasing function for the values given in this table.) To interpolate we can use the following format.

$$\cos \theta \qquad\qquad\qquad \theta$$

$$.0003 \left\{ \begin{matrix} .7501 \\ .7498 \\ \end{matrix} \right\} .0011 \qquad h \left\{ \begin{matrix} 41.4° \\ ? \\ \end{matrix} \right\} .1°$$
$$.7490 \qquad\qquad\qquad 41.5°$$

$$\frac{.0003}{.0011} = \frac{3}{11} = \frac{h}{.1}$$

$$11h = .3$$

$$h = .03°, \text{ to the nearest hundredth}$$

Therefore, $\theta = 41.4° + .03° = 41.43°$.

Radian Measure

EXAMPLE 6

Find cos 1.4556.

Locate 1.4556 in the column on the far right labeled **RADIANS** *at the bottom.* Then read across to the column labeled **COS** *at the bottom.* Therefore,

$$\cos 1.4556 = .1149.$$

EXAMPLE 7

Find x, to the nearest hundredth of a radian, if $\cos \theta = .0650$.

In the column labeled **COS** *at the bottom,* we find that .0650 falls between .0645 and .0663. Since .0650 is closer to .0645 we obtain $x = 1.5062$ by reading across in the far right column. Therefore, to the nearest hundredth of a radian, we obtain

$$x = 1.51.$$

Table A/Values of the Trigonometric Functions

RADIANS	DEGREES	SIN	COS	TAN	COT		
.0000	.0°	.0000	1.0000	.0000	—	90.0°	1.5708
.0017	.1°	.0017	1.0000	.0017	573.0	89.9°	1.5691
.0035	.2°	.0035	1.0000	.0035	286.5	89.8°	1.5673
.0052	.3°	.0052	1.0000	.0052	191.0	89.7°	1.5656
.0070	.4°	.0070	1.0000	.0070	143.2	89.6°	1.5638
.0087	.5°	.0087	1.0000	.0087	114.6	89.5°	1.5621
.0105	.6°	.0105	.9999	.0105	95.49	89.4°	1.5603
.0122	.7°	.0122	.9999	.0122	81.85	89.3°	1.5586
.0140	.8°	.0140	.9999	.0140	71.62	89.2°	1.5568
.0157	.9°	.0157	.9999	.0157	63.66	89.1°	1.5551
		COS	SIN	COT	TAN	DEGREES	RADIANS

Table A/Values of the Trigonometric Functions   (*Continued*)

RADIANS	DEGREES	SIN	COS	TAN	COT		
.0175	1.0°	.0175	.9998	.0175	57.29	89.0°	1.5533
.0192	1.1°	.0192	.9998	.0192	52.08	88.9°	1.5516
.0209	1.2°	.0209	.9998	.0209	47.74	88.8°	1.5499
.0227	1.3°	.0227	.9997	.0227	44.07	88.7°	1.5481
.0244	1.4°	.0244	.9997	.0244	40.92	88.6°	1.5464
.0262	1.5°	.0262	.9997	.0262	38.19	88.5°	1.5446
.0279	1.6°	.0279	.9996	.0279	35.80	88.4°	1.5429
.0297	1.7°	.0297	.9996	.0297	33.69	88.3°	1.5411
.0314	1.8°	.0314	.9995	.0314	31.82	88.2°	1.5394
.0332	1.9°	.0332	.9995	.0332	30.14	88.1°	1.5376
.0349	2.0°	.0349	.9994	.0349	28.64	88.0°	1.5359
.0367	2.1°	.0366	.9993	.0367	27.27	87.9°	1.5341
.0384	2.2°	.0384	.9993	.0384	26.03	87.8°	1.5324
.0401	2.3°	.0401	.9992	.0402	24.90	87.7°	1.5307
.0419	2.4°	.0419	.9991	.0419	23.86	87.6°	1.5289
.0436	2.5°	.0436	.9990	.0437	22.90	87.5°	1.5272
.0454	2.6°	.0454	.9990	.0454	22.02	87.4°	1.5254
.0471	2.7°	.0471	.9989	.0472	21.20	87.3°	1.5237
.0489	2.8°	.0488	.9988	.0489	20.45	87.2°	1.5219
.0506	2.9°	.0506	.9987	.0507	19.74	87.1°	1.5202
.0524	3.0°	.0523	.9986	.0524	19.08	87.0°	1.5184
.0541	3.1°	.0541	.9985	.0542	18.46	86.9°	1.5167
.0559	3.2°	.0558	.9984	.0559	17.89	86.8°	1.5149
.0576	3.3°	.0576	.9983	.0577	17.34	86.7°	1.5132
.0593	3.4°	.0593	.9982	.0594	16.83	86.6°	1.5115
.0611	3.5°	.0610	.9981	.0612	16.35	86.5°	1.5097
.0628	3.6°	.0628	.9980	.0629	15.89	86.4°	1.5080
.0646	3.7°	.0645	.9979	.0647	15.46	86.3°	1.5062
.0663	3.8°	.0663	.9978	.0664	15.06	86.2°	1.5045
.0681	3.9°	.0680	.9977	.0682	14.67	86.1°	1.5027
.0698	4.0°	.0698	.9976	.0699	14.30	86.0°	1.5010
.0716	4.1°	.0715	.9974	.0717	13.95	85.9°	1.4992
.0733	4.2°	.0732	.9973	.0734	13.62	85.8°	1.4975
.0750	4.3°	.0750	.9972	.0752	13.30	85.7°	1.4957
.0768	4.4°	.0767	.9971	.0769	13.00	85.6°	1.4940
.0785	4.5°	.0785	.9969	.0787	12.71	85.5°	1.4923
.0803	4.6°	.0802	.9968	.0805	12.43	85.4°	1.4905
.0820	4.7°	.0819	.9966	.0822	12.16	85.3°	1.4888
.0838	4.8°	.0837	.9965	.0840	11.91	85.2°	1.4870
.0855	4.9°	.0854	.9963·	.0857	11.66	85.1°	1.4853
		COS	SIN	COT	TAN	DEGREES	RADIANS

Table A/Values of the Trigonometric Functions (*Continued*)

RADIANS	DEGREES	SIN	COS	TAN	COT		
.0873	5.0°	.0872	.9962	.0875	11.43	85.0°	1.4835
.0890	5.1°	.0889	.9960	.0892	11.20	84.9°	1.4818
.0908	5.2°	.0906	.9959	.0910	10.99	84.8°	1.4800
.0925	5.3°	.0924	.9957	.0928	10.78	84.7°	1.4783
.0942	5.4°	.0941	.9956	.0945	10.58	84.6°	1.4765
.0960	5.5°	.0958	.9954	.0963	10.39	84.5°	1.4748
.0977	5.6°	.0976	.9952	.0981	10.20	84.4°	1.4731
.0995	5.7°	.0993	.9951	.0998	10.02	84.3°	1.4713
.1012	5.8°	.1011	.9949	.1016	9.845	84.2°	1.4696
.1030	5.9°	.1028	.9947	.1033	9.677	84.1°	1.4678
.1047	6.0°	.1045	.9945	.1051	9.514	84.0°	1.4661
.1065	6.1°	.1063	.9943	.1069	9.357	83.9°	1.4643
.1082	6.2°	.1080	.9942	.1086	9.205	83.8°	1.4626
.1100	6.3°	.1097	.9940	.1104	9.058	83.7°	1.4608
.1117	6.4°	.1115	.9938	.1122	8.915	83.6°	1.4591
.1134	6.5°	.1132	.9936	.1139	8.777	83.5°	1.4573
.1152	6.6°	.1149	.9934	.1157	8.643	83.4°	1.4556
.1169	6.7°	.1167	.9932	.1175	8.513	83.3°	1.4539
.1187	6.8°	.1184	.9930	.1192	8.386	83.2°	1.4521
.1204	6.9°	.1201	.9928	.1210	8.264	83.1°	1.4504
.1222	7.0°	.1219	.9925	.1228	8.144	83.0°	1.4486
.1239	7.1°	.1236	.9923	.1246	8.028	82.9°	1.4469
.1257	7.2°	.1253	.9921	.1263	7.916	82.8°	1.4451
.1274	7.3°	.1271	.9919	.1281	7.806	82.7°	1.4434
.1292	7.4°	.1288	.9917	.1299	7.700	82.6°	1.4416
.1309	7.5°	.1305	.9914	.1317	7.596	82.5°	1.4399
.1326	7.6°	.1323	.9912	.1334	7.495	82.4°	1.4382
.1344	7.7°	.1340	.9910	.1352	7.396	82.3°	1.4364
.1361	7.8°	.1357	.9907	.1370	7.300	82.2°	1.4347
.1379	7.9°	.1374	.9905	.1388	7.207	82.1°	1.4329
.1396	8.0°	.1392	.9903	.1405	7.115	82.0°	1.4312
.1414	8.1°	.1409	.9900	.1423	7.026	81.9°	1.4294
.1431	8.2°	.1426	.9898	.1441	6.940	81.8°	1.4277
.1449	8.3°	.1444	.9895	.1459	6.855	81.7°	1.4259
.1466	8.4°	.1461	.9893	.1477	6.772	81.6°	1.4242
.1484	8.5°	.1478	.9890	.1495	6.691	81.5°	1.4224
.1501	8.6°	.1495	.9888	.1512	6.612	81.4°	1.4207
.1518	8.7°	.1513	.9885	.1530	6.535	81.3°	1.4190
.1536	8.8°	.1530	.9882	.1548	6.460	81.2°	1.4172
.1553	8.9°	.1547	.9880	.1566	6.386	81.1°	1.4155
		COS	SIN	COT	TAN	DEGREES	RADIANS

Table A/Values of the Trigonometric Functions   (*Continued*)

RADIANS	DEGREES	SIN	COS	TAN	COT		
.1571	9.0°	.1564	.9877	.1584	6.314	81.0°	1.4137
.1588	9.1°	.1582	.9874	.1602	6.243	80.9°	1.4120
.1606	9.2°	.1599	.9871	.1620	6.174	80.8°	1.4102
.1623	9.3°	.1616	.9869	.1638	6.107	80.7°	1.4085
.1641	9.4°	.1633	.9866	.1655	6.041	80.6°	1.4067
.1658	9.5°	.1650	.9863	.1673	5.976	80.5°	1.4050
.1676	9.6°	.1668	.9860	.1691	5.912	80.4°	1.4032
.1693	9.7°	.1685	.9857	.1709	5.850	80.3°	1.4015
.1710	9.8°	.1702	.9854	.1727	5.789	80.2°	1.3998
.1728	9.9°	.1719	.9851	.1745	5.730	80.1°	1.3980
.1745	10.0°	.1736	.9848	.1763	5.671	80.0°	1.3963
.1763	10.1°	.1754	.9845	.1781	5.614	79.9°	1.3945
.1780	10.2°	.1771	.9842	.1799	5.558	79.8°	1.3928
.1798	10.3°	.1788	.9839	.1817	5.503	79.7°	1.3910
.1815	10.4°	.1805	.9836	.1835	5.449	79.6°	1.3893
.1833	10.5°	.1822	.9833	.1853	5.396	79.5°	1.3875
.1850	10.6°	.1840	.9829	.1871	5.343	79.4°	1.3858
.1868	10.7°	.1857	.9826	.1890	5.292	79.3°	1.3840
.1885	10.8°	.1874	.9823	.1908	5.242	79.2°	1.3823
.1902	10.9°	.1891	.9820	.1926	5.193	79.1°	1.3806
.1920	11.0°	.1908	.9816	.1944	5.145	79.0°	1.3788
.1937	11.1°	.1925	.9813	.1962	5.097	78.9°	1.3771
.1955	11.2°	.1942	.9810	.1980	5.050	78.8°	1.3753
.1972	11.3°	.1959	.9806	.1998	5.005	78.7°	1.3736
.1990	11.4°	.1977	.9803	.2016	4.959	78.6°	1.3718
.2007	11.5°	.1994	.9799	.2035	4.915	78.5°	1.3701
.2025	11.6°	.2011	.9796	.2053	4.872	78.4°	1.3683
.2042	11.7°	.2028	.9792	.2071	4.829	78.3°	1.3666
.2059	11.8°	.2045	.9789	.2089	4.787	78.2°	1.3648
.2077	11.9°	.2062	.9785	.2107	4.745	78.1°	1.3631
.2094	12.0°	.2079	.9781	.2126	4.705	78.0°	1.3614
.2112	12.1°	.2096	.9778	.2144	4.665	77.9°	1.3596
.2129	12.2°	.2113	.9774	.2162	4.625	77.8°	1.3579
.2147	12.3°	.2130	.9770	.2180	4.586	77.7°	1.3561
.2164	12.4°	.2147	.9767	.2199	4.548	77.6°	1.3544
.2182	12.5°	.2164	.9763	.2217	4.511	77.5°	1.3526
.2199	12.6°	.2181	.9759	.2235	4.474	77.4°	1.3509
.2217	12.7°	.2198	.9755	.2254	4.437	77.3°	1.3491
.2234	12.8°	.2215	.9751	.2272	4.402	77.2°	1.3474
.2251	12.9°	.2233	.9748	.2290	4.366	77.1°	1.3456
		COS	SIN	COT	TAN	DEGREES	RADIANS

Table A/Values of the Trigonometric Functions   (*Continued*)

RADIANS	DEGREES	SIN	COS	TAN	COT		
.2269	13.0°	.2250	.9744	.2309	4.331	77.0°	1.3439
.2286	13.1°	.2267	.9740	.2327	4.297	76.9°	1.3422
.2304	13.2°	.2284	.9736	.2345	4.264	76.8°	1.3404
.2321	13.3°	.2300	.9732	.2364	4.230	76.7°	1.3387
.2339	13.4°	.2317	.9728	.2382	4.198	76.6°	1.3369
.2356	13.5°	.2334	.9724	.2401	4.165	76.5°	1.3352
.2374	13.6°	.2351	.9720	.2419	4.134	76.4°	1.3334
.2391	13.7°	.2368	.9715	.2438	4.102	76.3°	1.3317
.2409	13.8°	.2385	.9711	.2456	4.071	76.2°	1.3299
.2426	13.9°	.2402	.9707	.2475	4.041	76.1°	1.3282
.2443	14.0°	.2419	.9703	.2493	4.011	76.0°	1.3265
.2461	14.1°	.2436	.9699	.2512	3.981	75.9°	1.3247
.2478	14.2°	.2453	.9694	.2530	3.952	75.8°	1.3230
.2496	14.3°	.2470	.9690	.2549	3.923	75.7°	1.3212
.2513	14.4°	.2487	.9686	.2568	3.895	75.6°	1.3195
.2531	14.5°	.2504	.9681	.2586	3.867	75.5°	1.3177
.2548	14.6°	.2521	.9677	.2605	3.839	75.4°	1.3160
.2566	14.7°	.2538	.9673	.2623	3.812	75.3°	1.3142
.2583	14.8°	.2554	.9668	.2642	3.785	75.2°	1.3125
.2601	14.9°	.2571	.9664	.2661	3.758	75.1°	1.3107
.2618	15.0°	.2588	.9659	.2679	3.732	75.0°	1.3090
.2635	15.1°	.2605	.9655	.2698	3.706	74.9°	1.3073
.2653	15.2°	.2622	.9650	.2717	3.681	74.8°	1.3055
.2670	15.3°	.2639	.9646	.2736	3.655	74.7°	1.3038
.2688	15.4°	.2656	.9641	.2754	3.630	74.6°	1.3020
.2705	15.5°	.2672	.9636	.2773	3.606	74.5°	1.3003
.2723	15.6°	.2689	.9632	.2792	3.582	74.4°	1.2985
.2740	15.7°	.2706	.9627	.2811	3.558	74.3°	1.2968
.2758	15.8°	.2723	.9622	.2830	3.534	74.2°	1.2950
.2775	15.9°	.2740	.9617	.2849	3.511	74.1°	1.2933
.2793	16.0°	.2756	.9613	.2867	3.487	74.0°	1.2915
.2810	16.1°	.2773	.9608	.2886	3.465	73.9°	1.2898
.2827	16.2°	.2790	.9603	.2905	3.442	73.8°	1.2881
.2845	16.3°	.2807	.9598	.2924	3.420	73.7°	1.2863
.2862	16.4°	.2823	.9593	.2943	3.398	73.6°	1.2846
.2880	16.5°	.2840	.9588	.2962	3.376	73.5°	1.2828
.2897	16.6°	.2857	.9583	.2981	3.354	73.4°	1.2811
.2915	16.7°	.2874	.9578	.3000	3.333	73.3°	1.2793
.2932	16.8°	.2890	.9573	.3019	3.312	73.2°	1.2776
.2950	16.9°	.2907	.9568	.3038	3.291	73.1°	1.2758
		COS	SIN	COT	TAN	DEGREES	RADIANS

Table A/Values of the Trigonometric Functions   (*Continued*)

RADIANS	DEGREES	SIN	COS	TAN	COT		
.2967	17.0°	.2924	.9563	.3057	3.271	73.0°	1.2741
.2985	17.1°	.2940	.9558	.3076	3.251	72.9°	1.2723
.3002	17.2°	.2957	.9553	.3096	3.230	72.8°	1.2706
.3019	17.3°	.2974	.9548	.3115	3.211	72.7°	1.2689
.3037	17.4°	.2990	.9542	.3134	3.191	72.6°	1.2671
.3054	17.5°	.3007	.9537	.3153	3.172	72.5°	1.2654
.3072	17.6°	.3024	.9532	.3172	3.152	72.4°	1.2636
.3089	17.7°	.3040	.9527	.3191	3.133	72.3°	1.2619
.3107	17.8°	.3057	.9521	.3211	3.115	72.2°	1.2601
.3124	17.9°	.3074	.9516	.3230	3.096	72.1°	1.2584
.3142	18.0°	.3090	.9511	.3249	3.078	72.0°	1.2566
.3159	18.1°	.3107	.9505	.3269	3.060	71.9°	1.2549
.3176	18.2°	.3123	.9500	.3288	3.042	71.8°	1.2531
.3194	18.3°	.3140	.9494	.3307	3.024	71.7°	1.2514
.3211	18.4°	.3156	.9489	.3327	3.006	71.6°	1.2497
.3229	18.5°	.3173	.9483	.3346	2.989	71.5°	1.2479
.3246	18.6°	.3190	.9478	.3365	2.971	71.4°	1.2462
.3264	18.7°	.3206	.9472	.3385	2.954	71.3°	1.2444
.3281	18.8°	.3223	.9466	.3404	2.937	71.2°	1.2427
.3299	18.9°	.3239	.9461	.3424	2.921	71.1°	1.2409
.3316	19.0°	.3256	.9455	.3443	2.904	71.0°	1.2392
.3334	19.1°	.3272	.9449	.3463	2.888	70.9°	1.2374
.3351	19.2°	.3289	.9444	.3482	2.872	70.8°	1.2357
.3368	19.3°	.3305	.9438	.3502	2.856	70.7°	1.2339
.3386	19.4°	.3322	.9432	.3522	2.840	70.6°	1.2322
.3403	19.5°	.3338	.9426	.3541	2.824	70.5°	1.2305
.3421	19.6°	.3355	.9421	.3561	2.808	70.4°	1.2287
.3438	19.7°	.3371	.9415	.3581	2.793	70.3°	1.2270
.3456	19.8°	.3387	.9409	.3600	2.778	70.2°	1.2252
.3473	19.9°	.3404	.9403	.3620	2.762	70.1°	1.2235
.3491	20.0°	.3420	.9397	.3640	2.747	70.0°	1.2217
.3508	20.1°	.3437	.9391	.3659	2.733	69.9°	1.2200
.3526	20.2°	.3453	.9385	.3679	2.718	69.8°	1.2182
.3543	20.3°	.3469	.9379	.3699	2.703	69.7°	1.2165
.3560	20.4°	.3486	.9373	.3719	2.689	69.6°	1.2147
.3578	20.5°	.3502	.9367	.3739	2.675	69.5°	1.2130
.3595	20.6°	.3518	.9361	.3759	2.660	69.4°	1.2113
.3613	20.7°	.3535	.9354	.3779	2.646	69.3°	1.2095
.3630	20.8°	.3551	.9348	.3799	2.633	69.2°	1.2078
.3648	20.9°	.3567	.9342	.3819	2.619	69.1°	1.2060
		COS	SIN	COT	TAN	DEGREES	RADIANS

Table A/Values of the Trigonometric Functions (*Continued*)

RADIANS	DEGREES	SIN	COS	TAN	COT		
.3665	21.0°	.3584	.9336	.3839	2.605	69.0°	1.2043
.3683	21.1°	.3600	.9330	.3859	2.592	68.9°	1.2025
.3700	21.2°	.3616	.9323	.3879	2.578	68.8°	1.2008
.3718	21.3°	.3633	.9317	.3899	2.565	68.7°	1.1990
.3735	21.4°	.3649	.9311	.3919	2.552	68.6°	1.1973
.3752	21.5°	.3665	.9304	.3939	2.539	68.5°	1.1956
.3770	21.6°	.3681	.9298	.3959	2.526	68.4°	1.1938
.3787	21.7°	.3697	.9291	.3979	2.513	68.3°	1.1921
.3805	21.8°	.3714	.9285	.4000	2.500	68.2°	1.1903
.3822	21.9°	.3730	.9278	.4020	2.488	68.1°	1.1886
.3840	22.0°	.3746	.9272	.4040	2.475	68.0°	1.1868
.3857	22.1°	.3762	.9265	.4061	2.463	67.9°	1.1851
.3875	22.2°	.3778	.9259	.4081	2.450	67.8°	1.1833
.3892	22.3°	.3795	.9252	.4101	2.438	67.7°	1.1816
.3910	22.4°	.3811	.9245	.4122	2.426	67.6°	1.1798
.3927	22.5°	.3827	.9239	.4142	2.414	67.5°	1.1781
.3944	22.6°	.3843	.9232	.4163	2.402	67.4°	1.1764
.3962	22.7°	.3859	.9225	.4183	2.391	67.3°	1.1746
.3979	22.8°	.3875	.9219	.4204	2.379	67.2°	1.1729
.3997	22.9°	.3891	.9212	.4224	2.367	67.1°	1.1711
.4014	23.0°	.3907	.9205	.4245	2.356	67.0°	1.1694
.4032	23.1°	.3923	.9198	.4265	2.344	66.9°	1.1676
.4049	23.2°	.3939	.9191	.4286	2.333	66.8°	1.1659
.4067	23.3°	.3955	.9184	.4307	2.322	66.7°	1.1641
.4084	23.4°	.3971	.9178	.4327	2.311	66.6°	1.1624
.4102	23.5°	.3987	.9171	.4348	2.300	66.5°	1.1606
.4119	23.6°	.4003	.9164	.4369	2.289	66.4°	1.1589
.4136	23.7°	.4019	.9157	.4390	2.278	66.3°	1.1572
.4154	23.8°	.4035	.9150	.4411	2.267	66.2°	1.1554
.4171	23.9°	.4051	.9143	.4431	2.257	66.1°	1.1537
.4189	24.0°	.4067	.9135	.4452	2.246	66.0°	1.1519
.4206	24.1°	.4083	.9128	.4473	2.236	65.9°	1.1502
.4224	24.2°	.4099	.9121	.4494	2.225	65.8°	1.1484
.4241	24.3°	.4115	.9114	.4515	2.215	65.7°	1.1467
.4259	24.4°	.4131	.9107	.4536	2.204	65.6°	1.1449
.4276	24.5°	.4147	.9100	.4557	2.194	65.5°	1.1432
.4294	24.6°	.4163	.9092	.4578	2.184	65.4°	1.1414
.4311	24.7°	.4179	.9085	.4599	2.174	65.3°	1.1397
.4328	24.8°	.4195	.9078	.4621	2.164	65.2°	1.1380
.4346	24.9°	.4210	.9070	.4642	2.154	65.1°	1.1362
		COS	SIN	COT	TAN	DEGREES	RADIANS

Table A/Values of the Trigonometric Functions   (*Continued*)

RADIANS	DEGREES	SIN	COS	TAN	COT		
.4363	25.0°	.4226	.9063	.4663	2.145	65.0°	1.1345
.4381	25.1°	.4242	.9056	.4684	2.135	64.9°	1.1327
.4398	25.2°	.4258	.9048	.4706	2.125	64.8°	1.1310
.4416	25.3°	.4274	.9041	.4727	2.116	64.7°	1.1292
.4433	25.4°	.4289	.9033	.4748	2.106	64.6°	1.1275
.4451	25.5°	.4305	.9026	.4770	2.097	64.5°	1.1257
.4468	25.6°	.4321	.9018	.4791	2.087	64.4°	1.1240
.4485	25.7°	.4337	.9011	.4813	2.078	64.3°	1.1222
.4503	25.8°	.4352	.9003	.4834	2.069	64.2°	1.1205
.4520	25.9°	.4368	.8996	.4856	2.059	64.1°	1.1188
.4538	26.0°	.4384	.8988	.4877	2.050	64.0°	1.1170
.4555	26.1°	.4399	.8980	.4899	2.041	63.9°	1.1153
.4573	26.2°	.4415	.8973	.4921	2.032	63.8°	1.1135
.4590	26.3°	.4431	.8965	.4942	2.023	63.7°	1.1118
.4608	26.4°	.4446	.8957	.4964	2.014	63.6°	1.1100
.4625	26.5°	.4462	.8949	.4986	2.006	63.5°	1.1083
.4643	26.6°	.4478	.8942	.5008	1.997	63.4°	1.1065
.4660	26.7°	.4493	.8934	.5029	1.988	63.3°	1.1048
.4677	26.8°	.4509	.8926	.5051	1.980	63.2°	1.1030
.4695	26.9°	.4524	.8918	.5073	1.971	63.1°	1.1013
.4712	27.0°	.4540	.8910	.5095	1.963	63.0°	1.0996
.4730	27.1°	.4555	.8902	.5117	1.954	62.9°	1.0978
.4747	27.2°	.4571	.8894	.5139	1.946	62.8°	1.0961
.4765	27.3°	.4586	.8886	.5161	1.937	62.7°	1.0943
.4782	27.4°	.4602	.8878	.5184	1.929	62.6°	1.0926
.4800	27.5°	.4617	.8870	.5206	1.921	62.5°	1.0908
.4817	27.6°	.4633	.8862	.5228	1.913	62.4°	1.0891
.4835	27.7°	.4648	.8854	.5250	1.905	62.3°	1.0873
.4852	27.8°	.4664	.8846	.5272	1.897	62.2°	1.0856
.4869	27.9°	.4679	.8838	.5295	1.889	62.1°	1.0838
.4887	28.0°	.4695	.8829	.5317	1.881	62.0°	1.0821
.4904	28.1°	.4710	.8821	.5340	1.873	61.9°	1.0804
.4922	28.2°	.4726	.8813	.5362	1.865	61.8°	1.0786
.4939	28.3°	.4741	.8805	.5384	1.857	61.7°	1.0769
.4957	28.4°	.4756	.8796	.5407	1.849	61.6°	1.0751
.4974	28.5°	.4772	.8788	.5430	1.842	61.5°	1.0734
.4992	28.6°	.4787	.8780	.5452	1.834	61.4°	1.0716
.5009	28.7°	.4802	.8771	.5475	1.827	61.3°	1.0699
.5027	28.8°	.4818	.8763	.5498	1.819	61.2°	1.0681
.5044	28.9°	.4833	.8755	.5520	1.811	61.1°	1.0664
		COS	SIN	COT	TAN	DEGREES	RADIANS

Table A/Values of the Trigonometric Functions (*Continued*)

RADIANS	DEGREES	SIN	COS	TAN	COT		
.5061	29.0°	.4848	.8746	.5543	1.804	61.0°	1.0647
.5079	29.1°	.4863	.8738	.5566	1.797	60.9°	1.0629
.5096	29.2°	.4879	.8729	.5589	1.789	60.8°	1.0612
.5114	29.3°	.4894	.8721	.5612	1.782	60.7°	1.0594
.5131	29.4°	.4909	.8712	.5635	1.775	60.6°	1.0577
.5149	29.5°	.4924	.8704	.5658	1.767	60.5°	1.0559
.5166	29.6°	.4939	.8695	.5681	1.760	60.4°	1.0542
.5184	29.7°	.4955	.8686	.5704	1.753	60.3°	1.0524
.5201	29.8°	.4970	.8678	.5727	1.746	60.2°	1.0507
.5219	29.9°	.4985	.8669	.5750	1.739	60.1°	1.0489
.5236	30.0°	.5000	.8660	.5774	1.732	60.0°	1.0472
.5253	30.1°	.5015	.8652	.5797	1.725	59.9°	1.0455
.5271	30.2°	.5030	.8643	.5820	1.718	59.8°	1.0437
.5288	30.3°	.5045	.8634	.5844	1.711	59.7°	1.0420
.5306	30.4°	.5060	.8625	.5867	1.704	59.6°	1.0402
.5323	30.5°	.5075	.8616	.5890	1.698	59.5°	1.0385
.5341	30.6°	.5090	.8607	.5914	1.691	59.4°	1.0367
.5358	30.7°	.5105	.8599	.5938	1.684	59.3°	1.0350
.5376	30.8°	.5120	.8590	.5961	1.678	59.2°	1.0332
.5393	30.9°	.5135	.8581	.5985	1.671	59.1°	1.0315
.5411	31.0°	.5150	.8572	.6009	1.664	59.0°	1.0297
.5428	31.1°	.5165	.8563	.6032	1.658	58.9°	1.0280
.5445	31.2°	.5180	.8554	.6056	1.651	58.8°	1.0263
.5463	31.3°	.5195	.8545	.6080	1.645	58.7°	1.0245
.5480	31.4°	.5210	.8536	.6104	1.638	58.6°	1.0228
.5498	31.5°	.5225	.8526	.6128	1.632	58.5°	1.0210
.5515	31.6°	.5240	.8517	.6152	1.625	58.4°	1.0193
.5533	31.7°	.5255	.8508	.6176	1.619	58.3°	1.0175
.5550	31.8°	.5270	.8499	.6200	1.613	58.2°	1.0158
.5568	31.9°	.5284	.8490	.6224	1.607	58.1°	1.0140
.5585	32.0°	.5299	.8480	.6249	1.600	58.0°	1.0123
.5603	32.1°	.5314	.8471	.6273	1.594	57.9°	1.0105
.5620	32.2°	.5329	.8462	.6297	1.588	57.8°	1.0088
.5637	32.3°	.5344	.8453	.6322	1.582	57.7°	1.0071
.5655	32.4°	.5358	.8443	.6346	1.576	57.6°	1.0053
.5672	32.5°	.5373	.8434	.6371	1.570	57.5°	1.0036
.5690	32.6°	.5388	.8425	.6395	1.564	57.4°	1.0018
.5707	32.7°	.5402	.8415	.6420	1.558	57.3°	1.0001
.5725	32.8°	.5417	.8406	.6445	1.552	57.2°	.9983
.5742	32.9°	.5432	.8396	.6469	1.546	57.1°	.9966
		COS	SIN	COT	TAN	DEGREES	RADIANS

Table A/Values of the Trigonometric Functions   (*Continued*)

RADIANS	DEGREES	SIN	COS	TAN	COT		
.5760	33.0°	.5446	.8387	.6494	1.540	57.0°	.9948
.5777	33.1°	.5461	.8377	.6519	1.534	56.9°	.9931
.5794	33.2°	.5476	.8368	.6544	1.528	56.8°	.9913
.5812	33.3°	.5490	.8358	.6569	1.522	56.7°	.9896
.5829	33.4°	.5505	.8348	.6594	1.517	56.6°	.9879
.5847	33.5°	.5519	.8339	.6619	1.511	56.5°	.9861
.5864	33.6°	.5534	.8329	.6644	1.505	56.4°	.9844
.5882	33.7°	.5548	.8320	.6669	1.499	56.3°	.9826
.5899	33.8°	.5563	.8310	.6694	1.494	56.2°	.9809
.5917	33.9°	.5577	.8300	.6720	1.488	56.1°	.9791
.5934	34.0°	.5592	.8290	.6745	1.483	56.0°	.9774
.5952	34.1°	.5606	.8281	.6771	1.477	55.9°	.9756
.5969	34.2°	.5621	.8271	.6796	1.471	55.8°	.9739
.5986	34.3°	.5635	.8261	.6822	1.466	55.7°	.9721
.6004	34.4°	.5650	.8251	.6847	1.460	55.6°	.9704
.6021	34.5°	.5664	.8241	.6873	1.455	55.5°	.9687
.6039	34.6°	.5678	.8231	.6899	1.450	55.4°	.9669
.6056	34.7°	.5693	.8221	.6924	1.444	55.3°	.9652
.6074	34.8°	.5707	.8211	.6950	1.439	55.2°	.9634
.6091	34.9°	.5721	.8202	.6976	1.433	55.1°	.9617
.6109	35.0°	.5736	.8192	.7002	1.428	55.0°	.9599
.6126	35.1°	.5750	.8181	.7028	1.423	54.9°	.9582
.6144	35.2°	.5764	.8171	.7054	1.418	54.8°	.9564
.6161	35.3°	.5779	.8161	.7080	1.412	54.7°	.9547
.6178	35.4°	.5793	.8151	.7107	1.407	54.6°	.9530
.6196	35.5°	.5807	.8141	.7133	1.402	54.5°	.9512
.6213	35.6°	.5821	.8131	.7159	1.397	54.4°	.9495
.6231	35.7°	.5835	.8121	.7186	1.392	54.3°	.9477
.6248	35.8°	.5850	.8111	.7212	1.387	54.2°	.9460
.6266	35.9°	.5864	.8100	.7239	1.381	54.1°	.9442
.6283	36.0°	.5878	.8090	.7265	1.376	54.0°	.9425
.6301	36.1°	.5892	.8080	.7292	1.371	53.9°	.9407
.6318	36.2°	.5906	.8070	.7319	1.366	53.8°	.9390
.6336	36.3°	.5920	.8059	.7346	1.361	53.7°	.9372
.6353	36.4°	.5934	.8049	.7373	1.356	53.6°	.9355
.6370	36.5°	.5948	.8039	.7400	1.351	53.5°	.9338
.6388	36.6°	.5962	.8028	.7427	1.347	53.4°	.9320
.6405	36.7°	.5976	.8018	.7454	1.342	53.3°	.9303
.6423	36.8°	.5990	.8007	.7481	1.337	53.2°	.9285
.6440	36.9°	.6004	.7997	.7508	1.332	53.1°	.9268
		COS	SIN	COT	TAN	DEGREES	RADIANS

Table A/Values of the Trigonometric Functions   (*Continued*)

RADIANS	DEGREES	SIN	COS	TAN	COT		
.6458	37.0°	.6018	.7986	.7536	1.327	53.0°	.9250
.6475	37.1°	.6032	.7976	.7563	1.322	52.9°	.9233
.6493	37.2°	.6046	.7965	.7590	1.317	52.8°	.9215
.6510	37.3°	.6060	.7955	.7618	1.313	52.7°	.9198
.6528	37.4°	.6074	.7944	.7646	1.308	52.6°	.9180
.6545	37.5°	.6088	.7934	.7673	1.303	52.5°	.9163
.6562	37.6°	.6101	.7923	.7701	1.299	52.4°	.9146
.6580	37.7°	.6115	.7912	.7729	1.294	52.3°	.9128
.6597	37.8°	.6129	.7902	.7757	1.289	52.2°	.9111
.6615	37.9°	.6143	.7891	.7785	1.285	52.1°	.9093
.6632	38.0°	.6157	.7880	.7813	1.280	52.0°	.9076
.6650	38.1°	.6170	.7869	.7841	1.275	51.9°	.9058
.6667	38.2°	.6184	.7859	.7869	1.271	51.8°	.9041
.6685	38.3°	.6198	.7848	.7898	1.266	51.7°	.9023
.6702	38.4°	.6211	.7837	.7926	1.262	51.6°	.9006
.6720	38.5°	.6225	.7826	.7954	1.257	51.5°	.8988
.6737	38.6°	.6239	.7815	.7983	1.253	51.4°	.8971
.6754	38.7°	.6252	.7804	.8012	1.248	51.3°	.8954
.6772	38.8°	.6266	.7793	.8040	1.244	51.2°	.8936
.6789	38.9°	.6280	.7782	.8069	1.239	51.1°	.8919
.6807	39.0°	.6293	.7771	.8098	1.235	51.0°	.8901
.6824	39.1°	.6307	.7760	.8127	1.230	50.9°	.8884
.6842	39.2°	.6320	.7749	.8156	1.226	50.8°	.8866
.6859	39.3°	.6334	.7738	.8185	1.222	50.7°	.8849
.6877	39.4°	.6347	.7727	.8214	1.217	50.6°	.8831
.6894	39.5°	.6361	.7716	.8243	1.213	50.5°	.8814
.6912	39.6°	.6374	.7705	.8273	1.209	50.4°	.8796
.6929	39.7°	.6388	.7694	.8302	1.205	50.3°	.8779
.6946	39.8°	.6401	.7683	.8332	1.200	50.2°	.8762
.6964	39.9°	.6414	.7672	.8361	1.196	50.1°	.8744
.6981	40.0°	.6428	.7660	.8391	1.192	50.0°	.8727
.6999	40.1°	.6441	.7649	.8421	1.188	49.9°	.8709
.7016	40.2°	.6455	.7638	.8451	1.183	49.8°	.8692
.7034	40.3°	.6468	.7627	.8481	1.179	49.7°	.8674
.7051	40.4°	.6481	.7615	.8511	1.175	49.6°	.8657
.7069	40.5°	.6494	.7604	.8541	1.171	49.5°	.8639
.7086	40.6°	.6508	.7593	.8571	1.167	49.4°	.8622
.7103	40.7°	.6521	.7581	.8601	1.163	49.3°	.8604
.7121	40.8°	.6534	.7570	.8632	1.159	49.2°	.8587
.7138	40.9°	.6547	.7559	.8662	1.154	49.1°	.8570
		COS	SIN	COT	TAN	DEGREES	RADIANS

Table A/Values of the Trigonometric Functions   (*Continued*)

RADIANS	DEGREES	SIN	COS	TAN	COT		
.7156	41.0°	.6561	.7547	.8693	1.150	49.0°	.8552
.7173	41.1°	.6574	.7536	.8724	1.146	48.9°	.8535
.7191	41.2°	.6587	.7524	.8754	1.142	48.8°	.8517
.7208	41.3°	.6600	.7513	.8785	1.138	48.7°	.8500
.7226	41.4°	.6613	.7501	.8816	1.134	48.6°	.8482
.7243	41.5°	.6626	.7490	.8847	1.130	48.5°	.8465
.7261	41.6°	.6639	.7478	.8878	1.126	48.4°	.8447
.7278	41.7°	.6652	.7466	.8910	1.122	48.3°	.8430
.7295	41.8°	.6665	.7455	.8941	1.118	48.2°	.8412
.7313	41.9°	.6678	.7443	.8972	1.115	48.1°	.8395
.7330	42.0°	.6691	.7431	.9004	1.111	48.0°	.8378
.7348	42.1°	.6704	.7420	.9036	1.107	47.9°	.8360
.7365	42.2°	.6717	.7408	.9067	1.103	47.8°	.8343
.7383	42.3°	.6730	.7396	.9099	1.099	47.7°	.8325
.7400	42.4°	.6743	.7385	.9131	1.095	47.6°	.8308
.7418	42.5°	.6756	.7373	.9163	1.091	47.5°	.8290
.7435	42.6°	.6769	.7361	.9195	1.087	47.4°	.8273
.7453	42.7°	.6782	.7349	.9228	1.084	47.3°	.8255
.7470	42.8°	.6794	.7337	.9260	1.080	47.2°	.8238
.7487	42.9°	.6807	.7325	.9293	1.076	47.1°	.8221
.7505	43.0°	.6820	.7314	.9325	1.072	47.0°	.8203
.7522	43.1°	.6833	.7302	.9358	1.069	46.9°	.8186
.7540	43.2°	.6845	.7290	.9391	1.065	46.8°	.8168
.7557	43.3°	.6858	.7278	.9424	1.061	46.7°	.8151
.7575	43.4°	.6871	.7266	.9457	1.057	46.6°	.8133
.7592	43.5°	.6884	.7254	.9490	1.054	46.5°	.8116
.7610	43.6°	.6896	.7242	.9523	1.050	46.4°	.8098
.7627	43.7°	.6909	.7230	.9556	1.046	46.3°	.8081
.7645	43.8°	.6921	.7218	.9590	1.043	46.2°	.8063
.7662	43.9°	.6934	.7206	.9623	1.039	46.1°	.8046
.7679	44.0°	.6947	.7193	.9657	1.036	46.0°	.8029
.7697	44.1°	.6959	.7181	.9691	1.032	45.9°	.8011
.7714	44.2°	.6972	.7169	.9725	1.028	45.8°	.7994
.7732	44.3°	.6984	.7157	.9759	1.025	45.7°	.7976
.7749	44.4°	.6997	.7145	.9793	1.021	45.6°	.7959
.7767	44.5°	.7009	.7133	.9827	1.018	45.5°	.7941
.7784	44.6°	.7022	.7120	.9861	1.014	45.4°	.7924
.7802	44.7°	.7034	.7108	.9896	1.011	45.3°	.7906
.7819	44.8°	.7046	.7096	.9930	1.007	45.2°	.7889
.7837	44.9°	.7059	.7083	.9965	1.003	45.1°	.7871
.7854	45.0°	.7071	.7071	1.0000	1.000	45.0°	.7854
		COS	SIN	COT	TAN	DEGREES	RADIANS

Appendix B

Table of Common Logarithms

Table B/Common Logarithms

N	0	1	2	3	4	5	6	7	8	9
1.0	.0000	.0043	.0086	.0128	.0170	.0212	.0253	.0294	.0334	.0374
1.1	.0414	.0453	.0492	.0531	.0569	.0607	.0645	.0682	.0719	.0755
1.2	.0792	.0828	.0864	.0899	.0934	.0969	.1004	.1038	.1072	.1106
1.3	.1139	.1173	.1206	.1239	.1271	.1303	.1335	.1367	.1399	.1430
1.4	.1461	.1492	.1523	.1553	.1584	.1614	.1644	.1673	.1703	.1732
1.5	.1761	.1790	.1818	.1847	.1875	.1903	.1931	.1959	.1987	.2014
1.6	.2041	.2068	.2095	.2122	.2148	.2175	.2201	.2227	.2253	.2279
1.7	.2304	.2330	.2355	.2380	.2405	.2430	.2455	.2480	.2504	.2529
1.8	.2553	.2577	.2601	.2625	.2648	.2672	.2695	.2718	.2742	.2765
1.9	.2788	.2810	.2833	.2856	.2878	.2900	.2923	.2945	.2967	.2989
2.0	.3010	.3032	.3054	.3075	.3096	.3118	.3139	.3160	.3181	.3201
2.1	.3222	.3243	.3263	.3284	.3304	.3324	.3345	.3365	.3385	.3404
2.2	.3424	.3444	.3464	.3483	.3502	.3522	.3541	.3560	.3579	.3598
2.3	.3617	.3636	.3655	.3674	.3692	.3711	.3729	.3747	.3766	.3784
2.4	.3802	.3820	.3838	.3856	.3874	.3892	.3909	.3927	.3945	.3962
2.5	.3979	.3997	.4014	.4031	.4048	.4065	.4082	.4099	.4116	.4133
2.6	.4150	.4166	.4183	.4200	.4216	.4232	.4249	.4265	.4281	.4298
2.7	.4314	.4330	.4346	.4362	.4378	.4393	.4409	.4425	.4440	.4456
2.8	.4472	.4487	.4502	.4518	.4533	.4548	.4564	.4579	.4594	.4609
2.9	.4624	.4639	.4654	.4669	.4683	.4698	.4713	.4728	.4742	.4757
3.0	.4771	.4786	.4800	.4814	.4829	.4843	.4857	.4871	.4886	.4900
3.1	.4914	.4928	.4942	.4955	.4969	.4983	.4997	.5011	.5024	.5038
3.2	.5051	.5065	.5079	.5092	.5105	.5119	.5132	.5145	.5159	.5172
3.3	.5185	.5198	.5211	.5224	.5237	.5250	.5263	.5276	.5289	.5302
3.4	.5315	.5328	.5340	.5353	.5366	.5378	.5391	.5403	.5416	.5428
3.5	.5441	.5453	.5465	.5478	.5490	.5502	.5514	.5527	.5539	.5551
3.6	.5563	.5575	.5587	.5599	.5611	.5623	.5635	.5647	.5658	.5670
3.7	.5682	.5694	.5705	.5717	.5729	.5740	.5752	.5763	.5775	.5786
3.8	.5798	.5809	.5821	.5832	.5843	.5855	.5866	.5877	.5888	.5899
3.9	.5911	.5922	.5933	.5944	.5955	.5966	.5977	.5988	.5999	.6010
4.0	.6021	.6031	.6042	.6053	.6064	.6075	.6085	.6096	.6107	.6117
4.1	.6128	.6138	.6149	.6160	.6170	.6180	.6191	.6201	.6212	.6222
4.2	.6232	.6243	.6253	.6263	.6274	.6284	.6294	.6304	.6314	.6325
4.3	.6335	.6345	.6355	.6365	.6375	.6385	.6395	.6405	.6415	.6425
4.4	.6435	.6444	.6454	.6464	.6474	.6484	.6493	.6503	.6513	.6522
4.5	.6532	.6542	.6551	.6561	.6571	.6580	.6590	.6599	.6609	6618
4.6	.6628	.6637	.6646	.6656	.6665	.6675	.6684	.6693	.6702	.6712
4.7	.6721	.6730	.6739	.6749	.6758	.6767	.6776	.6785	.6794	.6803
4.8	.6812	.6821	.6830	.6839	.6848	.6857	.6866	.6875	.6884	.6893
4.9	.6902	.6911	.6920	.6928	.6937	.6946	.6955	.6964	.6972	.6981

Table B/Common Logarithms   (*Continued*)

N	0	1	2	3	4	5	6	7	8	9
5.0	.6990	.6998	.7007	.7016	.7024	.7033	.7042	.7050	.7059	.7067
5.1	.7076	.7084	.7093	.7101	.7110	.7118	.7126	.7135	.7143	.7152
5.2	.7160	.7168	.7177	.7185	.7193	.7202	.7210	.7218	.7226	.7235
5.3	.7243	.7251	.7259	.7267	.7275	.7284	.7292	.7300	.7308	.7316
5.4	.7324	.7332	.7340	.7348	.7356	.7364	.7372	.7380	.7388	.7396
5.5	.7404	.7412	.7419	.7427	.7435	.7443	.7451	.7459	.7466	.7474
5.6	.7482	.7490	.7497	.7505	.7513	.7520	.7528	.7536	.7543	.7551
5.7	.7559	.7566	.7574	.7582	.7589	.7597	.7604	.7612	.7619	.7627
5.8	.7634	.7642	.7649	.7657	.7664	.7672	.7679	.7686	.7694	.7701
5.9	.7709	.7716	.7723	.7731	.7738	.7745	.7752	.7760	.7767	.7774
6.0	.7782	.7789	.7796	.7803	.7810	.7818	.7825	.7832	.7839	.7846
6.1	.7853	.7860	.7868	.7875	.7882	.7889	.7896	.7903	.7910	.7917
6.2	.7924	.7931	.7938	.7945	.7952	.7959	.7966	.7973	.7980	.7987
6.3	.7993	.8000	.8007	.8014	.8021	.8028	.8035	.8041	.8048	.8055
6.4	.8062	.8069	.8075	.8082	.8089	.8096	.8102	.8109	.8116	.8122
6.5	.8129	.8136	.8142	.8149	.8156	.8162	.8169	.8176	.8182	.8189
6.6	.8195	.8202	.8209	.8215	.8222	.8228	.8235	.8241	.8248	.8254
6.7	.8261	.8267	.8274	.8280	.8287	.8293	.8299	.8306	.8312	.8319
6.8	.8325	.8331	.8338	.8344	.8351	.8357	.8363	.8370	.8376	.8382
6.9	.8388	.8395	.8401	.8407	.8414	.8420	.8426	.8432	.8439	.8445
7.0	.8451	.8457	.8463	.8470	.8476	.8482	.8488	.8494	.8500	.8506
7.1	.8513	.8519	.8525	.8531	.8537	.8543	.8549	.8555	.8561	.8567
7.2	.8573	.8579	.8585	.8591	.8597	.8603	.8609	.8615	.8621	.8627
7.3	.8633	.8639	.8645	.8651	.8657	.8663	.8669	.8675	.8681	.8686
7.4	.8692	.8698	.8704	.8710	.8716	.8722	.8727	.8733	.8739	.8745
7.5	.8751	.8756	.8762	.8768	.8774	.8779	.8785	.8791	.8797	.8802
7.6	.8808	.8814	.8820	.8825	.8831	.8837	.8842	.8848	.8854	.8859
7.7	.8865	.8871	.8876	.8882	.8887	.8893	.8899	.8904	.8910	.8915
7.8	.8921	.8927	.8932	.8938	.8943	.8949	.8954	.8960	.8965	.8971
7.9	.8976	.8982	.8987	.8993	.8998	.9004	.9009	.9015	.9020	.9025
8.0	.9031	.9036	.9042	.9047	.9053	.9058	.9063	.9069	.9074	.9079
8.1	.9085	.9090	.9096	.9101	.9106	.9112	.9117	.9122	.9128	.9133
8.2	.9138	.9143	.9149	.9154	.9159	.9165	.9170	.9175	.9180	.9186
8.3	.9191	.9196	.9201	.9206	.9212	.9217	.9222	.9227	.9232	.9238
8.4	.9243	.9248	.9253	.9258	.9263	.9269	.9274	.9279	.9284	.9289
8.5	.9294	.9299	.9304	.9309	.9315	.9320	.9325	.9330	.9335	.9340
8.6	.9345	.9350	.9355	.9360	.9365	.9370	.9375	.9380	.9385	.9390
8.7	.9395	.9400	.9405	.9410	.9415	.9420	.9425	.9430	.9435	.9440
8.8	.9445	.9450	.9455	.9460	.9465	.9469	.9474	.9479	.9484	.9489
8.9	.9494	.9499	.9504	.9509	.9513	.9518	.9523	.9528	.9533	.9538
9.0	.9542	.9547	.9552	.9557	.9562	.9566	.9571	.9576	.9581	.9586
9.1	.9590	.9595	.9600	.9605	.9609	.9614	.9619	.9624	.9628	.9633
9.2	.9638	.9643	.9647	.9652	.9657	.9661	.9666	.9671	.9675	.9680
9.3	.9685	.9689	.9694	.9699	.9703	.9708	.9713	.9717	.9722	.9727
9.4	.9731	.9736	.9741	.9745	.9750	.9754	.9759	.9763	.9768	.9773

Table B/Common Logarithms   (*Continued*)

N	0	1	2	3	4	5	6	7	8	9
9.5	.9777	.9782	.9786	.9791	.9795	.9800	.9805	.9809	.9814	.9818
9.6	.9823	.9827	.9832	.9836	.9841	.9845	.9850	.9854	.9859	.9863
9.7	.9868	.9872	.9877	.9881	.9886	.9890	.9894	.9899	.9903	.9908
9.8	.9912	.9917	.9921	.9926	.9930	.9934	.9939	.9943	.9948	.9952
9.9	.9956	.9961	.9965	.9969	.9974	.9978	.9983	.9987	.9991	.9996

Answers to Odd-Numbered Exercises and All Review Problems

CHAPTER 1

Problem Set 1.1 (page 8)

1. True   **3.** False   **5.** False   **7.** True
9. False   **11.** 46   **13.** 0, -14, and 46
15. $\sqrt{5}, -\sqrt{2}$, and $-\pi$ **17.** 0 and -14 **19.** 18
21. 39   **23.** 29   **25.** 1 if $x > 0$ and -1 if $x < 0$
27. -1 if $x > 0$ and 1 if $x < 0$ **29.** I and III
31. II and III   **33.** 0   **35.** 10   **37.** $\sqrt{34}$
39. $4\sqrt{2}$ **41.** $2\sqrt{13}$ **43.** 10
45. Two sides are of length $2\sqrt{10}$.
47. The distance between $(3,1)$ and $(-2,6)$ equals the distance between $(3,1)$ and $(8,-4)$ which is $5\sqrt{2}$ units.
49. 0.71   **51.** 0.87

Problem Set 1.2 (page 16)

1. $D = \{1,2,3,4\}$, $R = \{4,6,12,17\}$; it is a function.
3. $D = \{0,1,-1\}$, $R = \{0,1,-1\}$; it is not a function.
5. $D = \{1\}$, $R = \{3,4,-1,-2\}$; it is not a function.
7. $D = \{\text{all reals}\}$, $R = \{\text{all reals}\}$; it is a function.
9. All reals   **11.** All reals except zero
13. All reals except -2 and 3
15. All reals except -2 and -3
17. All reals except -2 and 0
19. All reals except 3 and -3
21. $D = \{x \mid x \geq -1\}$ **23.** 6, 12, -3, $3a + 6$
25. $-4, -4, -12, -28$ **27.** 0, 2, $2\sqrt{3}$, 5
29. 1, 11, 11, 29   **31.** 3   **33.** $2a + h - 6$

35. $(f \circ g)(x) = 6x - 2$, $D = \{\text{all reals}\}$
     $(g \circ f)(x) = 6x - 1$, $D = \{\text{all reals}\}$
37. $(f \circ g)(x) = 10x + 2$, $D = \{\text{all reals}\}$
     $(g \circ f)(x) = 10x - 5$, $D = \{\text{all reals}\}$
39. $(f \circ g)(x) = 3x^2 + 7$, $D = \{\text{all reals}\}$
     $(g \circ f)(x) = 9x^2 + 24x + 17$, $D = \{\text{all reals}\}$
41. $(f \circ g)(x) = 3x^2 + 9x - 16$, $D = \{\text{all reals}\}$
     $(g \circ f)(x) = 9x^2 - 15x$, $D = \{\text{all reals}\}$
43. $(f \circ g)(x) = \dfrac{1}{2x + 7}$, $D = \{x \mid x \neq -\frac{7}{2}\}$
     $(g \circ f)(x) = \dfrac{7x + 2}{x}$, $D = \{x \mid x \neq 0\}$
45. $(f \circ g)(x) = \sqrt{3x - 3}$, $D = \{x \mid x \geq 1\}$
     $(g \circ f)(x) = 3\sqrt{x - 2} - 1$, $D = \{x \mid x \geq 2\}$
47. $(f \circ g)(x) = \dfrac{x}{2 - x}$, $D = \{x \mid x \neq 0 \text{ and } x \neq 2\}$
     $(g \circ f)(x) = 2x - 2$, $D = \{x \mid x \neq 1\}$
49. 455, 1505, 2255, 4505   **51.** $-1500, 600, 1000, 900$
53. 12.56, 28.26, 452.16, 907.46

Problem Set 1.3 (page 22)

1. 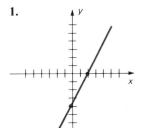   **3.**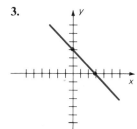

5.

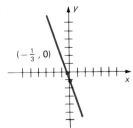

$\left(-\frac{1}{3}, 0\right)$

7.

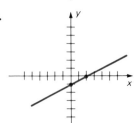

$(1, -4)$

29.

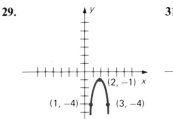

$(1, -4)$ $(2, -1)$ $(3, -4)$

31. (a)

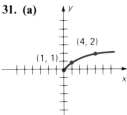

$(4, 2)$ $(1, 1)$

9.

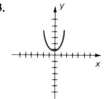

11.

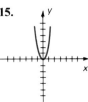

Problem Set 1.4 (page 29)

1. $52°$ **3.** $25°$ and $65°$ **5.** $20°$ and $160°$
7. $14.50°$ **9.** $22°18'$ **11.** $8.76°$
13. $45°19'12''$ **15.** $150.17°$ **17.** $9°7'48''$

19. $\dfrac{\pi}{18}$ **21.** $\dfrac{4\pi}{9}$ **23.** $\dfrac{5\pi}{6}$ **25.** $\dfrac{5\pi}{4}$ **27.** $\dfrac{\pi}{6}$

29. $-\dfrac{19\pi}{6}$ **31.** $20°$ **33.** $130°$ **35.** $240°$

37. $390°$ **39.** $-45°$ **41.** $-210°$ **43.** $114.6°$
45. $401.1°$ **47.** $-229.2°$ **49.** $.5$ **51.** $.3$
53. -4.4 **55.** 46.1 inches   **57.** 17.9 centimeters
59. $127.2°$ **61.** $412.5°$ **63.** $630°$
65. 3.8 inches   **67.** 2.8 revolutions   **69.** 7 feet
71. $4\sqrt{3}$ centimeters
73. $a = 6$ yards and $b = 6\sqrt{3}$ yards
75. $a = b = 5\sqrt{2}$ meters   **77.** 7.8 centimeters
79. 36.4 feet   **81.** 127.3 feet   **83.** $3\sqrt{3}$ centimeters
85. 10.8 centimeters

13.

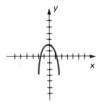

15.

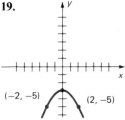

17.

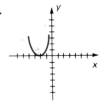

19.

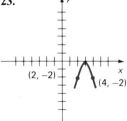

$(-2, -5)$ $(2, -5)$

Chapter 1 Review Problem Set (page 34)

1. 1 and 10 **2.** $0, 1$, and 10 **3.** $0, -1, 1$, and 10
4. 0 and -1 **5.** $0, \frac{5}{6}, -\frac{1}{2}, 0.14, 0.\overline{12}, -1, 1$, and 10

6. $\sqrt{7}, -\sqrt{3}, -\dfrac{\sqrt{2}}{2}$, and 2π **7.** $0, 1$, and 10

8. 1 and 10 **9.** 10 **10.** $\sqrt{2}$ **11.** $\sqrt{41}$
12. $\sqrt{74}$ **13.** All reals except 3
14. All reals except -4 and 4
15. All reals except 0 and 2 **16.** $D = \{x \mid x \geq \frac{3}{4}\}$
17. $8, 13, -22$ **18.** $-1, -3, 11$
19. $-13, -5, -20$ **20.** $6a + 3h - 2$
21. $(f \circ g)(x) = -12x - 15$ $D = \{\text{all reals}\}$
 $(g \circ f)(x) = -12x + 5$ $D = \{\text{all reals}\}$
22. $(f \circ g)(x) = 2x^2 - 1$ $D = \{\text{all reals}\}$
 $(g \circ f)(x) = 4x^2 + 4x$ $D = \{\text{all reals}\}$
23. $(f \circ g)(x) = \sqrt{5x - 1}$ $D = \{x \mid x \geq \frac{1}{5}\}$
 $(g \circ f)(x) = 5\sqrt{x + 1} - 2$ $D = \{x \mid x \geq -1\}$

21.

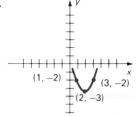

23.

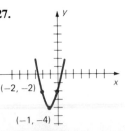

$(2, -2)$ $(4, -2)$

25.

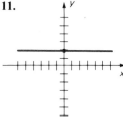

$(1, -2)$ $(3, -2)$ $(2, -3)$

27.

$(-2, -2)$ $(-1, -4)$

24. $(f \circ g)(x) = \dfrac{x - 3}{x - 1}$ $D = \{x \mid x \neq 3 \text{ and } x \neq 1\}$

$(g \circ f)(x) = \dfrac{2x + 2}{-3x - 2}$ $D = \{x \mid x \neq -1 \text{ and } x \neq -\frac{2}{3}\}$

25.   **26.**

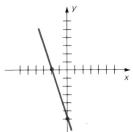

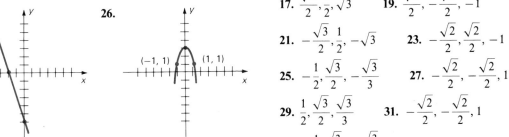

27.   **28.**

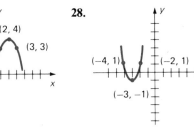

29. 35° and 55°   **30.** 35.28° and 82.26°
31. 93°21′ and 163°16′12″

32. (a) $\dfrac{7\pi}{3}$ (b) $\dfrac{19\pi}{6}$ (c) $-\dfrac{\pi}{4}$

33. (a) 210°   (b) −240°   (c) 765°
34. 71.2 centimeters   **35.** 31.8 inches   **36.** 25.6 feet
37. 32.5 meters   **38.** 43.3 feet

CHAPTER 2

Problem Set 2.1 (page 43)

For Problems 1–15, the answers are given in the order of sin θ, cos θ, tan θ, csc θ, sec θ, and cot θ.

1. $-\frac{4}{5}, \frac{3}{5}, -\frac{4}{3}, -\frac{5}{4}, \frac{5}{3}, -\frac{3}{4}$
3. $\frac{12}{13}, -\frac{5}{13}, -\frac{12}{5}, \frac{13}{12}, -\frac{13}{5}, -\frac{5}{12}$

5. $-\dfrac{\sqrt{2}}{2}, \dfrac{\sqrt{2}}{2}, -1, -\sqrt{2}, \sqrt{2}, -1$

7. $-\dfrac{3\sqrt{13}}{13}, -\dfrac{2\sqrt{13}}{13}, \dfrac{3}{2}, -\dfrac{\sqrt{13}}{3}, -\dfrac{\sqrt{13}}{2}, \dfrac{2}{3}$

9. $\dfrac{2\sqrt{5}}{5}, \dfrac{\sqrt{5}}{5}, 2, \dfrac{\sqrt{5}}{2}, \sqrt{5}, \dfrac{1}{2}$

11. $-\dfrac{\sqrt{10}}{10}, \dfrac{3\sqrt{10}}{10}, -\dfrac{1}{3}, -\sqrt{10}, \dfrac{\sqrt{10}}{3}, -3$

13. 1, 0, undefined, 1, undefined, 0
15. −1, 0, undefined, −1, undefined, 0

For Problems 17–33, the answers are given in the order of sin θ, cos θ, and tan θ.

17. $\dfrac{\sqrt{3}}{2}, \dfrac{1}{2}, \sqrt{3}$ **19.** $\dfrac{\sqrt{2}}{2}, -\dfrac{\sqrt{2}}{2}, -1$

21. $-\dfrac{\sqrt{3}}{2}, \dfrac{1}{2}, -\sqrt{3}$ **23.** $-\dfrac{\sqrt{2}}{2}, \dfrac{\sqrt{2}}{2}, -1$

25. $-\dfrac{1}{2}, \dfrac{\sqrt{3}}{2}, -\dfrac{\sqrt{3}}{3}$ **27.** $-\dfrac{\sqrt{2}}{2}, -\dfrac{\sqrt{2}}{2}, 1$

29. $\dfrac{1}{2}, \dfrac{\sqrt{3}}{2}, \dfrac{\sqrt{3}}{3}$ **31.** $-\dfrac{\sqrt{2}}{2}, -\dfrac{\sqrt{2}}{2}, 1$

33. $-\dfrac{1}{2}, \dfrac{\sqrt{3}}{2}, -\dfrac{\sqrt{3}}{3}$

35.

θ	RADIANS	SIN θ	COS θ
0°	0	0	1
30°	$\dfrac{\pi}{6}$	$\dfrac{1}{2}$	$\dfrac{\sqrt{3}}{2}$
45°	$\dfrac{\pi}{4}$	$\dfrac{\sqrt{2}}{2}$	$\dfrac{\sqrt{2}}{2}$
60°	$\dfrac{\pi}{3}$	$\dfrac{\sqrt{3}}{2}$	$\dfrac{1}{2}$
90°	$\dfrac{\pi}{2}$	1	0
180°	π	0	−1
270°	$\dfrac{3}{2}\pi$	−1	0

TAN θ	CSC θ	SEC θ	COT θ
0	undefined	1	undefined
$\dfrac{\sqrt{3}}{3}$	2	$\dfrac{2\sqrt{3}}{3}$	$\sqrt{3}$
1	$\sqrt{2}$	$\sqrt{2}$	1
$\sqrt{3}$	$\dfrac{2\sqrt{3}}{3}$	2	$\dfrac{\sqrt{3}}{3}$
undefined	1	undefined	0
0	undefined	−1	undefined
undefined	−1	undefined	0

37. $-\dfrac{\sqrt{2}}{2}$ **39.** $\dfrac{3\sqrt{10}}{10}$

41. $\sin\theta = -\frac{3}{5}$ and $\cot\theta = \frac{4}{3}$
43. $\sin\theta = \frac{7}{25}$ and $\sec\theta = \frac{25}{24}$ **45.** IV   **47.** II or IV
49. $\theta = 60°$ **51.** $\theta = 210°$ **53.** $\theta = 270°$

Problem Set 2.2 (page 50)

1. I   **3.** IV   **5.** III   **7.** II   **9.** $150°$

11. $240°$ **13.** $300°$ **15.** $240°$ **17.** $\dfrac{3\pi}{2}$

19. $\dfrac{7\pi}{6}$ **21.** $\dfrac{3\pi}{4}$ **23.** $85°$ **25.** $71.8°$

27. $73°$ **29.** $\dfrac{\pi}{4}$ **31.** $\dfrac{\pi}{3}$ **33.** $\dfrac{\pi}{3}$ **35.** $\dfrac{\sqrt{3}}{2}$

37. $\dfrac{\sqrt{3}}{2}$ **39.** $-\sqrt{3}$ **41.** $\sqrt{2}$ **43.** 2

45. $-\frac{1}{2}$ **47.** $\frac{1}{2}$ **49.** $-\sqrt{3}$ **51.** -1

53. $\dfrac{\sqrt{2}}{2}$ **55.** $\dfrac{\sqrt{3}}{2}$ **57.** $\sqrt{3}$ **59.** $-\dfrac{\sqrt{2}}{2}$

61. $\dfrac{\sqrt{3}}{3}$ **63.** $\frac{1}{2}$ **65.** undefined   **67.** .9659

69. 4.0108   **71.** .8607   **73.** $-.9135$
75. 1.0358   **77.** -3.9408 **79.** -1.6003

81. $-\frac{1}{2}$ **83.** $-\dfrac{2\sqrt{3}}{3}$ **85.** $-.1080$

87. -1.897 **89.** .8902   **91.** 2.1642
93. -1.4916

Problem Set 2.3 (page 57)

1. $12.9°$ **3.** $66.1°$ **5.** $18.6°$ **7.** $60.3°$
9. $31.2°$ **11.** $81.1°$ **13.** $71.2°$
15. $B = 53°$, $a = 11$, and $c = 18$
17. $A = 67°$, $a = 28$, and $c = 31$
19. $B = 23°$, $a = 24$, and $b = 10$
21. $c = 13$, $A = 23°$, and $B = 67°$
23. $a = 26$, $A = 66°$, and $B = 24°$
25. 23.0 feet   **27.** 92.1 miles meters   **29.** 1454 feet
31. 42 yards   **33.** 630 feet   **35.** 19.6 feet
37. 222 feet   **39.** $44.0°$ **41.** 74 feet
43. 20 feet   **45.** $29.5°$ **47.** 4044 miles
49. $23.2°$

Problem Set 2.4 (page 65)

1. 9.6 centimeters   **3.** 12.3 feet
5. 39.9 centimeters   **7.** $94.7°$ **9.** $28.8°$
11. $A = 33.1°$, $B = 128.7°$, and $C = 18.2°$

13. $b = 56.2$ feet, $A = 45.9°$, and $C = 34.1°$
15. $C = 90°$
17. Two sides of length 14.7 centimeters and two sides of length 21.8 centimeters
19. 19 miles   **21.** 46 meters   **23.** 318 feet
25. $77.4°$ **27.** 1479 miles   **29.** $66.7°$

Problem Set 2.5 (page 74)

1. $c = 13.8$ centimeters   **3.** $a = 16.8$ meters
5. $B = 102°$ and $b = 20.9$ centimeters
7. $B = 31°$ and $b = 52.0$ miles
9. $A = 56.9°$ and $C = 51.4°$
11. $b = 34.6$ feet and $c = 18.8$ feet
13. One triangle; $C = 33.4°$
15. No triangle determined
17. Two triangles; $C = 52.2°$, $B = 99.8°$, and $b = 39.9$ miles or $C = 127.8°$, $B = 24.2°$, and $b = 16.6$ miles
19. One triangle; $C = 90°$ and $b = 36.4$ centimeters
23. 48.7 feet   **25.** 79 feet   **27.** 229.6 feet
29. 168 miles or 43 miles   **31.** 3.2 feet
33. (a) 158 square meters   (c) 655 square feet
35. (a) 107 square centimeters   (c) 140 square inches

Problem Set 2.6 (page 84)

1. 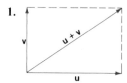   **3.**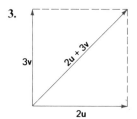

5. 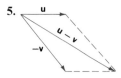   **7.**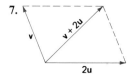

9.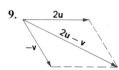

11. $9.1°$ and 253.2 miles per hour   **13.** N75.5°E
15. 90.1 pounds and $33.7°$
17. 35.3 kilometers per hour
19. 42.3 pounds and 15.4 pounds
21. 16.4 pounds and 11.5 pounds
23. 794.5 pounds   **25.** $44.0°$
27. 334.0 kilograms and $28.4°$

29. 93.4 miles and N71.5°W
31. 131.0 miles and N86.0°E

Chapter 2 Review Problem Set (page 87)

1. $\frac{4}{5}, -\frac{3}{5}, -\frac{4}{3}$ **2.** $-\frac{3\sqrt{10}}{10}, \frac{\sqrt{10}}{10}, -3$

3. $-\frac{2\sqrt{5}}{5}, -\frac{\sqrt{5}}{5}, 2$ **4.** $-\frac{12}{13}, -\frac{5}{13}, \frac{12}{5}$ **5.** $-\frac{\sqrt{3}}{2}$

6. $\frac{1}{2}$ **7.** $\sqrt{2}$ **8.** -2 **9.** -1 **10.** $-\frac{\sqrt{3}}{3}$

11. $\frac{\sqrt{3}}{2}$ **12.** $\frac{1}{2}$ **13.** undefined   **14.** undefined

15. -1 **16.** 1   **17.** II   **18.** IV   **19.** -5
20. $\frac{12}{13}$ **21.** $-.7986$ **22.** $-.4067$ **23.** 49.6°
24. 68.3°   **25.** 85.8°   **26.** 61.9°   **27.** 42.5 feet
28. 135 feet   **29.** 28.4 miles   **30.** 132.0 yards
31. Two   **32.** 84.5 feet   **33.** 118.2°
34. 518 miles   **35.** 257 yards   **36.** 83.3 feet
37. 34.5 pounds and 28.9 pounds   **38.** 1457.3 pounds
39. 286.4 pounds and 12.3°
40. 149.2 miles and N46.1°W

CHAPTER 3

Problem Set 3.1 (page 96)

The answers for Problems 1–11 are given in the order sin s, cos s, tan s, csc s, sec s, and cot s.

1. $\frac{\sqrt{2}}{2}, -\frac{\sqrt{2}}{2}, -1, \sqrt{2}, -\sqrt{2}, -1$

3. $-1, 0,$ undefined, $-1,$ undefined, -1

5. $-\frac{\sqrt{2}}{2}, \frac{\sqrt{2}}{2}, -1, -\sqrt{2}, \sqrt{2}, -1$

7. $-\frac{\sqrt{3}}{2}, -\frac{1}{2}, \sqrt{3}, -\frac{2\sqrt{3}}{3}, -2, \frac{\sqrt{3}}{3}$

9. $1, 0,$ undefined, $1,$ undefined, 0

11. $-\frac{\sqrt{2}}{2}, -\frac{\sqrt{2}}{2}, 1, -\sqrt{2}, -\sqrt{2}, 1$

13.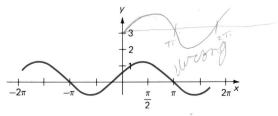

15.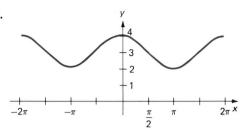
17.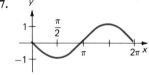
19.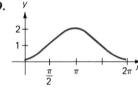
21.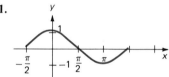
23.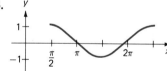
25.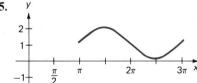
27.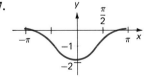
29.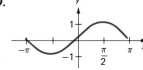

274   ANSWERS TO ODD-NUMBERED EXERCISES AND ALL REVIEW PROBLEMS

31.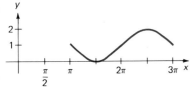

33.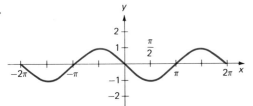

35.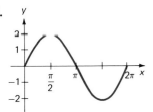

37.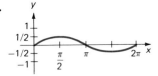

39.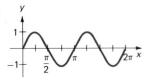

41.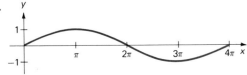

Problem Set 3.2 (page 102)

1.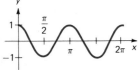

3.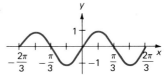

5.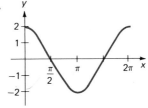

7.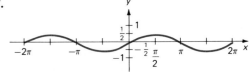

9.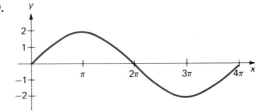

11.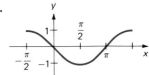

13.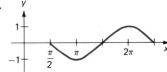

15.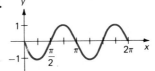

17.

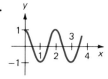

19. period: 2π
amplitude: 3

phase shift: $\dfrac{\pi}{2}$ to the left

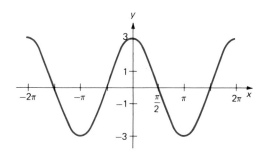

21. period: 2π
amplitude: 2
phase shift: π to the right

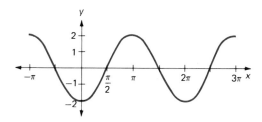

23. period: 2π
amplitude: $\frac{1}{2}$

phase shift: $\dfrac{\pi}{4}$ to the left

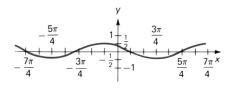

25. period: 2π
amplitude: 2
phase shift: π to the left

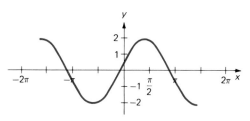

27. period: π
amplitude: 2
phase shift: π to the right

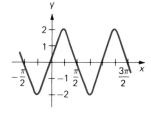

29. period: $\dfrac{2\pi}{3}$

amplitude: $\frac{1}{2}$
phase shift: π to the left

31. period: π

amplitude: 1

phase shift: $\dfrac{\pi}{2}$ to the right

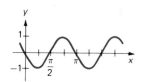

33. period: $\dfrac{2\pi}{3}$

amplitude: 2

phase shift: $\dfrac{\pi}{3}$ to the right

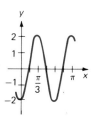

35. period: $\dfrac{2\pi}{3}$

amplitude: $\frac{1}{2}$

phase shift: $\dfrac{\pi}{3}$ to the right

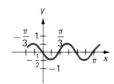

37. period: 4π
amplitude: 2
phase shift: π to the left

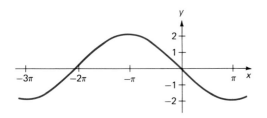

39. period: π
amplitude: 3

phase shift: $\dfrac{\pi}{2}$ to the left

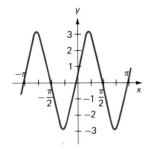

41.

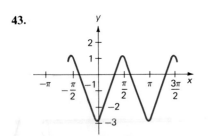

43.

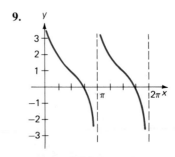

Problem Set 3.3 (page 106)

1.

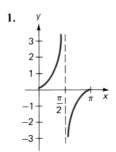

3.

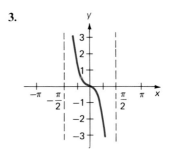

5.

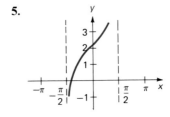

7.

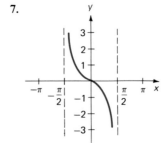

9.

11.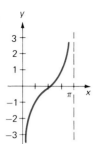

13.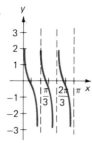

15.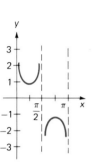

17.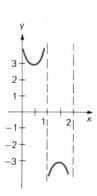

19.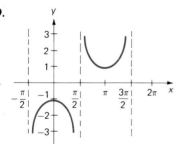

21.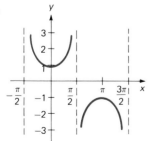

23. period: π

phase shift: $\dfrac{\pi}{2}$ to the left

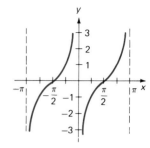

25. period: π

phase shift: $\dfrac{\pi}{4}$ to the right

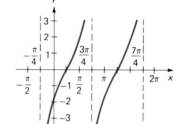

27. period: $\dfrac{\pi}{2}$

phase shift: $\dfrac{\pi}{2}$ to the left

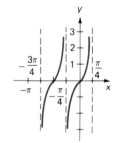

29. period: π

phase shift: $\dfrac{\pi}{2}$ to the right

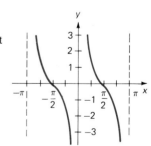

31. period: $\dfrac{\pi}{3}$

phase shift: $\dfrac{2\pi}{3}$ to the right

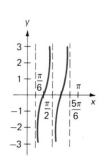

33. period: π

phase shift: $\dfrac{\pi}{4}$ to the right

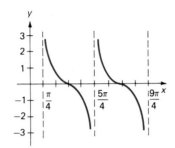

35. period: 2π

phase shift: $\dfrac{\pi}{2}$ to the left

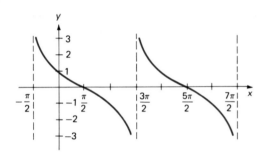

37. period: 2π

phase shift: $\dfrac{\pi}{2}$ to the right

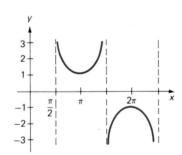

39. period: 2π

phase shift: $\dfrac{\pi}{4}$ to the left

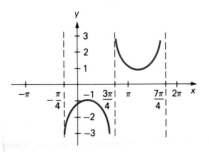

41. period: π

phase shift: $\dfrac{\pi}{2}$ to the left

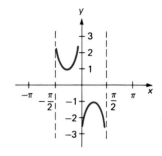

43. period: 2π

phase shift: π to the left

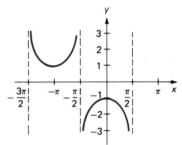

45. period: 2π

phase shift: $\dfrac{\pi}{4}$ to the right

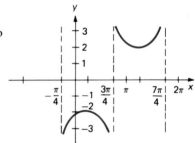

Problem Set 3.4 (page 114)

1.

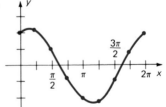

3.

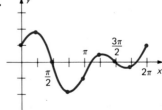

5.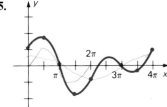

7.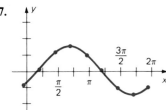

9.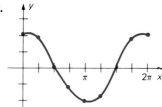

11.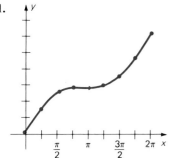

13. Function   **15.** Not a function   **17.** Function
19. One-to-one function
21. Not a one-to-one function
23. One-to-one function
25. Domain of f: $\{1, 2, 5\}$
   Range of f: $\{5, 9, 21\}$
   $f^{-1} = \{(5, 1), (9, 2), (21, 5)\}$
   Domain of f^{-1}: $\{5, 9, 21\}$
   Range of f^{-1}: $\{1, 2, 5\}$
27. Domain of f: $\{0, 2, -1, -2\}$
   Range of f: $\{0, 8, -1, -8\}$
   $f^{-1} = \{(0, 0), (8, 2), (-1, -1), (-8, -2)\}$
   Domain of f^{-1}: $\{0, 8, -1, -8\}$
   Range of f^{-1}: $\{0, 2, -1, -2\}$

29. $f^{-1}(x) = \dfrac{x - 3}{2}$ **31.** $f^{-1}(x) = 2x$

33. $f^{-1}(x) = \dfrac{x - 9}{5}$ **35.** $f^{-1}(x) = 3x + 12$

37. $f^{-1}(x) = \dfrac{3x - 15}{2}$ **39.** $f^{-1}(x) = \dfrac{x}{4}$

41. $f^{-1}(x) = \dfrac{x - 9}{2}$ **43.** $f^{-1}(x) = -\dfrac{3}{2}x$

45. $f^{-1}(x) = \dfrac{-x - 4}{3}$ **47.** $f^{-1}(x) = \dfrac{12x + 10}{9}$

49. $f^{-1}(x) = \dfrac{1}{6}x$ **51.** $f^{-1}(x) = -4x$

53. $f^{-1}(x) = \dfrac{x + 1}{2}$ **55.** $f^{-1}(x) = \dfrac{-x + 3}{4}$

57. $f^{-1}(x) = \sqrt{x}$,   where $x \geq 0$
59. Every nonconstant linear function is a one-to-one function.

Problem Set 3.5 (page 120)

1. $\dfrac{\pi}{4}$ **3.** $-\dfrac{\pi}{3}$ **5.** $\dfrac{\pi}{3}$ **7.** $\dfrac{\pi}{6}$ **9.** $\dfrac{\pi}{4}$

11. $\dfrac{\pi}{3}$ **13.** $-\dfrac{\pi}{2}$ **15.** 30°   **17.** 135°

19. 0°   **21.** −45°   **23.** 0.366   **25.** −1.154
27. 0.940   **29.** 2.000   **31.** 1.455   **33.** −0.196
35. 25.5°   **37.** −55.2°   **39.** 74.7°   **41.** 99.3°

43. 86.0°   **45.** −83.1°   **47.** $\dfrac{\sqrt{3}}{2}$ **49.** 0

51. 1   **53.** $\dfrac{\sqrt{3}}{2}$ **55.** $\sqrt{2}$ **57.** $\dfrac{3}{5}$ **59.** $\dfrac{3}{5}$

61. $-\dfrac{4}{3}$ **63.** $\dfrac{\sqrt{5}}{3}$ **65.** $-2\sqrt{2}$ **67.** $\dfrac{4\sqrt{7}}{7}$

69. $-\dfrac{3\sqrt{10}}{20}$

71.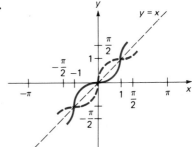

73.

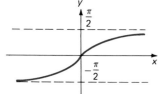

75.

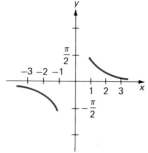

33. period of $\pi/2$, no amplitude, phase shift of $\pi/8$ to the left
34. period of $2\pi/3$, no amplitude, phase shift of $\pi/12$ to the right
35. period of $2\pi/3$, no amplitude, phase shift of 2 to the right
36. period of π, no amplitude, phase shift of $\pi/2$ to the right
37.

38.

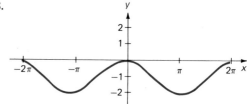

Chapter 3 Review Problem Set (page 124)

1. $-\frac{1}{2}$ **2.** 1 **3.** $\dfrac{\sqrt{2}}{2}$ **4.** $\frac{5}{13}$ **5.** $\dfrac{\sqrt{3}}{3}$

6. $-\dfrac{2\sqrt{13}}{13}$ **7.** $-\dfrac{\pi}{6}$ **8.** $\dfrac{5\pi}{6}$ **9.** $-\dfrac{\pi}{6}$

10. $-\dfrac{\pi}{3}$ **11.** $60°$ **12.** $120°$ **13.** $-45°$

14. $-30°$ **15.** 2.180 **16.** $-.807$
17. -1.557 **18.** 1.228 **19.** $75.6°$
20. $-24.2°$ **21.** $-83.8°$ **22.** $124.2°$
23. period of 2π, amplitude of 4, phase shift of $\pi/4$ units to the left
24. period of π, amplitude of 3, phase shift to $\pi/3$ units to the right
25. period of $\pi/2$, no amplitude, phase shift of π units to the left
26. period of $2\pi/3$, amplitude of 1, phase shift of $\pi/3$ units to the right
27. period of 2, amplitude of 2, phase shift of 1 unit to the left
28. period of 2, amplitude of 2, phase shift of $\frac{1}{2}$ unit to the right
29. period of $2\pi/3$, amplitude of 4, no phase shift
30. period of π, amplitude of 5, no phase shift
31. period of $\pi/3$, no amplitude, phase shift of $\pi/3$ to the right
32. period of $\pi/4$, no amplitude, phase shift of $\pi/4$ to the left

39.

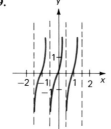

40.

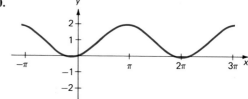

41.

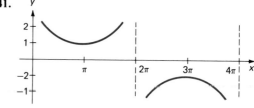

42.

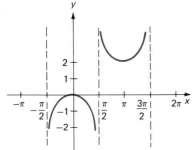

43.

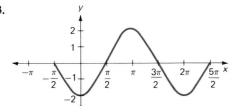

44.

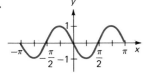

45.

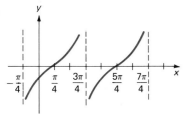

46.

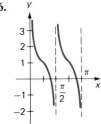

47.

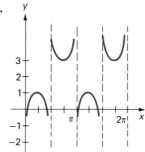

48.

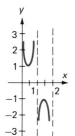

49.

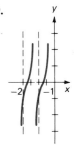

50.

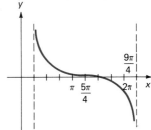

51.

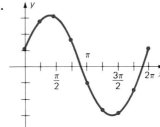

52.

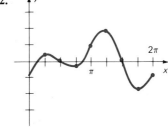

53. Yes   **54.** Yes   **55.** No   **56.** No

57. $f^{-1}(x) = \dfrac{x+1}{6}$ **58.** $f^{-1}(x) = \dfrac{3x-21}{2}$

59. $f^{-1}(x) = \dfrac{-35x-10}{21}$ **60.** $f^{-1}(x) = \dfrac{-9x+2}{45}$

CHAPTER 4

Problem Set 4.1 (page 135)

1. $\sec\theta = \dfrac{r}{x} = \dfrac{1}{\dfrac{x}{r}} = \cos\theta$ $\cot\theta = \dfrac{x}{y} = \dfrac{1}{\dfrac{y}{x}} = \dfrac{1}{\tan\theta}$

For Problems 3–11, the answers are arranged in the order:
$\sin\theta$, $\cos\theta$, $\tan\theta$, $\csc\theta$, $\sec\theta$, and $\cot\theta$ omitting the value given in the problem.

3. $\frac{3}{5}, \frac{4}{3}, \frac{5}{4}, \frac{5}{3}, \frac{3}{4}$ **5.** $-\frac{12}{13}, -\frac{5}{13}, -\frac{13}{12}, -\frac{13}{5}, \frac{5}{12}$

7. $-\frac{3}{5}, -\frac{4}{3}, \frac{5}{4}, -\frac{5}{3}, -\frac{3}{4}$

9. $-\dfrac{\sqrt{5}}{5}, \dfrac{2\sqrt{5}}{5}, -\sqrt{5}, \dfrac{\sqrt{5}}{2}, -2$

11. $\dfrac{2}{3}, -\dfrac{\sqrt{5}}{3}, -\dfrac{2\sqrt{5}}{5}, -\dfrac{3\sqrt{5}}{5}, -\dfrac{\sqrt{5}}{2}$ **13.** $\sin\theta$

15. $\cot x$ **17.** $-\tan^2 x$ **19.** 1   **21.** $\sec\theta$

23. $\cos x$

Problem Set 4.2 (page 141)

1. 60° and 120°   **3.** 180°   **5.** 120° and 300°

7. 270°   **9.** $\dfrac{4\pi}{3}$ and $\dfrac{5\pi}{3}$ **11.** 0

13. $\dfrac{\pi}{4}, \dfrac{7\pi}{6}, \dfrac{5\pi}{4}$, and $\dfrac{11\pi}{6}$ **15.** No solutions

17. $0, \pi, \dfrac{\pi}{6}, \dfrac{5\pi}{6}$ **19.** 0°, 45°, 180°, 225°

21. 45°, 90°, 135°, 225°, 270°, and 315°

23. $\dfrac{2\pi}{3}, \pi$, and $\dfrac{4\pi}{3}$ **25.** 60°, 180°, and 300°

27. $\dfrac{\pi}{3}, \pi$, and $\dfrac{5\pi}{3}$ **29.** $0, \dfrac{\pi}{2}$, and $\dfrac{3\pi}{2}$ **31.** 0 and $\dfrac{\pi}{2}$

33. $\dfrac{\pi}{4}, \dfrac{3\pi}{4}, \dfrac{5\pi}{4}$, and $\dfrac{7\pi}{4}$

For Problems 35–45, n is an integer.

35. $30° + n \cdot 360°$ and $330° + n \cdot 360°$

37. $\dfrac{4\pi}{3} + 2\pi n$ and $\dfrac{5\pi}{3} + 2\pi n$ **39.** $\dfrac{3\pi}{4} + \pi n$

41. $60° + n \cdot 180°$ and $120° + n \cdot 180°$

43. $\dfrac{\pi}{4} + \pi n$ and $\dfrac{\pi}{2} + \pi n$

45. $\dfrac{\pi}{6} + 2\pi n$ and $\dfrac{5\pi}{6} + 2\pi n$ and $\dfrac{3\pi}{2} + 2\pi n$

$\left(\text{All of these can be represented by the one expression}\right.$

$\left.\dfrac{\pi}{6} + \dfrac{2\pi n}{3}.\right)$

47. 347.5° and 192.5°   **49.** 130.0° and 230.0°
51. 287.7° and 107.7°   **53.** 19.5° and 160.5°
55. 48.2° and 311.8°   **57.** 15.5° and 164.5°
59. 3.93 and 5.49   **61.** 2.53 and 3.75
63. 2.23 and 5.37   **65.** .25 and 2.89
67. 3.74 and 5.68   **69.** .90 and 5.38

Problem Set 4.3 (page 149)

1. $\dfrac{\sqrt{6}-\sqrt{2}}{4}$ **3.** $2+\sqrt{3}$ **5.** $\dfrac{\sqrt{6}+\sqrt{2}}{4}$

7. $\dfrac{-\sqrt{6}-\sqrt{2}}{4}$ **9.** $2+\sqrt{3}$ **11.** $\dfrac{\sqrt{6}+\sqrt{2}}{4}$

13. $\dfrac{\sqrt{6}+\sqrt{2}}{4}$ **15.** $-\frac{77}{85}; -\frac{13}{84}$ **17.** $-\frac{36}{85}; \frac{13}{85}$

19. $\frac{416}{425}; -\frac{87}{425}$ **21.** $\frac{4}{5}$ **23.** $-\frac{84}{13}$ **25.** $\dfrac{9\sqrt{10}}{170}$

45. (a) $\frac{1}{2}\cos 5\theta + \frac{1}{2}\cos\theta$ (c) $\frac{1}{2}\cos 2\theta - \frac{1}{2}\cos 4\theta$
(e) $\sin 12\theta + \sin 6\theta$

Problem Set 4.4 (page 157)

1. $\frac{24}{25}, \frac{7}{25}, \frac{24}{7}$ **3.** $-\frac{120}{169}, -\frac{119}{169}, \frac{120}{119}$

5. $-\frac{336}{625}, \frac{527}{625}, -\frac{336}{527}$ **7.** $\frac{4}{5}, \frac{3}{5}, \frac{4}{3}$ **9.** $\dfrac{\sqrt{2-\sqrt{3}}}{2}$

11. $\sqrt{7-4\sqrt{3}}$ **13.** $1-\sqrt{2}$ **15.** $\dfrac{\sqrt{2-\sqrt{2}}}{2}$

17. $2+\sqrt{3}$ **19.** $-\dfrac{\sqrt{2-\sqrt{3}}}{2}$ **21.** $\dfrac{\sqrt{10}}{10}, \dfrac{3\sqrt{10}}{10}, \dfrac{1}{3}$

23. $\dfrac{2\sqrt{5}}{5}, -\dfrac{\sqrt{5}}{5}, -2$ **25.** $\dfrac{3\sqrt{13}}{13}, \dfrac{2\sqrt{13}}{13}, \dfrac{3}{2}$

27. $\dfrac{\sqrt{6}}{6}, -\dfrac{\sqrt{30}}{6}, -\dfrac{\sqrt{5}}{5}$ **29.** 0°, 60°, 180°, and 300°

31. $30°, 90°,$ and $150°$ **33.** $90°$ and $270°$
35. $0°$ and $240°$
37. $0°, 30°, 90°, 150°, 180°, 210°, 270°,$ and $330°$

39. $\dfrac{\pi}{6}, \dfrac{\pi}{2}, \dfrac{5\pi}{6},$ and $\dfrac{3\pi}{2}$ **41.** $\dfrac{7\pi}{6}, \dfrac{3\pi}{2},$ and $\dfrac{11\pi}{6}$

43. $0, \dfrac{\pi}{3},$ and $\dfrac{5\pi}{3}$ **45.** $\dfrac{\pi}{6}, \dfrac{\pi}{2}, \dfrac{5\pi}{6},$ and $\dfrac{3\pi}{2}$

47. $0, \dfrac{\pi}{2},$ and $\dfrac{3\pi}{2}$ **63.** $\cos 3\theta = 4\cos^3\theta - 3\cos\theta$

65. $\cos 6\theta = 32\cos^6\theta - 48\cos^4\theta + 18\cos^2\theta - 1$

Chapter 4 Review Problem Set (page 159)

1. $-\dfrac{5}{13}; \dfrac{13}{12}; -\dfrac{12}{5}$ **2.** $-\dfrac{63}{65}; -\dfrac{56}{65}; \dfrac{63}{16}$

3. $-\dfrac{120}{169}; \dfrac{119}{168}; -\dfrac{120}{119}$ **4.** $\dfrac{\sqrt{10}}{10}; -\dfrac{3\sqrt{10}}{10}; -\dfrac{1}{3}$

5. $\dfrac{\sqrt{6}-\sqrt{2}}{4}$ or $\dfrac{\sqrt{2-\sqrt{3}}}{2}$ **6.** $\dfrac{\sqrt{6}-\sqrt{2}}{4}$ or $\dfrac{\sqrt{2-\sqrt{3}}}{2}$

7. $\dfrac{\sqrt{6}+\sqrt{2}}{4}$ or $\dfrac{\sqrt{2+\sqrt{3}}}{2}$ **8.** $\sqrt{2}-1$ **9.** $\dfrac{24}{25}$

10. $\dfrac{33}{56}$ **11.** $0°, 90°,$ and $180°$ **12.** $30°$ and $150°$
13. $30°$ and $150°$ **14.** $0°, 60°, 180°,$ and $300°$
15. $120°, 180°,$ and $240°$ **16.** $60°$ and $300°$
17. $0°$ and $240°$ **18.** $0°, 60°, 180°,$ and $300°$

19. $\dfrac{7\pi}{6}$ and $\dfrac{11\pi}{6}$ **20.** $\dfrac{\pi}{4}, \dfrac{\pi}{2}, \dfrac{3\pi}{4}, \dfrac{5\pi}{4},$ and $\dfrac{7\pi}{4}$

21. $\dfrac{\pi}{6}, \dfrac{\pi}{3}, \dfrac{5\pi}{6},$ and $\dfrac{5\pi}{3}$ **22.** $\dfrac{\pi}{8}, \dfrac{5\pi}{8}, \dfrac{9\pi}{8},$ and $\dfrac{13\pi}{8}$

23. $\dfrac{\pi}{3}$ **24.** $0, \dfrac{\pi}{2}, \pi,$ and $\dfrac{3\pi}{2}$

25. $2\pi n$, where n is an integer
26. $45° + n \cdot 180°, 210° + n \cdot 360°,$ and $330° + n \cdot 360°$ where n is an integer
27. $220°$ and $158.0°$ **28.** $196.3°$ and $343.7°$
29. $.98, 1.81, 4.12,$ and 4.95 **30.** 2.24 and 4.04

CHAPTER 5

Problem Set 5.1 (page 168)

1. $13 + 8i$ **3.** $3 + 4i$ **5.** $-11 + i$
7. $-1 - 2i$ **9.** $-\dfrac{3}{20} + \dfrac{5}{12}i$ **11.** $\dfrac{7}{10} - \dfrac{11}{12}i$
13. $4 + 0i$ **15.** $3i$ **17.** $i\sqrt{19}$ **19.** $\dfrac{2}{3}i$
21. $2i\sqrt{2}$ **23.** $3i\sqrt{3}$ **25.** $3i\sqrt{6}$ **27.** $18i$

29. $12i\sqrt{2}$ **31.** -8 **33.** $-\sqrt{6}$ **35.** $-2\sqrt{5}$
37. $-2\sqrt{15}$ **39.** $-2\sqrt{14}$ **41.** 3 **43.** $\sqrt{6}$
45. -21 **47.** $8 + 12i$ **49.** $0 + 26i$
51. $53 - 26i$ **53.** $10 - 24i$ **55.** $-14 - 8i$
57. $-7 + 24i$ **59.** $-3 + 4i$ **61.** $113 + 0i$
63. $13 + 0i$ **65.** $-\dfrac{8}{13} + \dfrac{12}{13}i$ **67.** $1 - \dfrac{2}{3}i$
69. $0 - \dfrac{3}{2}i$ **71.** $\dfrac{22}{41} - \dfrac{7}{41}i$ **73.** $-1 + 2i$
75. $-\dfrac{17}{10} + \dfrac{1}{10}i$ **77.** $\dfrac{5}{13} - \dfrac{1}{13}i$ **79.** $\{-2 \pm i\sqrt{3}\}$

81. $\{-1 \pm 2i\}$ **83.** $\{\pm 2i\sqrt{6}\}$ **85.** $\left\{\dfrac{3 \pm i\sqrt{47}}{4}\right\}$

87. $\left\{\dfrac{-1 \pm i\sqrt{5}}{3}\right\}$

Problem Set 5.2 (page 173)

1. 5 **3.** 13 **5.** 5 **7.** $\sqrt{5}$ **9.** $\sqrt{13}$

11. 1 **13.** $2\sqrt{2}\left(\cos\dfrac{3\pi}{4} + i\sin\dfrac{3\pi}{-4}\right)$

15. $3\left(\cos\dfrac{3\pi}{2} + i\sin\dfrac{3\pi}{2}\right)$ **17.** $4\left(\cos\dfrac{11\pi}{6} + i\sin\dfrac{11\pi}{6}\right)$

19. $\sqrt{2}\left(\cos\dfrac{5\pi}{4} + i\sin\dfrac{5\pi}{4}\right)$ **21.** $2\left(\cos\dfrac{2\pi}{3} + i\sin\dfrac{2\pi}{3}\right)$

23. $5\sqrt{2}(\cos 315° + i\sin 315°)$
25. $2(\cos 180° + i\sin 180°)$ **27.** $2(\cos 30° + i\sin 30°)$
29. $8(\cos 240° + i\sin 240°)$
31. $12(\cos 330° + i\sin 330°)$
33. $\sqrt{13}(\cos 56.3° + i\sin 56.3°)$
35. $\sqrt{5}(\cos 206.6° + i\sin 206.6°)$
37. $\sqrt{17}(\cos 346.0° + i\sin 346.0°)$
39. $2\sqrt{5}(\cos 116.6° + i\sin 116.6°)$ **41.** $2\sqrt{3} + 2i$

43. $-\dfrac{3\sqrt{2}}{2} - \dfrac{3\sqrt{2}}{2}i$ **45.** $-1 - \sqrt{3}i$

47. $\dfrac{1}{3} - \dfrac{\sqrt{3}}{3}i$ **49.** $6 + 0i$ **51.** $-4 - 4\sqrt{3}i$

53. $60 + 60\sqrt{3}i$ **55.** $\dfrac{3}{2} - \dfrac{3\sqrt{3}}{2}i$ **57.** $10\sqrt{3} - 10i$

59. $0 - 24i$ **61.** $20(\cos 75° + i\sin 75°)$

63. $6(\cos 70° + i\sin 70°)$ **65.** $35\left(\cos\dfrac{11\pi}{10} + i\sin\dfrac{11\pi}{10}\right)$

69. $-\dfrac{2}{3} - \dfrac{2}{3}i$ **71.** $0 + i$ **73.** $-\sqrt{3} - i$

Problem Set 5.3 (page 179)

1. $-1024 + 0i$ **3.** $0 - 32i$ **5.** $-972 - 972i$

7. $-8 + 8\sqrt{3}i$ **9.** $\dfrac{1}{2} - \dfrac{\sqrt{3}}{2}i$ **11.** $-\dfrac{\sqrt{2}}{2} - \dfrac{\sqrt{2}}{2}i$

13. $8 + 8\sqrt{3}\,i$ **15.** $-\dfrac{\sqrt{2}}{2} - \dfrac{\sqrt{2}}{2}\,i$

17. $2 + 0i, -1 + \sqrt{3}\,i,$ and $-1 - \sqrt{3}\,i$
19. $-2\sqrt{2} + 2\sqrt{2}\,i$ and $2\sqrt{2} - 2\sqrt{2}\,i$
21. $\sqrt{3} + i, -1 + \sqrt{3}\,i, -\sqrt{3} - i,$ and $1 - \sqrt{3}\,i$
23. $\sqrt[10]{2}(\cos\theta + i\sin\theta)$ where $\theta = 9°, 81°, 153°, 225°,$ and $297°$

25. $1 + 0i, \dfrac{1}{2} + \dfrac{\sqrt{3}}{2}\,i, -\dfrac{1}{2} + \dfrac{\sqrt{3}}{2}\,i, -1 + 0i, -\dfrac{1}{2} - \dfrac{\sqrt{3}}{2}\,i,$ and $\dfrac{1}{2} - \dfrac{\sqrt{3}}{2}\,i$

27. $\sqrt[3]{2}(\cos\theta + i\sin\theta)$ where $\theta = 40°, 160°,$ and $280°$
29. $\sqrt[5]{2}(\cos\theta + i\sin\theta)$ where $\theta = 27°, 99°, 171°, 243°,$ and $315°$

31. $\dfrac{3\sqrt{3}}{2} + \dfrac{3}{2}\,i$ and $-\dfrac{3\sqrt{3}}{2} - \dfrac{3}{2}\,i$

Problem Set 5.4 (page 185)

The points in Problems 1, 3, 5, 7, 9, and 11 are located on the following figure.

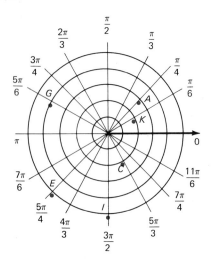

13. $\left(\dfrac{3\sqrt{3}}{2}, \dfrac{3}{2}\right)$ **15.** $(2\sqrt{2}, 2\sqrt{2})$ **17.** $(1, \sqrt{3})$

19. $\left(-\dfrac{3}{2}, -\dfrac{3\sqrt{3}}{2}\right)$ **21.** $(-2, -2\sqrt{3})$

23. $\left(-\dfrac{1}{2}, -\dfrac{\sqrt{3}}{2}\right)$ **25.** $\left(-\dfrac{7\sqrt{2}}{2}, -\dfrac{7\sqrt{2}}{2}\right)$

27. $(\sqrt{3}, 1)$ **29.** $(0, -3)$ **31.** $\left(2, \dfrac{3\pi}{4}\right)$

33. $\left(5, \dfrac{7\pi}{6}\right)$ **35.** $\left(6, \dfrac{5\pi}{3}\right)$ **37.** $(4, \pi)$

39. $\left(3, \dfrac{\pi}{6}\right)$ **41.** $\left(-2, \dfrac{5\pi}{4}\right)$ **43.** $\left(-4, \dfrac{2\pi}{3}\right)$

45. $\left(-5, \dfrac{11\pi}{6}\right)$ **47.** $(\sqrt{13}, 33.7°)$ **49.** $(5, 143.1°)$

51. $(\sqrt{17}, 194.0°)$ **53.** $r\sin\theta = 2$
55. $r(3\cos\theta - 2\sin\theta) = 4$ **57.** $\tan\theta = 1$
59. $r = 8\cos\theta$ **61.** $r = 1 - \cos\theta$
63. $r = 4\tan\theta\sec\theta$ **65.** $y = -4$
67. $x^2 + y^2 - 3x = 0$ **69.** $x^2 + y^2 - 2x - 3y = 0$
71. $y + 4x = 5$ **73.** $3x^2 + 4y^2 + 8x - 16 = 0$
75. $x^2 + y^2 - 3y = 2\sqrt{x^2 + y^2}$

77.

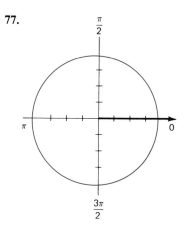

79.   **81.**

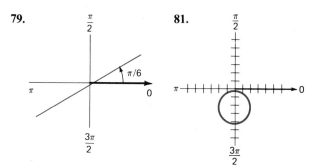

83.

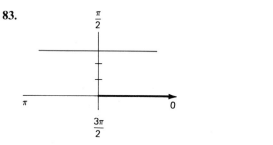

85.

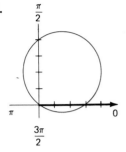

87.

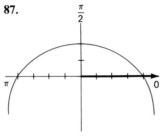

89.

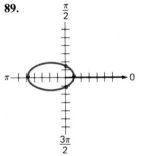

91.

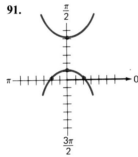

Problem Set 5.5 (page 190)

1. Polar axis   **3.** $\frac{\pi}{2}$-axis   **5.** Polar axis

7. $\frac{\pi}{2}$-axis   **9.** None   **11.** Polar axis

13. Polar axis   **15.** Pole

17.

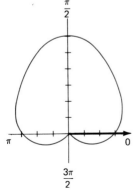

19.

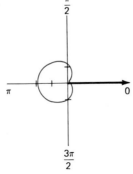

21.

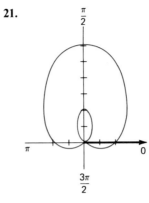

23.

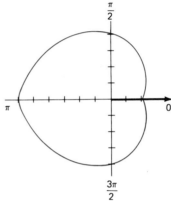

25.

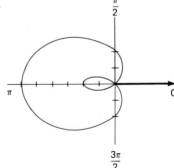

27.

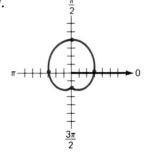

29.   **31.**

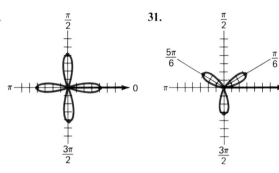

Problem Set 5.6 (page 194)

1. 5   **3.** $\sqrt{10}$ **5.** $2\sqrt{10}$ **7.** $2\sqrt{13}$

9. $\langle 4, 7 \rangle$; $\langle -2, -3 \rangle$; $\langle 15, 26 \rangle$; $\langle -13, -21 \rangle$

11. $\langle -5, 3 \rangle$; $\langle 3, -9 \rangle$; $\langle -16, 15 \rangle$; $\langle -3, -36 \rangle$

13. $\langle 3, -1 \rangle$; $\langle 11, -1 \rangle$; $\langle 5, -3 \rangle$; $\langle 34, -2 \rangle$

15. $\langle -5, -10 \rangle$; $\langle -1, -2 \rangle$; $\langle -17, -34 \rangle$; $\langle 4, 8 \rangle$

17. $\langle 2, 6 \rangle$ **19.** $\langle -4, 3 \rangle$ **21.** $\langle 0, 9 \rangle$

23. $\langle 3, -6 \rangle$ **25.** 10   **27.** 16   **29.** 19

31. Yes   **33.** No   **35.** Yes   **37.** 49.8°

39. 31.3°   **41.** 157.4°   **43.** 75.0°   **45.** 102.5°

33.

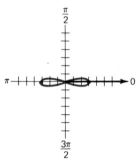

Chapter 5 Review Problem Set (page 198)

1. $-11 - 6i$ **2.** $-1 - 2i$ **3.** $1 - 2i$

4. $21 + 0i$ **5.** $26 - 7i$ **6.** $-25 + 15i$

7. $-14 - 12i$ **8.** $29 + 0i$ **9.** $0 - \frac{5}{3}i$

10. $-\frac{6}{25} + \frac{17}{25}i$ **11.** $0 + i$ **12.** $-\frac{12}{29} - \frac{30}{29}i$

13. $10i$ **14.** $2i\sqrt{10}$ **15.** $16i\sqrt{5}$ **16.** -12

17. $-4\sqrt{3}$ **18.** $2\sqrt{2}$ **19.** $2\sqrt{5}$

20. $2\left(\cos\dfrac{11\pi}{6} + i\sin\dfrac{11\pi}{6} \right)$ **21.** $6(\cos 225° + i\sin 225°)$

22. $0 - 5i$ **23.** $4 - 4\sqrt{3}\,i$ **24.** $16 + 0i$

25. $-512 + 512\sqrt{3}\,i$ **26.** $\dfrac{\sqrt{2}}{2} + \dfrac{\sqrt{2}}{2}i$

27. $0 + 3i,\ -\dfrac{3\sqrt{3}}{2} - \dfrac{3}{2}i$, and $\dfrac{3\sqrt{3}}{2} - \dfrac{3}{2}i$

28. $\dfrac{\sqrt{6}}{2} + \dfrac{\sqrt{2}}{2}i,\ -\dfrac{\sqrt{2}}{2} + \dfrac{\sqrt{6}}{2}i,\ -\dfrac{\sqrt{6}}{2} - \dfrac{\sqrt{2}}{2}i$, and

$\dfrac{\sqrt{2}}{2} - \dfrac{\sqrt{6}}{2}i$

29. $-2\sqrt{3} + 2i$ and $2\sqrt{3} - 2i$

30.   **31.**

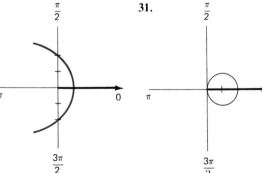

35.   **37.**

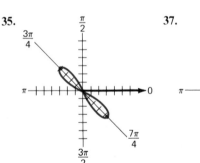
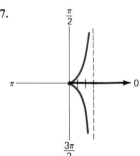

39.

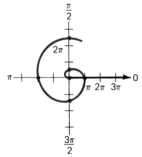

32.

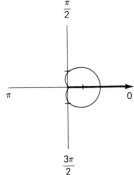

33.

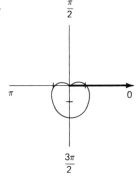

34.

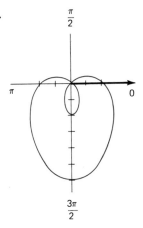

35.

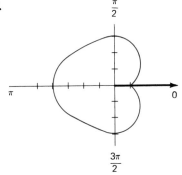

36. $x^2 + y^2 + x = \sqrt{x^2 + y^2}$, cardioid
37. $r = -3 \sec \theta \tan \theta$, parabola
38. $x^2 + y^2 - 3y = 0$, circle
39. $r = 2 + 3 \sin \theta$, limaçon
40. (a) $3\sqrt{2}$ (b) $2\sqrt{5}$
41. (a) $\langle 2, -6 \rangle, \langle -25, -2 \rangle, \langle -9, 16 \rangle$
  (b) $\langle 2, 12 \rangle, \langle 24, 18 \rangle, \langle -2, -30 \rangle$
42. (a) $45.0°$ (b) $133.7°$

CHAPTER 6

Problem Set 6.1 (page 206)

1. $\frac{1}{8}$ **3.** $-\frac{1}{1000}$ **5.** 27 **7.** 4 **9.** $-\frac{27}{8}$
11. 1 **13.** $\frac{16}{25}$ **15.** 4 **17.** $\frac{1}{100}$ or .01
19. $\frac{1}{100000}$ or .00001 **21.** 81 **23.** $\frac{1}{16}$ **25.** $\frac{3}{4}$
27. $\frac{256}{25}$ **29.** $\frac{16}{25}$ **31.** $\frac{64}{81}$ **33.** 64
35. $\frac{1}{100000}$ or .00001 **37.** $\frac{17}{72}$ **39.** $\frac{1}{6}$ **41.** $\frac{48}{19}$
43. 7 **45.** 8 **47.** -4 **49.** 2 **51.** 64

53. .001 **55.** $\frac{1}{32}$ **57.** 2 **59.** $\dfrac{6}{x^3 y}$

61. $\dfrac{6}{a^2 y^3}$ **63.** $\dfrac{4x^3}{y^5}$ **65.** $-\dfrac{5}{a^2 b}$ **67.** $\dfrac{1}{4x^2 y^4}$

69. $15x^{7/12}$ **71.** $y^{5/12}$ **73.** $64x^{3/4} y^{3/2}$

75. $4x^{4/15}$ **77.** $\dfrac{7}{a^{1/12}}$ **79.** $\dfrac{16x^{4/3}}{81y}$ **81.** $\dfrac{y^{3/2}}{x}$

83. $8a^{9/2} x^2$ **85.** $\{3\}$ **87.** $\{7\}$ **89.** $\{4\}$
91. $\{2\}$ **93.** $\{-2\}$ **95.** $\{\frac{5}{3}\}$ **97.** $\{\frac{3}{2}\}$
99. $\{\frac{4}{9}\}$

Problem Set 6.2 (page 214)

1.

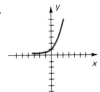

3.

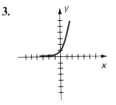

5.

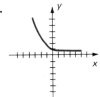

7.

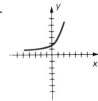

9.

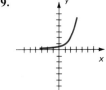

11.

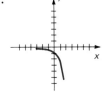

13.

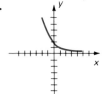

15.

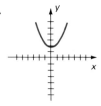

17.

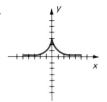

19.

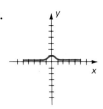

21.

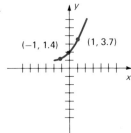

23.

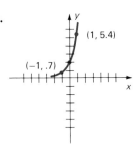

25.

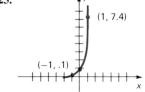

27. (a) \$.67   **(c)** \$2.31   **(e)** \$12,623
(g) \$803
29. \$384.66   **31.** \$480.31   **33.** \$2479.35
35. \$1816.70   **37.** \$1356.59   **39.** \$745.88
41. \$2174.40   **43.** \$4416.52

45.

	8%	10%	12%	14%
COMPOUNDED ANNUALLY	\$2159	2594	3106	3707
COMPOUNDED SEMIANNUALLY	2191	2653	3207	3870
COMPOUNDED QUARTERLY	2208	2685	3262	3959
COMPOUNDED MONTHLY	2220	2707	3300	4022
COMPOUNDED CONTINUOUSLY	2225	2718	3320	4055

47.

	8%	10%	12%	14%
5 YEARS	\$1492	1649	1822	2014
10 YEARS	2225	2718	3320	4055
15 YEARS	3320	4481	6049	8164
20 YEARS	4952	7388	11020	16440
25 YEARS	7388	12179	20079	33103

49. 8243; 22,405; 100,396   **51.** 203; 27; .5
53. (a) 82,887   **(c)** 96,299   **55.** 500; 210
57. $-\frac{1}{25}$

Problem Set 6.3 (page 223)

1. $\log_3 9 = 2$ **3.** $\log_5 125 = 3$ **5.** $\log_2(\frac{1}{16}) = -4$
7. $\log_{10}.01 = -2$ **9.** $2^6 = 64$ **11.** $10^{-1} = .1$
13. $2^{-4} = \frac{1}{16}$ **15.** 2   **17.** -1 **19.** 1
21. $\frac{1}{2}$ **23.** $\frac{1}{2}$ **25.** $-\frac{1}{8}$ **27.** 7   **29.** 0
31. $\{25\}$ **33.** $\{32\}$ **35.** $\{9\}$ **37.** $\{1\}$
39. 1.1461   **41.** .6020   **43.** 2.5353   **45.** .1505
47. 1.1268   **49.** 1.4471   **51.** 1.9912
53. 2.3010   **55.** 3.1461   **57.** $\log_b x + \log_b y + \log_b z$
59. $2\log_b x + 3\log_b y$ **61.** $\frac{1}{2}\log_b x + \frac{1}{2}\log_b y$

63. $\frac{1}{2}\log_b x - \frac{1}{2}\log_b y$ **65.** $\log_b\left(\dfrac{xy}{z}\right)$ **67.** $\log_b\left(\dfrac{x}{yz}\right)$

69. $\log_b(x\sqrt{y})$ **71.** $\log_b\left(\dfrac{x^2\sqrt{x-1}}{(2x+5)^4}\right)$ **73.** $\{2\}$

75. $\{2\}$ **77.** $\{\frac{2}{9}\}$ **79.** $\{6\}$

Problem Set 6.4 (page 232)

1.

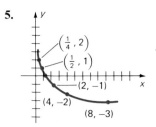

3.

5.

7. Same graph as Problem 1
9. Same graph as Problem 5

11.

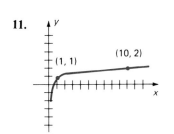

13.

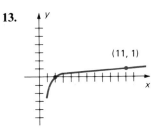

15. .9754   **17.** 1.5393   **19.** 3.6741
21. −.2132   **23.** −2.3279   **25.** .9405
27. 1.7551   **29.** 2.8500   **31.** .6920 + (−1)
33. .7226 + (−3)   **35.** 5.8825   **37.** 33.597
39. 8580.3   **41.** 3.5620   **43.** .71581
45. .0022172   **47.** 3.55   **49.** 30.4   **51.** 983
53. 640000   **55.** .542   **57.** .00731   **59.** .179
61. .6931   **63.** 3.0634   **65.** 6.0210
67. −1.1394   **69.** −2.6381   **71.** −7.1309
73. 1.3083   **75.** 4.3567   **77.** 6.4297
79. −.8675   **81.** −4.7677

Problem Set 6.5 (page 240)

1. {3.17}   **3.** {1.95}   **5.** {1.81}   **7.** {1.41}
9. {1.41}   **11.** {3.10}   **13.** {1.82}   **15.** {7.84}
17. {10.32}   **19.** {2}   **21.** {$\frac{29}{8}$}

23. $\left\{\dfrac{-1 + \sqrt{65}}{2}\right\}$ **25.** {$\sqrt{2}$}   **27.** {6}

29. {1,100}   **31.** 2.402   **33.** .461   **35.** 2.657
37. 1.211   **39.** 7.9 years   **41.** 12.2 years
43. 11.8%   **45.** 6.6 years   **47.** 1.5 hours
49. 34.7 years   **51.** .17   **55.** {1.13}
57. $x = \ln(y + \sqrt{y^2 + 1})$

Problem Set 6.6 (page 247)

1. .6362   **3.** 2.1103   **5.** −1 + .9426
7. −3 + .7148   **9.** 2.945   **11.** 155.8
13. .3181   **15.** 20,930   **17.** 5.749
19. 1,549,000   **21.** 614.1   **23.** .9614
25. 13.79   **27.** 15.91

Chapter 6 Review Problem Set (page 250)

1. $\frac{16}{9}$ **2.** −27   **3.** 36   **4.** $\frac{9}{4}$ **5.** 32
6. −125   **7.** 81   **8.** 3   **9.** −$\frac{4}{3}$ **10.** 3
11. −2   **12.** $\frac{1}{3}$ **13.** $\frac{1}{4}$ **14.** −5   **15.** 1

16. 12   **17.** −12y^3 **18.** $\dfrac{3x^3}{y^4}$ **19.** $\dfrac{a^2}{2b^3}$

20. $\dfrac{a^2 b^{14}}{81}$ **21.** −21$x^{11/12}$ **22.** 25ax^2

23. {5}   **24.** {$\frac{1}{9}$}   **25.** {$\frac{7}{2}$}   **26.** {3.40}
27. {8}   **28.** {$\frac{1}{11}$}   **29.** {1.95}   **30.** {1.41}
31. {1.56}   **32.** {20}   **33.** {10^{100}}   **34.** {2}
35. {$\frac{11}{2}$}   **36.** {0}   **37.** .3680   **38.** 1.3222
39. 1.4313   **40.** .5634
41. (a) $\log_b x - 2\log_b y$ (b) $\frac{1}{4}\log_b x + \frac{1}{2}\log_b y$
        (c) $\frac{1}{2}\log_b x - 3\log_b y$

42. (a) $\log_b x^3 y^2$ (b) $\log_b\left(\dfrac{\sqrt{y}}{x^4}\right)$ (c) $\log_b\left(\dfrac{\sqrt{xy}}{z^2}\right)$

43. 1.585   **44.** .631   **45.** 3.789   **46.** −2.120
47.

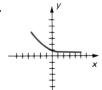

48.

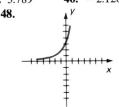

49.

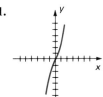

50.

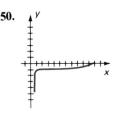

51.

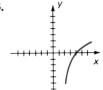

52.

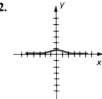

53.

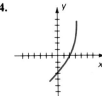

54.

55. $2219.91   **56.** $4797.55   **57.** $15,999.31
58. approximately 5.3 years
59. approximately 12.1 years   **60.** approximately 8.7%
61. 61,070; 67,493; 74,591   **62.** approximately 4.8 hours

Index